The World Map, 1300–1492

Published in association with the Center for American Places
Sante Fe, New Mexico, and Staunton, Virginia
www.americanplaces.org

The World Map, 1300–1492

The Persistence of Tradition and Transformation

EVELYN EDSON

The Johns Hopkins University Press
Baltimore

© 2007 The Johns Hopkins University Press

All rights reserved. Published 2007

Printed in the United States of America on acid-free paper

2 4 6 8 9 7 5 3

The Johns Hopkins University Press

2715 North Charles Street

Baltimore, Maryland 21218-4363

www.press.jhu.edu

Library of Congress Cataloging-in-Publication Data

Edson, Evelyn.

The world map, 1300–1492 : the persistence of tradition and transformation / Evelyn Edson.

p. cm.

Includes bibliographical references and index.

ISBN-13: 978-0-8018-8589-1 (hardcover : alk. paper)

ISBN-10: 0-8018-8589-2 (hardcover : alk. paper)

1. Geography, Medieval.   2. Geography—History—15th century.   3. Cartography—History.

4. Early maps.   I. Title.

GA221.E38 2007

910.9′023—dc22   2006030067

A catalog record for this book is available from the British Library.

CONTENTS

The European discovery of the Americas has been considered a great watershed in the history of maps. Until then, there were only the three known continents of the Old World to include, and suddenly it was necessary to map both sides of a round world with vast additional territories. The emphasis in this book is not to report on the change that discovery made but to argue that before it took place, mapmakers, thinking of space in a new way, were practiced in opening their imaginations and their maps to new lands. The results of the exploration of the African coast, the Atlantic islands, and the northern countries of Europe had already been incorporated onto maps. In the East, knowledge was imperfect, but by 1500 Europeans had a sense of the great space and powerful civilizations in Asia. Marco Polo had returned home talking of millions: millions of people, millions of ducats. The world map began to stretch to include what he had seen.

The change from the encyclopedic tradition of the medieval mappamundi, or world map, was not without its costs. Viewing space merely as distance to be traveled eliminated many of its most interesting qualities. A place of importance in the past, now destroyed or diminished; a place where a famous or holy person had lived or was buried; a place that was home to some exotic product or creature—all these considerations had guided medieval map painters in the construction of their world pictures. What was the reason for the change? Certainly, people had always traveled through space and were concerned about distance and direction; why did these considerations now become paramount? One reason was the growing aggressiveness of European civilization toward the rest of the world. The voyages of the Portuguese down the African coast were not a pure scientific enterprise but a quest for profits from gold, slaves, ivory, and other products that could be sold at home. How long it took to make a journey, avoiding maritime hazards, was extremely important to the income of the prince-patron. Sailors in the Mediterranean had learned their trade through apprenticeship, but there was no master pilot on the African coast. Information passed from one voyage to another, including accurate maps, became increasingly vital.

The move toward mapping measured space took place without a dramatic change in mapping technology. Tools such as the compass and the astrolabe had been known since the twelfth century at least, but the instruments were not finely tuned enough to avoid gross errors. The king of Portugal convened a group of scientists to study the problems of astronomical navigation in 1484, but the results were hardly encouraging. Driven by the demand for better observations, small incremental improvements in cartographic techniques over the next two centuries made it possible to produce more precise maps.

Looking at the maps of the sixteenth century, one is impressed to observe how quickly new spaces were mapped, however imperfectly at first. Was America part of Asia as Columbus had believed? Did the Pacific Ocean lie immediately on the other side of the Appalachian Mountains? After Magellan's ships returned from their epic voyage—alas, without their captain—Europeans began to realize how very large the Pacific was. It was the opinion of Antonio Pigafetta, a gentleman passenger, that the distance was so great that no one would ever try to cross it again. Behind the maps of the sixteenth century lies another story—the change in geographical conceptions, mapmaking, and map usage that took place in the late Middle Ages. Pietro Vesconte, Cresques Abraham, Fra Mauro, Andrea Bianco and Henricus Martellus Germanus were among the medieval cartographers who began to reshape the image of the world before Columbus sailed west. This book tells their story.

Like most writers, I have many people to thank. First on the list are my many colleagues in the field of the history of cartography. The notes show the excellent books and articles that aided me in my research, and I can only hope that I have done justice to the careful and thoughtful scholarship of their authors. In addition to published work, there are less formal sources of information, from speeches delivered at conferences, to pointed questions that follow lectures, to conversations over a glass of wine, to interchanges by e-mail, telephone, and good old snail mail. How kindly my fellow scholars have replied to my queries, suggested further reading, sent me offprints of their articles, and generally set me straight! Of particular note is the biennial International Conference on the History of Cartography, which never fails to provide exciting intellectual exchanges. I dare not start naming names, as the list would be too long—you know who you are!

It has been my privilege in the course of my research to visit a number of the great libraries of the world, true treasure houses, staffed with helpful, intelligent people, many of whom are scholars in their own right. Paradise may have disappeared from the late medieval world map, but on my personal map, it is a library, as beautiful as the Beinecke Library at Yale, surrounded by gardens like the Huntington, equipped with more manuscripts than the Vatican Library, and staffed by people as kind and helpful as those at the British Library (to name a few). I give special thanks to my two home libraries, the University of Virginia Library, which has generously extended borrowing privileges to its neighbors, and the Piedmont Virginia Community College Library, whose librarians have heroically assisted in tracking down obscure, foreign-language items through Interlibrary Loan. I also want to mention my first serious library, the Library of Congress, where I harassed the staff with my requests when I was still a high school student. Walking into the beautifully restored dome room today is always an experience of coming home.

Community college teachers teach five classes a semester and are required to hold ten office hours a week. That schedule makes grants and released time particularly welcome. I have been most fortunate to receive two of the grand prizes in the academic lottery: a fellowship from the National Endowment for the Humanities in 1999, which enabled me to do most of the research for this book, and a fellowship from the American Council of Learned Societies, which gave me a year in which to write it. While in England in 1999, I spent six months as a Fellow of the Senior Common Room at Merton College, a rare and wonderful experience at a place that was an important center of late medieval science. My thanks go to Sarah Bendall, then librarian at Merton, who sponsored me. My college generously supplemented the sabbatical grants that I received, and the College Foundation awarded me a grant that enabled me to spend a week in the fall of 2003 working at the Beinecke Library at Yale University. I have also been the grateful recipient of a month's fellowship at the Newberry Library in Chicago, where I studied astrolabes and maps. I do hope my work repays the confidence all these fine organizations have had in me.

I would also like to thank my Medieval Latin Reading Group—David Larrick, Marvin Colker, and Matt Clay—who have helped with translations and willingly read large chunks of Vincent of Beauvais, not to mention all of *The Golden Legend*, in most congenial weekly sessions.

My warmest appreciation goes also to my other friends and family, whose emotional support has been of overwhelming importance. My mother, Margery; my children, Ben, Meredith, and Pete; my fellow communards, Toots and Tom; my dear friends, thank you. And thanks most of all to Andy—Best Beloved, this book is for you.

The World Map, 1300–1492

Andrea Bianco's Three Maps

"Andrea Biancho de Veneciis me fecit, M CCCC XXXVI" (Andrea Biancho of Venice made me, 1436) is written on the title page of an atlas.[1] Turning the page, the reader admires the beautifully drawn charts of the nautical atlas, made to show navigation routes in the Mediterranean and Black Seas and along the Atlantic and North Sea coasts. Such atlases had been made in Venice for more than a century to be used by its sailors and merchants, who sent the Venetian fleet all over the known world. The last of the charts in the book is a "whole portolan," showing at one glance the region depicted piece by piece in the preceding pages (fig. I.1). It is the world familiar to western Europeans of the fifteenth century, reaching from Norway to the Nile, and from the Canary Islands to the eastern shore of the Black Sea.

There are two more maps in Bianco's collection, both world maps, each constructed according to different principles. One is drawn after the instructions of Claudius Ptolemy, whose *Geographia* had recently been translated into Latin and made available in western Europe (fig. I.2). The other is a mappamundi, or medieval world map, with east at the top, decorated with pictures of Adam and Eve, kings, dog-headed men, and dragons (fig. I.3). Bianco does not write about the maps but merely places them side by side, and thus he illustrates the great cartographic question of the fifteenth century: how should the world be mapped?

Bianco himself was a practical man, a sailor and galley captain. A Venetian living in the parish of St. Jerome near the old foundry, he was born sometime in the last two decades of the fourteenth century. In 1417 he made a will, leaving a little money to say masses for his soul. It was not unusual for a sailor to make an early will, given the hazards of his life. Another will follows in 1435, and from then on there are constant references to his career in the city records. Over fifty entries name him in connection with voyages to Tana on the Black Sea, Flanders, Beirut, Alexandria, Romania (the Byzantine Empire), Aigues-Mortes

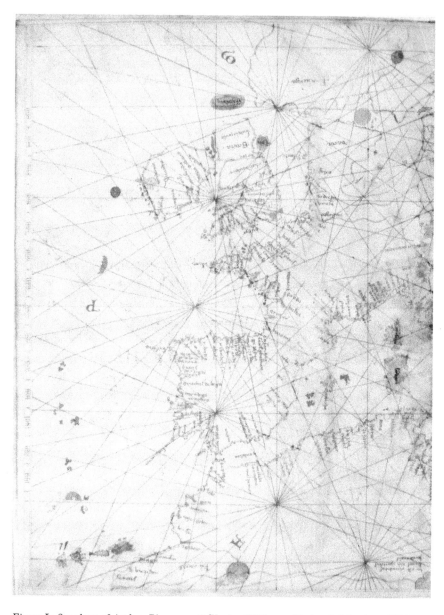

Figure I.1. Sea chart of Andrea Bianco, 1436 (Venice, Biblioteca Nazionale Marciana, MS It. Z,76, fol. 7). Andrea Bianco made this sea chart, which appears in an atlas with a set of charts mapping smaller parts of this region, as well as the two world maps in figs. I.2 and I.3. This chart shows the British Isles and the Baltic coast in the north, the Black Sea in the east, and the coast of Africa as far south as Cape Rosso. The atlas also includes instructions for navigating by the compass. Courtesy of the Biblioteca Nazionale Marciana.

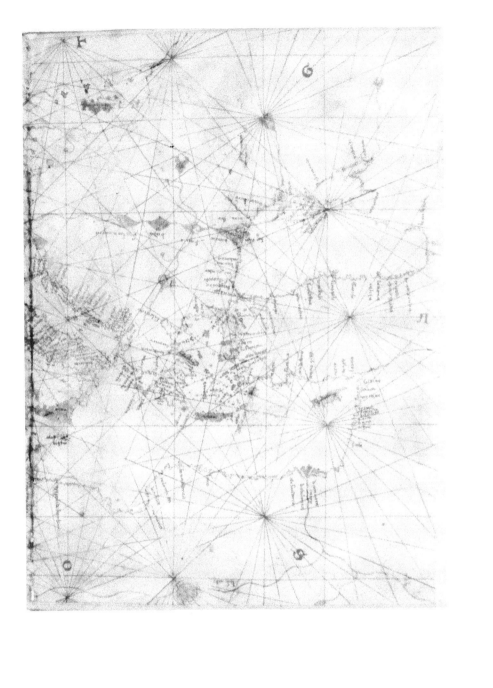

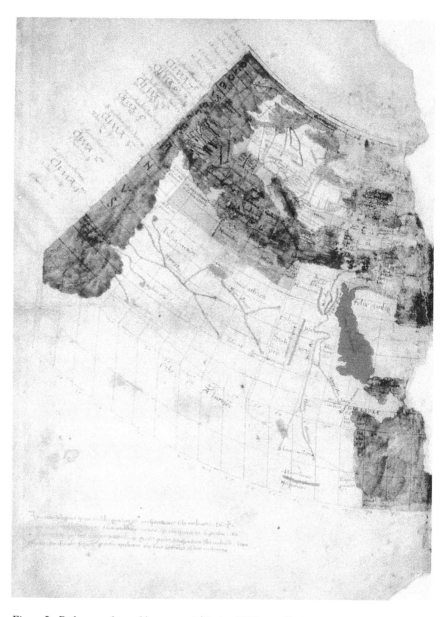

Figure I.2. Ptolemy-style world map, 1436 (Venice, Biblioteca Nazionale Marciana, MS It. Z, 76, fol. 9). In a different hand than the sea charts, this map was probably copied for Bianco. It is one of the earliest known world maps made according to Ptolemy's instructions. Note the shape of northern Britain, the enclosed Indian Ocean with its very large island of Taprobana, and the Red Sea. Courtesy of the Biblioteca Nazionale Marciana.

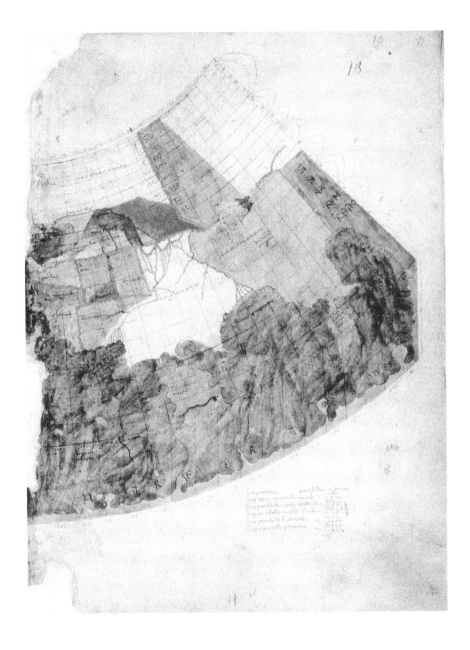

Figure I.3. Mappamundi, 1436 (Venice, Biblioteca Nazionale Marciana, MS It. Z,76, fol. 8). Oriented to the east, the mappamundi shows paradise and the four rivers that flow from it. The Indian Ocean is open to the east, and Africa extends to form its southern shore. The two poles are marked with semicircles, and the Atlantic Ocean is greatly enlarged to show the recently discovered islands of the Azores. Courtesy of the Biblioteca Nazionale Marciana.

in southern France, Cyprus, and "Barbaria" or Algeria. In 1448 he was in London, where he made another sea chart while waiting for his fleet to go through customs and reload its cargo. Here he calls himself "comito di galia" or ship's commander. He is described as "comitus" and "admiraltus" in the Venetian records, and there has been some discussion about his exact status, whether he was commander of the entire fleet or second in command, as these titles changed their meaning in the course of the century.[2]

The last years of Bianco's life found him again in Venice. Despite his wills, he had lived a long life, and now, in retirement, he was named as the collaborator with Fra Mauro on a great world map of midcentury, receiving payment from the monastery of San Michele da Murano for mapmaking supplies. The conflict implicit in his 1436 atlas is made explicit in the many paragraphs of text that adorn the world map, discussing the pros and cons of various methods of depicting the world. It is a struggle between the authority of the mighty classical past, the religious orthodoxy of the medieval mappaemundi, and the practical experience of sailors, "persons worthy of trust who have seen with their own eyes" (see chapter 6).[3]

Bianco opens his 1436 atlas with a page of instructions for "those who wish to sail by the compass." He depicts a beautiful compass rose of the type that appears on many medieval sea charts and a little figure holding a compass. A set of trigonometric tables, accompanied by an explanation, show how to calculate and set a course by the compass, taking into account the ship's speed and the deviations caused by contrary winds. It was a complex business. We find similar tables in mariners' handbooks, although probably only captains and pilots would have understood how they worked.[4]

Once one has mastered the difficult art of navigation in theory, one is ready to set sail. The pages that follow Bianco's instructions contain the detailed maps, beginning with one of the Black Sea. There the Italian cities had important trade depots, which received goods coming overland from the East. Venice's city, Tana, was located on the river Don (or Tanais) just east of the Crimea. From here, goods such as silk, alum, slaves, and furs could be transshipped to the markets of Europe. The earliest surviving marine charts date from the late thirteenth century and show the Mediterranean and Black Seas. Eventually, they were expanded to include the Atlantic coasts of Africa and Europe, as well as the northern seas. We do not know when the first one was made or even exactly how it was constructed, and speculation on this subject has been rife. Looking at them today, one cannot help but be struck by the contrast between their sober, (relative) accuracy and the imaginative, colorful mappaemundi of the medieval scholastic tradition. Giving mostly the contours of the coast and listing the ports of call, they have very little detail in the interior: no legends, no Biblical sites, no monsters (see chapter 2).

The world according to Ptolemy in Bianco's atlas is one of the earliest world maps we have in this format. Ptolemy's *Geography* had been lost to the West for centuries, though it seems to have been known to the Arabs and the Byzantines. Brought to Italy in the late fourteenth century, it was translated into Latin in Florence in 1406. Maps were then constructed according to Ptolemy's instructions. It is possible that Bianco had access to a copy of Ptolemy with maps, belonging to the Florentine Palla Strozzi, who in 1430 took his book with him into exile to Padua, close to Venice. The Ptolemy map is drawn in a different hand from the rest of the atlas, so perhaps someone was commissioned to copy it for Bianco (see chapter 5).

Ptolemy's *Geography* and the maps made from it burst upon the western world at a most receptive moment. Travelers of the later Middle Ages had brought back fascinating information about the worlds of the far East, northern Europe, and Africa. Mapmakers tried to incorporate their findings into their work but increasingly found it difficult to reconcile them with traditional geography and to fit it all within the medieval frame. What Ptolemy provided was a technique, using latitude and longitude to pinpoint the location of cities and other landmarks, which enabled one to construct an accurate map from tables alone. The problem was, of course, the data. Ptolemy provided locations for over eight thousand places, but many of the names were obsolete and some of the data erroneous. The resulting map was somewhat skewed, especially to the medieval eye accustomed to marine charts. Before the ink was dry on the first set of Ptolemy maps, they were being criticized, corrected and supplemented according to the geographical knowledge of the day.[5] Soon new maps, or Tabulae Modernae, were added to Ptolemaic atlases, beginning in 1427 with a map of Scandinavia, an area unknown to the classical world.[6]

Ptolemy's work challenged certain specific ideas of medieval geography. His idea that the Indian Ocean was a closed body of water, was to be disproved by Portuguese sailors near the end of the fifteenth century, when they rounded the Cape of Good Hope and sailed into the Indian Ocean. About the Caspian Sea, however, he was right. Medieval geographers, following Isidore, had described it as a gulf of the northern ocean, but Ptolemy insisted that it was an inland sea.[7] In the thirteenth century, explorers William Rubruck and Marco Polo reported that the Caspian was landlocked, and over the next century, the Caspian was to move around or even appear twice on maps until the question was finally settled by exploration.

Most important were Ptolemy's techniques for projecting an image of the spherical world onto a flat chart and making a true measured map. The marine charts had already begun to use a compass for direction and probably dead reckoning for distance, but they made no accommodation for the curvature of the earth. Ptolemy used astronomical observations, locating latitudes in relation to the position of the sun. Longitude was more difficult to calculate, as it depended on accurate timekeeping. Observing the exact time of a lunar eclipse in different locations was a good method of determining longitude, and, since eclipses were of intense interest to astrologers, these data were more accessible than one might expect. However, the inaccuracy of the timepieces of the day made longitude figures approximate at best. As for estimating longitude at sea, a clock that would keep accurate time on a rolling ship would not be invented until the eighteenth century. Ptolemy's was a great system, inspiring to contemplate, but it would be several centuries before it could be fully implemented.

The third of Bianco's world maps is a mappamundi of the medieval type. Placed in a circular frame, unlike the sea chart's rectangle and Ptolemy's fan shape, the map is oriented with east and paradise at the top and shows the three continents of the known world surrounded by a green ocean. The whole is encircled by a dark blue rim, studded with stars, representing the heavens (see chapter 1). The marine charts have affected the configuration of familiar coasts, and it is relatively easy to pick out Spain, the Mediterranean, and the British Isles, although details are suppressed by the small scale of the map. Once beyond the area of the usual navigational chart, the geography becomes vaguer. Africa stretches far to the east, and the Indian Ocean, filled with islands, is a long narrow gulf, oriented east-west. The Red Sea, helpfully colored red, runs to the west, then makes a right-angle turn to the north, but comes nowhere near the Mediterranean. In the far East the continent of Asia and a large island break into the celestial frame, implying that the known world has now extended beyond the hemisphere and cannot be contained within a circle. Or it may be an expression of the idea that the Garden of Eden, inaccessible to mankind, was located on a mountain that reached to the sphere of the moon.

In contrast to his nautical charts and his copy of Ptolemy's world map, the Bianco mappamundi is richly decorated with pictures and writing. Kings sit enthroned in royal tents, accompanied by their retinues, and cities are marked with little drawings of churches or castles. Christian theology is made visible, as Adam and Eve are shown in paradise with the tree of forbidden fruit, and in another picture God instructs Adam. The four rivers flow from paradise to water the globe. Near the center we find the Virgin Mary and child being adored by the Magi, and at the River Jordan, the baptism of Christ. Noah's Ark is shown in Armenia, and Mt. Sinai in Arabia. A fully caparisoned ship sails down the coast of West Africa toward the south, where a tribe of dog-faced men march under a banner. In the far south the continent of Africa is cut by a half-circle, suggesting the polar circle, and here are two dragons, a double-tailed merman, and a hanging man, possibly Judas. In the corresponding circle to the north are several human figures, and we are told that it is so cold there that all people are born savages. In more familiar lands we see exotic but real beasts such as elephants and camels. Scattered over the map are place-names like Cathay, Samarkand, and India in

Asia, while closer to home we find Paris, and the king of the French, Norway, Sweden, England, and Ireland. Place-names in Africa are mostly in the north, except for Ethiopia, a region that traditionally covered a lot of ground. In the far eastern extension of the African continent can be found Prester John, the Christian king of phenomenal virtue and wealth, who was the source of legends and the object of searches for centuries.

The mappamundi supplements the marine chart, which was all coasts and names, and shows us some of the less tangible aspects of the world, such as the sites important to the history of the Christian religion. It mixes ancient history (Alexander the Great) with modern place-names, such as Basra and Cairo, including a reference to Murad II, who ascended to the Turkish throne in 1422. The mappamundi was probably copied freehand from a model and embellished according to the author's tastes. There is little evidence that it was constructed according to any measure, except that it is divided into eight pie sections, like a rudimentary wind rose. Each radial line is marked at the edge with a letter standing for the Italian name of one of the eight directional winds used by sailors. Some of Bianco's contemporaries in Germany were experimenting with structuring a circular map using a longitude scale around the rim and a series of concentric circles for latitude.[8] Bianco, however, shows no scale, nor is the circular border divided into degrees.

Although some features of the three types of map affect the others, Bianco's choice is most interesting for the stark differences among them. He shows the world in three distinct ways—does he perceive them to be contradictory or complementary? Can one combine the practical needs of travelers with the science of the classical world and the profound philosophical and religious sense of the medieval scholastics? Twenty years later, on Fra Mauro's map, we see the mappamundi and the sea chart pulled together into a single work, accompanied by a long argument with Ptolemy pursued in bits of text all over the map. But the debate was opened in the 1436 atlas by Andrea Bianco placing three maps side by side.

CHAPTER ONE

The World View of the Mappamundi in the Thirteenth Century

In the late thirteenth century a precious gift was presented to Hereford Cathedral by Richard of Haldingham, a newly appointed cleric. It was an enormous map of the world, covered with colored pictures, place-names, and descriptive legends. Set in a frame surmounted by a dramatic scene of the Last Judgment, the map put the physical world in a spiritual context as the theater of human history in a universe created (and to be destroyed) by God.[1]

Classical Origins

The Hereford Cathedral map is the best surviving example of a group of artifacts which were once more numerous. The mappaemundi of the thirteenth century pulled together the world view of the High Middle Ages and presented it in a spatial format, incorporating history, geography, botany, zoology, ethnology, and theology into one harmonious and dazzling whole. The medieval conception of world geography was developed from a classical foundation made up of the speculative geography of the Greeks and the more practical, world-conquering, administrative knowledge of the Romans. Accidents of survival apparently determined which authors should remain influential and which should disappear. For example, the work of Macrobius, a fifth-century writer, was preserved perhaps because of its digressions into numerology and morality, while the geography of Ptolemy seems to have gone underground in the West for most of the Middle Ages.[2] Certain ideas, such as the sphericity of the Earth, made their way into the digests of knowledge assembled in the late empire, and these found a place in medieval libraries and influenced later geographical texts and maps. Some Roman map forms survived as well, if we are to trust the fidelity of manuscripts of classical works copied in the ninth century. Among these was europa & affrica   De. Asia & eius partibus   Ca·iii·

Asia ex noie
cuiusdã mu/
lieris est ap/
pellata· que apud anti/
quos imperiu orientis
tenuit. Hec in tercia or
bis parte disposita· ab
oriente ortu solis·a me
ridie·oceão·ab occiduo
nostro mari finitur· a
septentrione meothide
lacu & tanai fluuio ter
minatur.Habet autem
prouincias multas et re
giones·quaru breuiter nomina et situs expediam·sumpto initio
a paradiso Paradisus est locus in orientis partibus constitu/
tus·cuius vocabulum ex greco in latinum vertitur ortus.Porro
hebraice eden dicitur·quod in nostra lingua delicie interpretat·
quod verumqʒ iunctum facit ortum deliciarum·est enim omni
genere ligni & pomiferarum arborum consitus habens· etiam
lignum vite.Non ibi frigus· non estus· sed perpetua aeris tem/
peries·e cuius medio fons prorumpens·totum nemus irrigat· di
uiditurqʒ in quatuor nascentia flumina.Cuius loci post pecca/
tum hominis aditus interclusus est.Septus est eni vndiqʒ rom
phea flammea·id est muro igneo accinctus· ita ut eius cũ caelo
pene iungatur incendium. Cherubin quoqʒ id est angelorum
presidium arcendis spiritibus malis super romphee flagrantiã
ordinatum est·ut homines flamme·angelos vero malos angeli
boni submoueãt·ne cui carni vel spiritui transgressionis aditus
paradisi pateat.   India vocata ab indo flumine· quo ex parte
occidentali clauditur.Hec a meridiano mari porrecta vsqʒ ad
ortum solis·& a septentrione vsqʒ ad montem caucasum perue/
nit·habens gentes multas & oppida· insulam quoqʒ taprobane
gemmis & elephantibus refertam. Crisam & argiram auro ar/
gentoqʒ fecundas·vtilem quoqʒ arboribus foliis nunqm caren
tibus.Habet & flumina gangen & nidan & idaspen illustran/
tes indos.Terra indie fauonio spiritu saluberrima. In anno bis

Figure 1.1. T-O world map (from Isidore of Seville, *Etymologiae* [Augsburg, 1472]. 6.4 cm / 2.5″ dia). It is interesting that the earliest world map to appear in print should be this very traditional T-O style diagram, following the style of map found in many manuscripts of Isidore's popular work. Oriented to the east, the map shows the continents as the domains allotted to Noah's three sons. The "T" water boundary is labeled as the Mediterranean Sea, while most of these maps have the southern arm as the Nile and the northern arm as the River Don or Tanais. A band of ocean surrounds the land. Courtesy of the Rare Books Division, Library of Congress.

Figure 1.2. Zone map (from Macrobius, *Commentary on the Dream of Scipio* [Cologne, 1521].
4.5″ diam.). Unlike the T-O, the zone map goes beyond the known world to sketch out a continent in the southern hemisphere, here noted as "temperate" in climate and "the Antipodes,
unknown to us." A band of the sea, labeled *perusta* (hot), separates the two landmasses. The
map is oriented to the north, and in this version reverses the arrangement of places—perhaps
a hazard of the printing process—with Africa and the Mediterranean in the east, and India in
the west. To the far north are the Riphaean mountains and the island of Thile (Thule). Courtesy of Special Collections, University of Virginia Library.

the T-O diagram, a simple circular sketch of the world divided into three continents by a T-shape representing three water boundaries—the Nile, the River
Don (or Tanais), and the Black Sea (fig. 1.1).[3] Another survival was the zonal
diagram, which showed one face of the spherical earth, divided horizontally into
five zones—two frigid at the poles, two temperate, and one torrid at the equator
(fig. 1.2). The inhabited world occupied the northern temperate zone of one side

of the earth, and the hypothesis, inherited from Crates of Mallos of the second century BCE, was that similar landforms appeared in each of the other three quarters of the globe. Medieval authors copied these models as illustrations for their books, occasionally adding contemporary details.

Understanding the extent of the classical heritage is difficult, as no original world map survives from the ancient world. Numerous classical geographical texts, references, and digests were available to medieval scholars, but the relationship of text to map is always doubtful. A recent thesis by the German scholar Brigitte Englisch suggests that medieval maps were structured on a geometrical frame, expressing the order and number of the world as created by God. She studies maps from the eighth to the twelfth centuries, concluding that they used ancient geographical terms and concepts but reordered their world picture into a true medieval creation. By the thirteenth century, she concludes, the heyday of the large mappaemundi, the depiction of orderly space had been subordinated to other, broader goals.[4]

Among medieval maps, the one with the most authentic ancient pedigree is a twelfth-century copy of a fourth-century copy of a first-century original. Called the Peutinger Table or Chart, it is an itinerary map, showing roads radiating out from Rome and extending throughout the empire. Missing is the western sheet, which would have shown Britain and Spain, but the map extends as far east as China and shows the furthest limit of Alexander's conquests. The format is long and narrow (6.75 meters by 34 centimeters), and the itinerary form produces a number of distortions in geographical shapes. Although it purports to assist the traveler, the map is not drawn to scale. Its most outstanding feature is its practical bent—one can easily imagine the messengers of the Empire setting out along its delineated roads, stopping at bathhouses at the end of the day and stocking up at the granaries, helpfully marked. Although in form and character the Tabula is about as far from a medieval mappamundi as possible, it may have been altered in the process of copying and recopying.[5] These changes may have been by error or by design. For example, the version we now have includes Mt. Sinai and a text on the wanderings of the Israelites in the desert.[6] A recent article by Emily Albu proposes that the Tabula was constructed in the Carolingian period from ancient itineraries as part of Charlemagne's project to link his empire with the glory of Rome.[7]

The Medieval Mappamundi

By the end of the twelfth century, the Middle Ages had developed its own map, the mappamundi, which would present the entire history and philosophy of the human race organized within a geographical framework. All the geographical "facts" of the classical heritage were now transformed to give them a Christian meaning. For example, the traditional eastern orientation of the map now gained new significance from the presence of the Garden of Eden in the east. While not indifferent to spatial relationships, distances, and scale, mappaemundi were constructed according to metaphysical principles. For example, the small territory in the Middle East known as the holy land was expanded to make room for the many spiritually significant events and holy places to be found there, while the great sweep of northern and eastern Asia was generally abbreviated for lack of knowledge. A recent study by Alessandro Scafi has suggested a distinguishing characteristic of the mappaemundi—they included the earthly paradise, a place not geographically accessible in the usual sense.[8] When the earthly paradise no longer appears on the world map, we have a new mapping tradition, more devoted to the physical measure of space than to its transcendent theological meaning. The element of time, so important on the mappamundi, disappeared in favor of a world in the present tense. Paradise became an event of the past, problematic in the present, and no longer suitable for inclusion on a world map.

The term *mappamundi*, while used loosely in the Middle Ages to mean any world map and sometimes merely a list of places, will here be reserved for these detailed, theologically and historically conceived maps.[9] Most of them were quite large—the Hereford mappamundi is five feet in diameter—and were made to be mounted on the wall. The smallest, two Psalter maps of the 1260s, are tucked into a tiny prayer book, but the wealth of detail, sketched in with a fine brush, suggest that one of these was a copy of much larger original, perhaps the one painted on the walls of Westminster Palace.[10]

The large size of most mappaemundi made it possible to present a great deal of information, but their size was also their downfall. Not easily portable nor storable, mappaemundi were destroyed by fire, floods, war, religious quarrels, and housecleaning. The vellum was cut up for book bindings (the Aslake map) or turned so that the reverse side could be used for other works (the Evesham map). Some, frescoed on walls, were whitewashed over in succeeding centuries as more modern maps made them look obsolete (Chalivoy-Milon). The most recent casualty, the Ebstorf map, three-and-a-half meters square, was destroyed by the Allied bombing of Hanover in 1943. Others have come down to us in fragments, faded and water spotted. Only the Hereford Cathedral map survives entire and in relatively good condition to bear witness to that confident age of European history (fig. 1.3).[11]

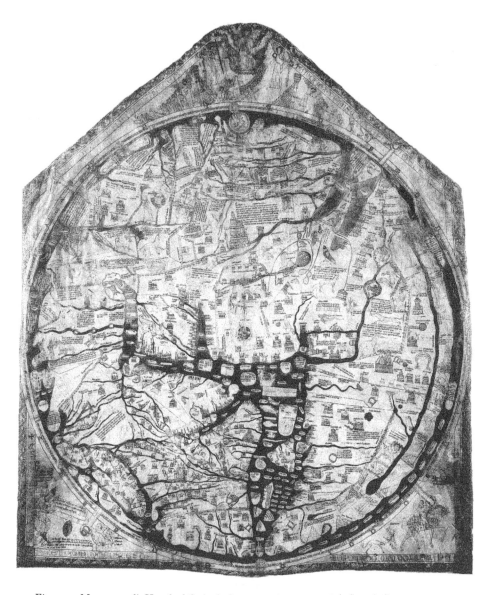

Figure 1.3. Mappamundi, Hereford Cathedral, 1290. 1.58 × 1.33 m / 5′ 2″ × 4′ 4″). Drawn on a
single ox hide, the Hereford Cathedral mappamundi is in the best condition of any of the few
surviving works of this type. Christ in Judgment presides over the map, paradise is the small
circle at the top (east), and the Red Sea and Persian Gulf make up the two-pronged body
of water at the top right. The Mediterranean stretches from the west or bottom of the map,
where the two Pillars of Hercules can be seen. Britain and Ireland are the three large islands
at the lower left. Courtesy of the Dean and the Chapter of Hereford Cathedral.

Shape of the Medieval World

What did the thirteenth century world look like to the mapmaker and map reader of Hereford? To begin with, the world was a sphere, and the mappamundi echoed that shape in the round form of the inhabited world it portrayed. Actually, complained the thirteenth-century English chronicler and mapmaker Matthew Paris, the inhabited world should be depicted as a chlamys, or spread-out cape.[12] But the medieval convention was to show it as the *orbis terrarum* or circle of lands, with the three continents surrounded by a narrow band of ocean. The flat appearance of these maps has misled some modern observers, who insist that medieval people believed in a flat earth.[13] The problem of representing a three-dimensional sphere on a plane surface still poses difficulties for mapmakers.

On the mappamundi the inhabited world was divided into three continents or "parts," Asia, Europe and Africa, as it was in the T-O diagram, with the traditional boundaries between them.[14] The continents were usually described as the inheritances of Noah's three sons, a division which took place in Genesis 10 but is expanded in the Book of Jubilees.[15] The Hereford map makes only passing references to this division, but on many smaller medieval world maps the sons are displayed prominently (see fig. 1.1). On a T-O diagram the continents were shown schematically, but on the larger maps the rivers meander and the borders are irregular. On the Hereford mappamundi, the Nile comes from the east, turning abruptly to the north to reach the Mediterranean Sea.

The threefold division of the world was overlaid by a fourfold division, based on the cardinal directions, represented by the four major winds at the four corners of the earth. The winds, personified by squatting gargoyles on the Hereford map, were expanded in number to twelve, each wind accompanied by a brief description of its various names and characteristics. For example, the southwesterly wind, Africus, was also called Libs, from Libya. It produced thunderstorms with lots of rain and lightning. The north wind, called Aquilo or Boreas, was cold and dry. It did not disperse the clouds but held in the rain. Thus, in addition to providing a sense of direction, the winds gave information on the weather.

Outside the three continents, it had been suggested by the Greek cosmographers that there might be a fourth continent in the southern hemisphere, as well as landmasses in the unseen western hemisphere. Maps in the manuscripts of St. Beatus's eighth-century commentary on the Apocalypse showed a body of land to the south which seems to be this fourth, Australian continent.[16] Zonal maps often showed a southern continent (see fig. 1.2), but such a body was usually excluded from the narrower compass of the mappamundi, which showed only the *ecumene,* or the known, inhabited world. It is possible that the band of monstrous races, shown on the southern fringe of the Hereford and Psalter maps and separated from the rest of the world by a narrow band of water, is a vestige of this imagined continent.[17] Some impossibly remote and inaccessible place was a good venue for these creatures of the imagination (fig. 1.4).

Measurement and Scale

The mappamundi, while striving for a just proportion and correct layout of lands, used no consistent scale of measurement. Its spatial organization is dynamic, representing the significance of places rather than their size or distance. The prime example is the holy land, which contains so many important sites crammed into a small geographical space. On the Hereford map this area gets disproportionate space not only because there was a lot to show but also to emphasize its importance. Less well-known or interesting regions were correspondingly shrunk in size. Elsewhere, places were sometimes dislocated to make room for pictures or lengthy descriptive legends. The Hereford mapmaker was not indifferent to measures of space. Indeed, the map contains several quotations from Pliny's *Natural History* on the dimensions of islands and distances between places. For example, on the size of France: "From the Rhine to the Pyrenees, and from the Ocean to the Cevennes and Iura mountains, excluding Narbonensian Gaul, [the territory] is in length 330 passuum and in width, 318."[18] However, dimensions are not given in all cases, and no attempt is made to represent them proportionally on the map. The mapmaker does concern himself with relative place; that is, places are defined relative to the positions of other places—north of, across from, far away—without more definite measures of distance. Why medieval mapmakers were not more interested in scale is a challenging question.[19] Distances, sometimes given in terms of days' journeys, do appear in written itineraries and travel accounts, and on some regional maps, for example, Matthew Paris's thirteenth-century map of Palestine.[20] Makers of world maps did not have all the needed data or the tools to get the data for their enormous project. Besides, a map drawn purely to scale would have lacked space for some most important features, such as depictions of animals, historic characters, and pilgrimage sites. The idea of a grid map, based on latitude and longitude, was proposed in the thirteenth century by Roger Bacon, but if he made such a map, it does not survive.[21] A grid-based map implies that all points on the surface of the earth are of equal importance, a concept that did not harmonize with the hierarchical world view of the mappaemundi.

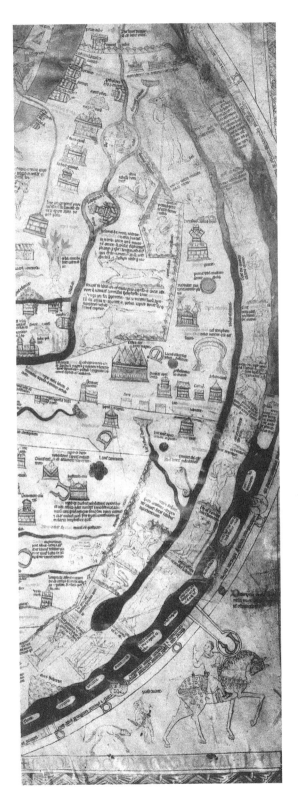

Figure 1.4. Detail of the Hereford mappamundi, the far south. The right or southern edge of the map contains a lineup of the monstrous races, from the "Gens sine auribus" (people without ears) at the top to the four-eyed people of Ethiopia and the Gangines "with whom there is no friendship." The Nile runs from east to west, splitting to form the delta at the left of the picture. A number of other monsters are shown in Africa, including a mandrake (middle left), a unicorn, and a centaur. Courtesy of the Dean and the Chapter of Hereford Cathedral.

Orientation

Until the fourteenth century nearly all European world maps were oriented to the east. East was the direction of sunrise, a symbol of beginnings and the dawn of light, but it was also the location of the earthly paradise, site of the creation of the human race. On the Hereford map the Garden of Eden appears on the top as an island surrounded by a wall and a circle of fire. Inside, Adam and Eve are shown eating the forbidden fruit by the fountain of the four rivers. On other maps, paradise is shown as a castle on a mountain, surrounded by impenetrable fortifications, emphasizing its inaccessibility. A series of maps made to accompany Beatus's eighth-century commentary on the Apocalypse shows Adam and Eve in the Garden, clutching their fig leaves and looking reproachfully at the snake. The choice of this moment indicates that the fall has already taken place, and hence human history has begun. The Hereford map shows Adam and Eve a second time. Now they are outside of paradise, where they are being pursued by an angel with a flaming sword. Thus East is not only a sacred direction but the direction of the beginning of the world and of time.

If the Garden of Eden is the point farthest east, the Straits of Gibraltar or the Pillars of Hercules are the farthest west. Although by 1300 ships were regularly going through this passage, Gibraltar was conventionally held to be the limit of human travel. In the *Inferno* Dante describes Ulysses' journey through this strait into the south Atlantic, where he perishes, as "a fool's flight."[22] To the north and south there were no standard markings, although some legendary travelers, such as Alexander the Great and the god Bacchus, were said to have set up pillars or altars to mark the extent of their journeys. Some medieval maps stress these landmarks or limits, along with warnings of extremes of heat and cold and fearsome monsters in the northern and southern fringes of the inhabited world.[23]

Centering the Map

While a sphere has no center on its surface, the flat, circular image of the medieval map did have a center. The center was not usually stressed until the thirteenth century, when a new emphasis on Jerusalem, probably due to the European involvement in the Crusades, put that holy city in the middle. Jerusalem is central on the Hereford, Ebstorf, and Psalter maps, all from the mid-thirteenth century. The placement is justified by a reference to the Book of Ezekiel 5:5, where the prophet reports that God told him, "I have placed Jerusalem in the midst of the nations and with all the lands around it."[24] However, Ezekiel is speaking of a visionary Jerusalem, not the actual city, and earlier medieval geographers had tended to follow Isidore of Seville, who described Jerusalem as "umbilicus regionis totius," the center of the whole region, that is, of Judaea.[25] In a recent study of Jerusalem on medieval maps, Ingrid Baumgärtner has pointed out that descriptions of the city as the center of the earth appear more frequently in written accounts than on maps, and it was only after the loss of Jerusalem in 1244 that it became relatively common to place the city in the center of world maps.[26] Overall, it seems likely that the new emphasis is spiritual rather than physical. The loss of Jerusalem was a great source of grief and guilt to Western Christendom, and the real city seems to have been transformed in the imagination into a shimmering vision of heavenly perfection, now out of reach. As the mappamundi tradition began to lose its hold on the European mind, later mapmakers, trying to represent the earth in a more physically accurate mode, felt called upon to explain the displacement of Jerusalem from the center of their maps.[27]

Framing the Map

Outside the circle of the Hereford map proper are pictures and texts that further develop the meaning of the whole. It is interesting that these texts are written in Norman French, while the inscriptions on the map proper are in Latin. Above is a painting of the Last Judgment with Christ enthroned and the damned and saved to his left and right, respectively. Below Christ, the Virgin Mary bares her breasts, reminding her son of how she nurtured him and urging compassion on the unfortunate sinners.[28] The Judgment is not only a warning to the heedless but also an indication of the ephemeral nature of worldly reality as laid out on the map. The world will come to an end, just as it was once created. This message is reinforced by the letters M-O-R-S (death), which appear one by one in small medallions around the rim of the map. The physical life of a human being, like the world itself, is a fleeting thing. On the fringes of the contemporary Duchy of Cornwall mappamundi, a series of pictures show the various ages of man progressing toward death and, hopefully, transmutation into an angelic form.[29]

A slightly different message is borne by the Ebstorf map, where the world is dramatically superimposed on the physical body of Christ. Here, we are to remember Christ's dual role, how he became a man and died a human death as a sacrifice for sinners, but also how, as God, he created the earth. The text surrounding the image of Christ reinforces his world-encompassing qualities. An alpha and an omega appear beside his head in the east. His hands and feet, displaying their wounds, appear at the north, south, and west. The right hand is labeled "Dextera Domini feci(t virtutem)" (The right hand of the Lord does mighty works). At the left hand is the text "Terram palmo concludit" (He holds the world in the palm of his hand). At the feet are the words "Usque ad finem fortiter, suaviter disponens omnia," part of a verse from the Book of Wisdom 8:1. ([Wisdom which you have produced from the mouth of the Highest] reaches powerfully from one end of the earth to the other, ordering all things well.)[30] These passages emphasize Christ's embrace of the earth in terms of both time and space.

Several other medieval world maps are drawn with Christ embracing or presiding over the world. On the thirteenth-century English Psalter map, Christ, flanked by incense-bearing angels, stands above the world and blesses it. In the version of the map on the reverse side of the folio, Christ embraces the earth. His feet, shown below, trample two dragons or wyverns, symbols of evil.[31] This depiction has the same theme of world enclosure, but only the Ebstorf map shows the superimposition of the earth on the body of the crucified Christ.

In the lower left-hand corner of the Hereford map is a text describing the survey of the earth supposedly instigated by Julius Caesar. Next to the inscription appears Augustus Caesar ordering the three surveyors to go out and do their work; we may assume that Julius failed to complete the job. Augustus, wearing the papal triple crown, holds up a document with an impressive hanging seal, thus lending imperial authority to the mission and the map. In the lower right-hand corner is an enigmatic scene of a huntsman with a greyhound and a man on horseback, accompanied by the words "passe avant" (go forward). Numerous ingenious suggestions have been made to explicate this scene, which might have some personal or local meaning for the mapmaker, but no agreement has been reached.[32]

Time on the Map

The Hereford mappamundi contains much more than geographical elements, although these features (continents, rivers, mountains, seas, islands) are the structure on which the rest is organized. The Last Judgment at the top is only the culmination of the religious story told in the representation of people and places all over the map. Old Testament history is laid out, from Noah's Ark, to the Tower of Babel, the granaries of Joseph, and the circuitous route of the Israelites from Egypt through the desert to the Promised Land. An image of the crucified Christ is drawn above Jerusalem, and the locations of the events of his life are pictured, such as Bethlehem, Nazareth, Cana, and the Sea of Galilee. The life of Paul is illustrated, from his birthplace in Tarsus and his conversion in Damascus through the various places where he preached the Gospel. The spread of Christianity is further chronicled by a portrait of Augustine in his cathedral church of Hippo and the monastery of St. Anthony in the Egyptian desert. The city of Rome is represented by an elaborate city symbol and described as "caput mundi tenet orbis frena" (the head of the world holds the reins of the globe). Given the antiquarian nature of the map, this phrase could refer to its past glories as capital of the Empire as well as to its present status as head of the Catholic Church and, theoretically, the Holy Roman Empire. Pilgrimage sites include Mont St. Michel, Santiago de Compostela, Sabaria in Hungary, and numerous locations in the holy land. The map was thus organized for the reader or student of sacred history who wished to imagine a pilgrimage or to arrange his meditations in spatial terms.

Christians believed that Christ would come again and that the world would end when all people had been converted. The Hereford map does not carry out this theme quite as completely as did the maps in the Beatus Apocalypse series, which showed each of the apostles, preaching throughout the world.[33] On the Hereford map the cathedral of Santiago de Compostela, erected at the outermost western edge, was a sign that the Last Judgment was imminent. Another theory, that of the Four Empires, suggested that world history was a succession of empires, originating in the east with the Babylonian Empire and moving west through Persia, Macedon, and Rome. Thus history moved from east to west toward the world's end, conceived in terms of space and time. The era of the Roman Empire was extended in its existence as the Byzantine Empire in the east and the Holy Roman Empire in the west, and so put off the fateful day.

Secular history appears on the Hereford map as well. The city of Troy, the places conquered or visited by Alexander the Great, and the Roman provinces and roads are present. In general, the makers of mappaemundi showed greater interest in ancient than in modern history, but a few contemporary sites and place-names appear, such as the unnamed "Saracen city" on the western bank of the Jordan River. Prague, Venice, and Worms, all important cities founded in the Middle Ages, are represented, as is Flanders, a key region in the medieval wool trade. In Britain the castles built by Edward I in his campaign to control Wales are represented by Conwy, built in 1277, and Caernarvon, begun in 1283.[34] In addition to some ancient itineraries, such as the voyages of Jason and Odysseus and the journeys of St. Paul, there are several modern routes drawn in, showing the roads followed by merchants on their way to the great medieval fairs.[35]

Political borders are shown only occasionally, such as the Franco-Burgundian border, whose location has interested scholars as evidence for dating the map.[36]

The Hereford map was made for long-term possession, and the shifting state of political entities would have rendered it quickly out of date. To what extent the concept of borders existed at all in the Middle Ages is unclear, as feudal loyalties continually crossed "national" lines, and national identity was trumped by the concept of supranational Christian unity.[37] The mapmaker preferred to show unchanging geographical areas, or even the Roman provinces, which had stood the test of time. A map based on current political categories would have been a sadly limited document in the thirteenth century. An attempt to compromise was made by Gervasius of Tilbury in his universal history (1215), when he wrote: "Nor should the reader ascribe to ignorance or mendacity the fact that the names we give are sometimes different from those known in our time, since at times we have paid homage to the past, while at other times we have had to fall in with spoken usage."[38]

Exotic Peoples and Animals

Outside Europe and the regions immediately surrounding the Mediterranean, place-names are fewer. Instead we find names of peoples or tribes, information on their appearance and customs, exotic animals, and scientific observations on climate and geography. Far northeastern Asia is "more horrible than one can imagine. The cold is intolerable with a constant blasting wind from the mountains."[39] Although strange creatures are scattered in many places on the map, they are more seriously concentrated on the southern rim of Africa, where they are lined up in an orderly fashion. These monsters were no invention of the Middle Ages but had been handed down from classical times through the respectable *Natural History* of Pliny (first century CE) and the less respectable but hugely entertaining work of Solinus (third century CE). Their pedigree extends at least as far back as Herodotus, and he professed to have taken them from still older sources. The range of abnormalities of the human form includes absence of features, such as no ears, no head, no mouth, no nose, or an excess, such as the organs of both sexes, multiple eyes, extremely long ears, and an upper lip that can be extended over the head.[40] Strange customs were also noted, such as a practice of the Philli, who test the legitimacy of their children by putting them in a pit of poisonous snakes. Bastards die at once, while legitimate children cannot be harmed.[41] Elsewhere on the map are cannibals and people who are excessively warlike, as well as those who are unusually peaceful and law-abiding. The Hyperboreans, for example, live a blessed life without discord or disease in the northern mists, and when they feel they have lived long enough, they cast themselves into the sea from a high rock. The sea, our mapmaker notes, borrowing from Solinus, they consider the best of tombs.[42]

Much of the liveliness of the map comes from the illustrations of the monsters, both human and animal. Nearly all the animals are found in the medieval bestiaries, which were descended from *Physiologus,* a Greek work of the third century CE. Christians had embellished this book of wonders with moral lessons. The pelican, for example, fed its young with its own blood and was a symbol of the sacrifice of Christ. The map presents the picture, along with the text, "I am pelican. For my chicks I rend my breast," but does not spell out the Christian allegory, which a contemporary reader was likely to understand.[43] The humans also provided moral lessons, either from their horridness or their virtue. As for the people with physical abnormalities, their existence made clear the variety and wonder of God's creation. In fact, they were often referred to not as monsters but as marvels or "mirabilia." Some, such as the Cynocephali or dog-heads, were capable of conversion, even sainthood.[44]

The original models for the pictures on the map are difficult to date. Similar illustrations appear on other medieval manuscripts, especially bestiaries. Works such as *The Marvels of the East,* based on a putative letter from Alexander to his teacher Aristotle, assemble pictures and brief descriptions of the various monstrous races of humanity to be found in the Orient.[45] Similar images can be found in cathedral sculpture. On the doors of the west façade of Modena Cathedral are carved a manticore, a cannibal, and a hermaphrodite, all represented on the Hereford map as well. Naomi Kline cites the use of model books, handbooks for artists which showed them how to draw animals, especially unfamiliar ones. It is possible that the originals for some of these depictions reach back to before medieval times.[46]

Legends and Labels

The 1,081 texts on the map are mostly brief and to the point. Most are merely names, but some places are given an additional phrase, such as "Enos, civitas antiquissima" or "Celdara (Kildare), the city of St. Bridget." Of the forty-four legends that consist of more than a dozen words, about half give dimensions and distances or are geographical descriptions. An example is the tag attached to the island of Taprobana, modern-day Sri Lanka: "Taphana, an island of India, lying toward the southeast where the Indian Ocean begins, has two summers and two winters every year, and the flowers bloom twice. Though the farther part of it is full of elephants and dragons, it has ten cities."[47] This description comes from Isidore's *Etymologies,* although a similar list of characteristics was common knowledge and could be found in any medieval geographical book. On the map a drawing of two dragons reinforces the textual account.

Often the inscriptions merely repeat what we see. The fauns, for example, are men that are half horse, as the drawing shows. Sometimes they clarify the picture: the manticores have three sets of teeth, gleaming eyes the color of blood, and a hissing voice.[48] Sometimes they add details which cannot be shown. We can see that the Anthropophagi have dismembered a human being and are eating body parts, but the legend goes on to tell us that at the time of the Antichrist they will come forth and inflict persecution on the entire world.[49] Exotic animals, such as the rhinoceros, the *monocerotus* (unicorn), the *bonnacon,* the tiger, and the *eale,* require fuller descriptions. We read, for example, that the monocerotus, can be captured by a virgin.

Sometimes the mapmaker is inspired to give information about the historical or religious significance of a place, such as Babilonia, where he tells us about its foundation and the great extent of its walls. Nearby a large image of the tower of Babel is labeled but not explained. Occasionally, there is no picture but a text alone. The story of the seven sleepers is written near the Baltic Sea: "It is not known how long they have been lying there, but as far as one can tell from their dress, they are believed to have been Romans."[50]

Sources for the Map

The Hereford Cathedral mappamundi drew from a number of ancient and contemporary sources. When weighing the evidence of experience against that of authority, authority usually won. The classical world had handed down to its successors a miscellaneous collection of geographical material: lists of names, provincial divisions, characteristics of climate and peoples, vivid anecdotes.[51] Much of this information was pulled together by Pliny and the juicy bits extracted by Solinus, who is cited a dozen times on the map, the most frequently named source. It is probable that the mapmaker knew these authors not from their own work but from the numerous geographical digests that circulated in the Middle Ages. Some of these were no more than lists of names, but others were more detailed.[52]

The Bible was of course the ultimate authority but was rather vague about geography, particularly outside the holy land. Such expressions as "the four corners of the earth" (Rev. 7:1) or "the waters above and under the earth" (Gen. 1:6–7) furrowed brows for centuries. Another convention of medieval geography, from Genesis 2:10–14, was that four rivers originating in the Garden of

Eden watered the entire world. Two of these rivers were the Tigris and Euphrates, which were well enough known, but the other two, the Phison and Gyon, were more difficult to identify. These were usually translated as the Ganges and the Nile.[53] How these rivers got from Eden to their appropriate locations was a problem, particularly for the Nile, which appeared to come from the south, not the east. One suggestion was that the rivers went underground and reappeared, and this configuration is sometimes drawn on mappaemundi. Thus the rivers maintained a vital connection between the off-limits earthly paradise and the human world.[54] When Gian Lorenzo Bernini designed his Four Rivers Fountain in Rome in the sixteenth century, he modernized the concept, replacing the Euphrates and Tigris with the Rio de la Plata from South America and the Danube from Europe. The Ganges and the Nile remained.

Among the early Church fathers, Saint Jerome (c. 340–420) was the most inclined to geography. In the late fourth century he reworked Eusebius's *Onomasticon*, a compendium of place-names used in the Bible, and entitled it *Liber de Situ et Nominibus Locorum Hebraicorum.* Jerome lived in the holy land during the last fifty years of his life and was literate in Hebrew as well as Greek and Latin. Thus he was well positioned to compile a geographical dictionary of the scriptures. Though there is a tantalizing reference to a map in the introduction to the *Liber,* the only maps to appear in connection with it are from the twelfth century and present numerous problems of correspondence.[55]

Another foundation of medieval knowledge was Paulus Orosius, who wrote a history *Seven Books Against the Pagans,* in the early fifth century.[56] An associate of St. Augustine, he was commissioned to write a book showing that the Christians were by no means responsible for the catastrophes of world history (as had been charged) but that many more disasters had taken place under pagan rule. He opened his book with an overview of the physical and political world, which was often copied separately. He is not cited directly on the Hereford map but is given credit for the whole work in the lower right-hand corner in an inscription which reads, "Descripcio orosii de ornesta mundi-sicut interius ostenditur" (Orosius' account of the *Ornesta* of the world, as shown within). "Ornesta," usually "Ormista," is the collapsed version of "Orosii mundi istoria" (Orosius's world history).[57] His theme, "the manifestation of divine providence in the vicissitudes of human history, eventually leading to the triumph of the true faith," was certainly compatible with the visual presentation on the Hereford mappamundi.[58]

Making use of Jerome and Orosius, Isidore of Seville composed his *Etymologies* or *Origins,* a universal encyclopedia that included several chapters on geography. Isidore, Bishop of Seville from 600–636, had access to a classical library and the diligence to cull from it the basics of geographical literacy, or what passed for such in the Middle Ages. His major accomplishment was to integrate classical and Biblical geography and to make geography a tool for understanding the Bible.[59] His work was a standard reference work for medieval libraries, and hundreds of manuscripts survive, as well as numerous copies of different sections of it. He is cited by name only once on the Hereford mappamundi, as the source of the story about the unicorn being captured by a virgin, but his influence appears behind many of the names and much of the organization of the map.

Writers of medieval geography books drew heavily on Orosius and Isidore, distilling their already brief descriptions into still more abbreviated works. A recent discovery of a most interesting text suggests that mapmakers might have worked from a specialized geographical digest made specifically to guide the mapmaking hand. The *Expositio Mappe Mundi,* unearthed by the French scholar Patrick Gautier Dalché in two German manuscripts of the fifteenth century, originated in the late twelfth century in Yorkshire.[60] References to "above," "near to," "across from" make it clear the writer is referring to a visual representation of the world, but it is unclear whether he is describing a map already made or giving directions for the creation of a map. There is an extremely high correlation between the legends on the Hereford mappamundi and those in the *Expositio:* 437 (out of 1091) are almost identical. This correspondence is even more impressive when one considers that the *Expositio* does not cover the entire world, omitting much of Asia, southern Africa, and northern Europe, either because part of the text was lost or because it was never finished.

The *Expositio* also describes illustrations very similar to those on the Hereford mappamundi, though the map contains many more than are described. For example, one illustration is described thus: "Postea pinguntur duo rustici hominis membra vorantes et intitulantur Antropafagi" (after this are painted two savage men devouring human limbs and labeled Anthropophagi). This exact scene can be found in northwestern Asia on the Hereford map.[61]

The medieval deference to classical authorities was in part justifiable humility. The conquests of Alexander, for example, were believed to have extended far beyond the world generally known in the thirteenth century. In Roger Bacon's *Opus Majus* (c. 1268), he comments that modern knowledge of Africa is inferior to that of the ancients, and he turns to Sallust's *Jugurthine War,* a history of the first century BCE, for his information. However, this situation was changing. While Bacon was writing, he had the opportunity to interview Friar William Rubruck, who had visited the court of the Tartars in northeast Asia in 1253 and had written a book about it. Bacon uses Rubruck's testimony to correct the view that the Caspian Sea was a gulf of the northern ocean.[62] Having traveled along the northern and western coasts, the friar reported that it was landlocked and that many lands lay to the north between the sea and the northern ocean. Bacon also consulted the work of Giovanni di Pian di Carpini (1246) on his travels in Asia. Medieval mapmakers often incorporated local knowledge, even on the most rigorously theological maps. The Ebstorf map features the small town of Ebstorf and the graves of the local martyrs, while the Hereford map has an excellent depiction of the area around Lincoln, where we assume it was made. Hereford and the River Wye were added later, perhaps after the map arrived at its final destination.

Roger Bacon is often cited as a man ahead of his time, in that he proposed the construction of a map using coordinates of longitude and latitude. If he actually made a model, it does not survive. He was also interested in eyewitness accounts. Even among antique authorities, he was careful to privilege those who saw with their own eyes. In other ways, however, Bacon was a clear contemporary of the Hereford map. Gaza, he tells us, is a populous city, according to information derived from St. Jerome, who lived in the fourth century. Even more, he sees the purpose of geographical knowledge as primarily spiritual, to enrich one's study of the Bible, and it is not surprising that he devotes a disproportionate amount of space to the holy land, as does the Hereford map. The factual knowledge of places led directly to interpretation in moral and allegorical terms. For example, he says that the river Jordan represents the world, for it flows into the Dead Sea, which represents Hell. The city of Jerusalem is interpreted morally as the holy soul that possesses peace of heart. Allegorically it is the Church Militant, while anagogically it is the Church Triumphant. These interpretations must be based on sound geographical knowledge but transcend physical reality in the elevation of a human mind to the highest spiritual levels. His geographical text makes explicit what the Hereford Cathedral map can only suggest by its form—that the world and all its wonders are set in a theological context. Place indeed has meaning.[63]

The Map in the Cathedral

An inscription below the Hereford mappamundi asks all who "have or hear or read or see" it to pray for Richard of Haldingham or Lafford, its maker or donor. How did people "read" the map? Today in its somewhat faded condition, many of the inscriptions are nearly impossible to make out through the glass in the necessarily dim conditions of the room in which it is mounted.[64] Even when it was new, its very size would have made it difficult to see all of it at once. In addition, its language (Latin) and some of the shorthand references and garbled or misspelled words would have been mysterious, even to a person living in the Middle Ages. Most tourists to Hereford today look at it reverently for a few minutes and pass on—would medieval people have behaved differently? The inscription suggests that the map might have been the subject of informative lectures or sermons, at which nonreaders might hear it expounded upon. In this way, the morals of beasts, the wonder of foreign peoples, and the great size and variety of the world would have been more fully appreciated.

The Hereford map, like most other mappaemundi, was associated with a religious institution, in this case a cathedral. Even before it came to Hereford, it was at the cathedral in Lincoln, where it or possibly its original model was made. At one time it was summarily assumed that the map had served as an altarpiece, partly because it was mounted in a wooden frame with flanking doors, similar to other altarpieces. Marcia Kupfer has convincingly scouted this idea, as the content of the Hereford map, the world in all its mundane variety, was too secular to be placed over the altar; when mappaemundi were painted on church walls, they were placed in the sections of the church frequented by the laity.[65] Recent work suggests that its original placement may have been near the tomb of the saintly Thomas Cantilupe, a former bishop of Hereford, who died in 1282 and was buried in the north transept in 1287 to the accompaniment of a flurry of miracles. The mappamundi may have served as an additional attraction for pilgrims.[66] The first historical references to the Hereford map are from the seventeenth century and have the map displayed in the former library, later the Lady Chapel. This location suggests that the map might have been used for study and teaching, to illustrate academic lectures or the narratives of unillustrated historical works, much as a globe or wall map functions in a modern library. The connection of mappaemundi with religious establishments is in part due to the association of scholarship with the clergy, the literate class of medieval society. Cathedrals and monasteries were centers of learning as well as worship, and so it is not surprising that maps were produced and housed in these places.[67] We also know of mappaemundi, now lost, which were on display in secular buildings. One of these, commissioned by Henry III and painted on a wall in the Palace of Westminster around 1230, perished in a fire in the mid-thirteenth century.

Uses of the Map

The mappamundi was an encyclopedia, and, like the encyclopedias on CD-ROMs, the form necessitated abridgement. The advantage of the form was that one could see the world all at once, the medieval equivalent of a photograph of the earth from space, or a "God's eye view." The map clarified the structure and relationship of geographical places. It would not have been of much use to a practical traveler, who would need more detailed information, but it provided an excellent overview of one's place in the world and in relation to world events. The Ebstorf map contains an inscription noting that the map was "of no small use to those travelers who read it, by pointing out the direction they should follow and allowing them, in a manner deserving the most gratitude, to choose at will how they should contemplate these monuments and travel these routes."[68] The Ebstorf map was certainly much too large to take on any journey, but we can imagine devout viewers looking longingly at the mappamundi's spiritual places of the holy land, imagining or even planning a voyage, or perhaps thinking of loved ones, merchants, pilgrims, or crusaders who were traveling in these faraway places. It is striking that the *Expositio Mappe Mundi*, the text that shows so much correspondence to the Hereford mappamundi, is bound with texts of clear use to travelers: *Liber Nautarum* on ships and weather and *De Viis Maris* describing navigation routes from northern England to the holy land and the west coast of India.[69]

A Philosophical Picture

On a less practical, more philosophical level, the map presents a unified view of the world in terms of time and space. We see the beginning in the Garden of Eden, which is also at the beginning or top of the map. As human history moves down the map from east to west, it reaches the Crucifixion in the center, the turning point. The walls of Jerusalem, below this scene, are shown in a flattened perspective that makes the city look like a cogwheel, a pivot on which the world really would turn. On the Ebstorf map the Resurrection is shown in this central spot, emphasizing the redemption of the human race. Then both worldly power and revelation move to the west, through the Roman Empire and its conversion to Christianity to the great kingdoms of northern Europe. The building of the cathedral of Santiago de Compostela, at the very ends of the earth in western Spain, symbolized the completion of the conversion of the world, after which God would bring it all to an end. Here one returns to the Last Judgment, which crowns the whole scene.

The mappaemundi presented a richly satisfying world picture that summed up the knowledge and culture of the High Middle Ages in an encyclopedic form similar to the great summas of the thirteenth century, such as those by St. Thomas Aquinas and Vincent of Beauvais. Its message went far beyond the purely physical representation of space that we assume to be the function of a map today. Instead, the *meaning* of space was its ambitious program. We see the same sweeping conception in the astronomical diagrams of the same period. Not content merely to show the earth with the planets, sun, and moon revolving around it, they include the ascending levels of the heavens as far as God Himself, as well as correspondences of time, the intervals of musical harmony, the qualities of matter, the humors of the human body, and the stages of human life. There seems to be no limit to the layers of meaning in the medieval universe.

Looking at the mappaemundi, we are impressed by the cultural consensus it expresses, particularly when we consider the contentious phase that was to follow. In the next century, maps were to be made which raised questions blithely passed over by the Hereford mapmaker. By the time Fra Mauro made his ambitious world map in 1450, a running debate had taken over the legends of the map, questioning the entire medieval world picture, while trying to cling to the elements that had made it most satisfying.[70]

CHAPTER TWO

Marine Charts and Sailing Directions

At the same time that monumental mappaemundi were being painted or posted on palace and cathedral walls, a very different kind of map was being made in medieval Europe, a sea chart that showed the coasts of the Mediterranean and Black Seas and the nearby Atlantic Ocean in startlingly accurate outline. These maps were not oriented to the east, as was common in mappaemundi, and were sparsely decorated. There were few inland features, such as mountain ranges and provincial boundaries, and place-names were almost exclusively those of coastal towns. Mouths of rivers were drawn but not their courses upstream. The plethora of anthropological, historical, mythological, zoological, and botanical information with which the mappaemundi abounded was absent. Geographical details were purely utilitarian, such as indications of shallows or other navigational hazards. For the modern eye, used to thinking of maps as functional and accustomed to the physically accurate geographical shapes, the late medieval marine charts seem a great improvement on the hazy geographic outlines and vague spatial relationships of the mappaemundi. E. L. Stevenson described them as "the first modern, scientific maps."[1]

Pisa Chart

The oldest marine chart to survive is the so-called Pisa Chart, made probably in Genoa around 1275–1300 (fig. 2.1).[2] Another, the Cortona chart, is dated around the mid-fourteenth century but appears to be based on an earlier original.[3] Still another from this era, made by Giovanni de Carignano of Genoa, was destroyed during World War II.[4] These three charts, roughly contemporary, form a new body of cartographic material around 1300. Because of their highly developed form and content, it is surmised that they cannot be the first of these charts to have been made but developed out of an older tradition. There are a

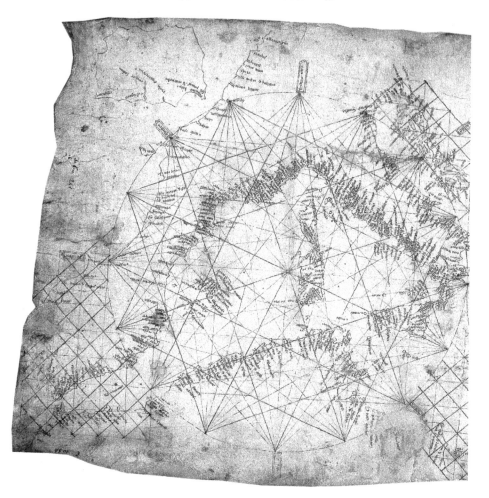

Figure 2.1. Pisa Chart, c. 1275 (Paris, Bibliothéque Nationale de France, Cartes et Plans, Rés. Ge. B 1118. 50 × 104 cm / 20″ × 40″). This is the oldest known marine chart, but its completeness and relative accuracy lead us to assume that it was not the first. The neck of the parchment is at the east. The map shows the Mediterranean, with Sardinia, the very large island left of center, and a thick-ankled Italy. The Atlantic coasts of France and Spain are lightly sketched, and the island of Britain is a poorly formed blob at the upper left. The damaged area would have shown the Black Sea. Courtesy of the Bibliothèque Nationale de France.

greater number of later surviving marine charts, with thirty charts and atlases from the fourteenth century and 150 from the fifteenth.[5]

The most striking feature of the early sea charts is their recognizable (to us) geographical forms, especially in the Mediterranean and Black Seas. The forms are somewhat simplified; for example, the spaces between headlands are repre-

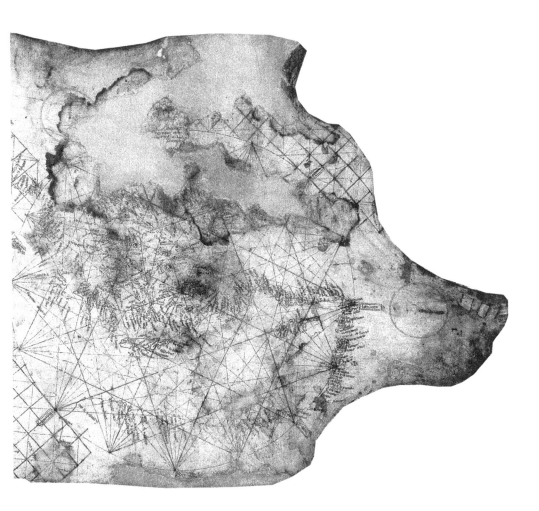

sented as simple arcs, a characteristic that becomes even more marked on later charts, where the coasts have definite, scalloped edges. Also interesting is that the Mediterranean, the length of which was overestimated by Ptolemy (61°), is shown very close to its correct longitudinal extent (41°). The Pisa Chart is not perfect—Italy is too thick in the ankle and the northern regions are sketched in rather roughly—but these areas were destined to be improved by later chart makers. Some of the standard features of charts appear in this earliest example, such as, the place-names (of which there are 927) are written on the land so as not to obscure the outline of the coasts. Usually reproduced with north on top for the comfort of modern viewers, these maps have no real top. Instead, the place-names are written so that one reads them in sequence while rotating the map or moving around it. Another revolutionary feature of the Pisa Chart is a mile scale, a line indicating five hundred miles, subdivided into smaller units. There is considerable debate about the actual length of the "portolan mile," as a great variety of distance units were used in Europe and the Mediterranean at this time. Modern scholars now generally agree that it equaled 1.25 kilometers, but there are inconsistencies and some doubts remain.[6]

Overlying the geographical forms of the Pisa chart are two large, tangential circles or wind roses, with one center just south of the island of Rhodes and the other on the west coast of Sardinia. Sixteen lines radiate from the center to sixteen points marked on the circumference, from which radiate eight lines or courses, producing a spiderweb effect. Eight of the sixteen points on the perimeter are labeled with the names of the eight principal winds used by navigators. The other points represent the half-winds. The radiating lines, called rhumb lines, were for the purposes of plotting a course at sea, but they may have also been used for construction of the chart. Tests with manuscript charts in the British Library and marine charts in Vienna and Lyons indicate that the rhumb line network was usually drawn first, before the coast was sketched in.[7] Outside the great circles, on the edges of the chart, are several rectangular grid structures. It is not clear what these were for, but perhaps they were a copying device to insure accuracy. Not until the sixteenth century would latitude and longitude scales and lines appear on European marine charts.[8] Sailors continued to navigate along rhumb lines until the introduction of the Mercator projection with its horizontal and vertical lines.[9]

The Pisa Chart is called so because it was for a long time in the collection of a family from Pisa. Usually, it is considered to be of Genoese origin because of evidence of an early mapmaking industry there and because one of the earliest references to the use of a map at sea (1270) takes place on board a Genoese ship. The dating of the chart to after 1244 is based on the establishment of the southern French port of Aigues-Mortes by Louis IX on that date.[10] It is generally dated before 1290 because the city of Acre on the eastern Mediterranean coast is marked with a cross. Acre fell to the Egyptian army in 1291, thereby passing permanently out of Christian hands, but dating medieval maps by a political event is doubtful at best. Later on, the island of Rhodes continued to be marked on maps with a Hospitallers' cross for a century after it had fallen to Turkish forces in 1523. A study of the flags bedecking numerous later sea charts have also been found to be a disappointing guide to dating.[11] Tony Campbell found that Christian conquests were quickly recognized and recorded, while cities conquered by the Muslims continued to be marked with a Christian flag, perhaps a symbol of hope rather than political reality.

The characteristics sea charts retained throughout their medieval career included accurate geographical forms, the wind rose with its radiating lines in multiples of eight (usually 32) (fig. 2.2) an established mile scale, and a series of common practices, such as the color coding of different features and the writing of the names on the land in a rotating direction. The advent of the Pisa Chart raises more questions than it answers. For example, the rhumb lines of the wind rose were apparently designed for use with the magnetic compass, yet evidence for the routine use of the compass at sea at that time is sketchy. Even more puzzling is the question of how the maps were constructed. Who surveyed all these coasts with such accuracy and what tools were used? And how was all this material compiled into a single chart? B. R. Motzo thinks that several regional charts were cobbled together, but this idea still requires that they be reduced to a common scale, not a simple task in an era of wildly unstandardized units.[12] One piece of evidence for Motzo's suggestion is that the Black Sea on early charts (this area is damaged on the Pisa Chart) is drawn to a different scale than the Mediterranean and continues to be drawn so for at least a century. It is too large in relation to the Mediterranean, which suggests that two charts of different scales have been put together. In 1409 Albertin de Virga drew a chart with a separate mile scale for the Black Sea, indicating his awareness of the problem.[13] Another indication that several sections were tacked together is that Alexandria and the Straits of Gibraltar are consistently drawn as though they were on the same latitude when, in fact, they are 5° apart. This gives the entire sea a slightly skewed appearance to our eyes.

The Portolan

The same data that appears on the early charts also shows up in contemporary portolans, or books of sailing directions, ancestors of the pilot books still used today.[14] These were written works giving the distance between different points on the coast and directions to follow from one place to another. They also included navigational information such as prevailing winds, fresh water supplies, hazards, and landmarks. The term *portolan* is used here for works of this type, while *marine chart* or *sea chart* is for the maps. These terms were used very loosely in the late Middle Ages and modern scholars have continued the practice. The Italian term for the written portolan was *compasso*, which creates confusion for English speakers because the magnetic needle also made its appearance at this time.[15] To compound the problem, *compasso* could mean sea chart and or dividers, the drawing aid, as well.

The oldest portolan extant is the *Compasso de Navegare*, from the mid-thirteenth century. The oldest surviving manuscript, which is the subject of a criti-

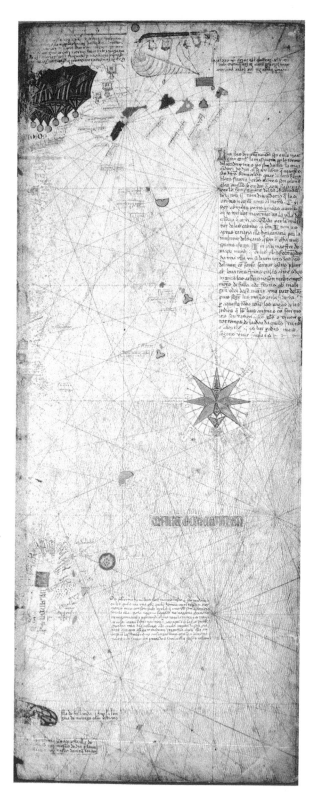

Figure 2.2. Compass wind rose (from the Catalan Atlas, c. 1375. Paris, Bibliothèque Nationale, MS Esp. 30, Panel 3a). The compass rose was a decorative feature of the later sea charts. They show the winds in multiples of eight, and the rhumb lines that stretch from their centers are lines of navigation. Most charts have multiple compass roses in order to allow pilots to follow a variety of routes. The appearance of a wind rose on the world map of the Catalan Atlas shows its derivation from the sea chart. Courtesy of the Bibliothèque Nationale de France.

cal edition by B. R. Motzo, is dated 1296 but is plainly a copy.[16] The *Compasso* takes us on a tour of Mediterranean and adjacent Atlantic and Black Sea coastlines, beginning at Cape St. Vincent in modern Portugal and proceeding in a clockwise direction, giving names of ports, the distances between them, and the direction in which one must sail. It includes other details of significance to navigators: warnings of shoals and reefs, descriptions of sea currents, and landing conditions for some harbors.[17] It places a special emphasis on landmarks visible from the sea, such as towers, churches, and mountains.

Sample entries read: "Licia [Laodicea] has a harbor with a chain, and the entrance is from the northwest. At the north entrance of the harbor is a tower. This tower, called Gloriata, is a good harbor, [entered] from the north. From Licia to Valenia, 15 miles going south." "Tyre is a harbor with several reefs to the northwest, so everyone ought to be directed out toward the south."[18] It is a document made to be used by sailors and, as such, is closely linked with the marine charts that appeared about the same time. Indeed, Motzo believes that the same hand produced the famous Pisa Chart and this portolan, and he attributes this feat ("one of the greatest monuments of human ingenuity") to the thirteenth-century mathematician Leonardo of Pisa or one of his followers.[19] Motzo's theory is based on the sophisticated trigonometry required in constructing the marine chart, but he has no other evidence to support his hypothesis.[20] As far as the Pisa Chart being the product of the *Compasso*'s author, Tony Campbell has observed that 40 percent of the names on the chart do not appear in the text. In addition, the language on the chart is a mixture of dialects, while the *Compasso* is written in a consistent form of northern Italian.[21] It thus is unlikely that one person was the author of both.

Where did the data compiled in the *Compasso* come from? Packed with information about distances between hundreds of landmarks and specific navigational hazards, it must have emerged from the collective experience of sea captains and pilots voyaging in the Mediterranean and adjacent seas. Motzo proposes that the author of the *Compasso* put together a number of "partial portolans," which covered local voyages. An example is the manuscript giving instructions for sailing from Acre to Venice.[22] Gautier Dalché dates this fragment in the mid-thirteenth century and thinks it was a memorandum copied for personal use by a merchant. Another document, a scroll dated 1300, contains a list of distances and directions for major geographical points from Venice to Alexandria.[23] A recent study of fifteenth-century navigational memorandum books shows other examples of material of this type.[24]

A peculiar feature in the *Compasso* are the traverses, or *pilleghi*, lengthy lines from one coastal feature to another, such as a point on the southern coast of France to a point in north Africa.[25] These might be interpreted as long-distance sailing routes, except they often ignore obstacles in the way, such as large islands.[26] In pilot books they appear to refer to the contrast between coasting routes and those across the open sea, more or less. The traverses could also be a way of orienting the seagoer within the larger context: for example, when in Alexandria, Venice is so many miles to the northwest, Acre so many miles to the east-northeast. It has also been suggested that the traverses are tools for the construction of a map, enabling the mapmaker to lay out a framework for his image. Another possibility is that they are taken from a map: in other words, the map came first and the portolan after. The question of whether the portolan or the map came first has yet to be resolved. Some years ago Jonathan Lanman attempted to construct, with moderate success, a chart using the data in the portolan.[27] It is clear, at any rate, that the authors of each, if not the same person, are using the same or similar data. In sum, the characteristics of the portolans are the crucial dimensions of distance and direction, the latter being given in terms of the medieval wind rose, using vernacular names for the winds. Portolans also included the traverses, relatively lengthy open sea routes, with distances and directions given. In addition, they presented useful facts about navigational hazards, landmarks, and havens, making their practical quality clear.

Origins

Ever since the portolans and marine charts began to be studied, some have suggested that they are survivals of classical antiquity. Such wide-ranging knowledge and experience of travel seems more likely to be a product of imperial Greco-Roman culture than the regionally fragmented Middle Ages. Surely they came out of the extensive Greek experience at sea and the Roman skill in surveying and administration. In addition, the early charts are sketchy on the western and northern coasts of Europe but perfect in presenting the Mediterranean and Black Sea area, the sailing domain of the ancient world. Thus a Mediterranean origin seems likely. However, no sea charts survive from classical antiquity, and textual references to maps are all too vague. The one promising map candidate, the Peutinger Chart, is strictly a land map, and its peculiar spatial distortions make it an unlikely precursor to the more accurate medieval charts. As for written sailing instructions, we do have land and sea itineraries from the ancient world, although none of them mention a map. Fragments of maritime itineraries or *periploi* survive from as early as the fifth or fourth century BCE (the Periplus of Scylax of Caryanda), but they tend to be literary works, containing historical or mythological details more appropriate to a travelogue than a practical sailing tool. The first-century Periplus of the Erthryean Sea describes the route down the Red Sea to India, with information of primary interest to a merchant, such as products, prices, and social and political conditions. For example, "Beyond Tabai, after a 400-stade sail along a peninsula toward which, moreover, the current sets, comes another port of trade, Opônê [modern Hafun], and it too offers a market for the aforementioned [olives, clothing, grain, wine, tin, etc.]. Its products for the most part are: cassia, aroma, moto, better-quality slaves, the greater number of which go to Egypt; tortoise shell in great quantity and finer than any other."[28] Navigational information is occasionally included (distances, difficult approaches to a harbor, landmarks) but not consistently. Such documents were probably written for, and by, the passengers on board rather than the sailors.

Other *periploi*, such as the maritime portion of the Antonine Itinerary (third century CE), are really just lists of places, giving only distances, not vital information about direction or navigational hazards like rocks, shoals, or contrary winds. The Antonine Itinerary, prepared for a journey of the emperor, probably Caracalla (213–214 CE), was a planning document rather than something one could use to find one's way at sea.[29] The Stadiasmus, or Circumnavigation of the Great Sea, is the survivor of antiquity most like a medieval portolan. Dating perhaps to the fourth century CE, it survives in fragments in a single tenth-century manuscript. One entry reads: "From Alexandria, sailing westward to Chersonessus—there is a harbor for smaller vessels, 70 stadia. From Chersonessus to Plinthine—there is a roadstead, the place lacks a harbor, 90 stadia."[30]

It is quite possible that the ancient world produced other documents like this, which have not been preserved. Whether they were handed down to the Middle Ages is unknown, but their use would have been limited, as conditions and routes of sea travel had changed since the fall of Rome. Place-names, settlement patterns, and trade routes were altered, as well as units of measurement. Another change was that the Greeks and Romans used a twelve-wind system, which we continue to see reproduced in medieval books of cosmology and on mappaemundi such as the one at Hereford Cathedral. The wind rose of the medieval navigators divided the circle into eight winds and their subdivisions. Only the eight principal winds were named. The others were described as being between two winds (the half-winds, e.g., "between greco and levante") or a quarter between the half and principal winds (e.g., "a quarter from greco toward levante"). The fifteenth-century author of a text on navigation claimed that the modern thirty-two-wind system made possible more "subtle" navigation than the classical twelve-wind division.[31] The portolans and charts also use vernacular rather than classical names for the winds. These changes would have made it difficult to translate any ancient document of this type into a tool suitable for medieval use.[32]

The Greeks were notable for their work in scientific geography, which dated back to the Ionian philosophers of the sixth century BCE and had its finest flowering in the work of Claudius Ptolemy (second century CE). However, these works were not for the practical navigator but for scholars and scientists.[33] Later, when Ptolemy's *Geography* was reintroduced to Europe in the fifteenth century, its contradictions with the forms of the sea charts would raise questions.

If there were any written transmission of charts and pilot books from the ancient world, one place to look for it would be in the Byzantine Empire. While surviving Byzantine navigational charts are later and derived from the Italian, there is one intriguing piece of textual evidence.[34] In the *Alexiad* the princess Anna Comnena reports on the poor success of an imperial general, trying to prevent the "Kelts" (that is, the Norman crusaders under Bohemond) from landing on the Dalmatian coast in 1107. The emperor, irritated at his minion's incompetence "drew him a map of the coasts of Lombardy and Illyricum, with the harbours on either side. This he sent to him (Isaac), adding written instructions. He advised him where to moor his ships and from what place to set sail if the wind was favourable, in order to attack the Kelts at sea."[35] If we can take Anna's description at face value, this would be the earliest description of a chart and accompanying portolan.[36]

The charts and portolans seem to have been, for lack of other evidence, a medieval European creation, appearing some time before 1300, and building on a substantial experience of sea travel. Theories that they came from China or the Arab world are interesting but lack solid documentary support and seem to be derived mostly from a low estimation of medieval European intelligence. The most ambitious and promising Arab source is the geographical text and maps of al-Idrīsī (d. 1154), compiled at the court of Roger II in Palermo. The king, deploring the state of geographical knowledge, instigated a survey of the world. The material was collected and published in graphic and verbal form by al-Idrīsī. The resulting world map is largely derived from Ptolemy and contains errors not seen in the marine charts. For example, the Mediterranean occupies half the inhabited world longitudinally, while Africa extends far to the east. Much of the text repeats the same information and misinformation available to the Christian world, though of course al-Idrīsī is better informed on distant parts of the Muslim world.[37] There are some portolan-like sections of his text, but they might have been derived from European sources since, as Gautier Dalché points out, his most specific information about directions and distances between ports is for Italy and the Mediterranean islands.[38] Al-Idrīsī was not translated into Latin until the seventeenth century, so it is unlikely that his work had any direct influence on chart makers.

Fuat Sezgin has tried to locate the source of Arabic influence even further back, at the court of the caliph al-Ma'mūn in Baghdad in the early ninth century, at which time a flourishing scientific and mathematical community was engaged in translating and studying the works of the Greeks.[39] A world map and several regional maps were made based on some kind of grid system. The oldest manuscript version of this map, Sezgin dates to the fourteenth century, although Gerald R. Tibbetts says it is a sixteenth-century copy.[40] Even Sezgin admits that the copy shows signs of distortion in the course of being copied and recopied. The grid, whether based on climates or latitudes and longitudes, is an interesting development, but since the sea charts did not use a grid but a circle of rhumb lines, there is no clear connection. Also Arabic maps tended to use abstract forms for geographical shapes, while itineraries were a series of circles linked by lines. Sezgin is more plausible when he cites Arabic influences on later European world maps, which incorporated material from the sea charts but also added information that might have come from Arabic nautical experience. (See chapter 3.)

A great deal of energy has also been spent trying to determine which European nationality created these "monuments of human ingenuity." At present, the Italian maritime cities, which produced the earliest documents and had extensive long-distance sailing experience, seem the most likely candidates, but the concept of nationality is a slippery one in the medieval period.[41] Sailors seem to have moved freely from place to place and had contact with a wide range of people of varying origins: Genoese sailors regularly served as admirals of the Portuguese fleet, a substantial Arabic community dwelt in Pisa in the twelfth century, and there were communities of Italian merchants established in cities throughout the Mediterranean. Making up the data for the maps and texts was probably the work of many anonymous navigators, including those from Arab countries, the Mediterranean islands, Greece, Italy, and France, as well as Catalans and Jews of all nationalities. How it was all put together in one coherent production remains, however, unknown.

Liber de Existencia Riveriarum

The French scholar Patrick Gautier Dalché has recently discovered a manuscript in the British Library which he dates between 1160 and 1200 and describes as a "precocious portolan."[42] The work is entitled *Liber de Existencia Riveriarum et Forma Maris Nostri Mediterranei* (Book of the position of the coasts and the form of our sea, the Mediterranean) and seems to have been made in Pisa. Its early date and its apparent link to an accompanying map (now lost) pushes the origin of chart/portolan back a full century. The *Liber* is a curious hybrid document, having many of the characteristics of a portolan (a coastal itinerary with distances and directions between objectives) but also including information about some inland sites and giving historical and religious information, particularly in the holy land, which we think of as more typical of the mappamundi. We know almost nothing about the author except that he says he has had experience at sea ("vidi et peregravi"—I have seen and traveled) and describes a journey he took from Alexandria to the Gulf of Aden.[43] The author tells us in his introduction that the work is being made at the request of a "canon of the cathedral of Pisa," who complained that his map (apparently a sea chart) was unsatisfactory, sacrificing *veritas* (truth) for *novitas* (novelty). The truth the mapmaker was sacrificing was the rich theological and historical content of the mappamundi, and the novelty was the facts-only approach of the sea chart. The author of the *Liber*, who writes in Latin, agrees to add details in his text which will satisfy the complaining canon. For example, on the problematic course of the Nile, he writes, "The river Nile has its origin from a mountain of lower Mauritania in Africa that is near the ocean. The Punic books affirm this; also we have received this information from King Juba. It is soon absorbed into the earth and proceeds underground to flow out again on the shore of the Red Sea, going around Ethiopia. It comes out through Egypt, and is divided into six branches at the sea, as we have said above, and enters the Mediterranean near Alexandria."[44] Elsewhere the author supplements the physical facts with Biblical references, such as: "From Caesarea to Mt. Carmel, 20 miles to the head of the gulf of Acre. About this mountain it says in the Song of Songs, 'Your head crowns you like Mt. Carmel,' [Song 7:5] and on this mountain Elijah met to converse with Elisha."[45]

The combination of map and text in the *Liber* shows that highly specific locational information was available in the twelfth century, if not yet in the form of the later portolan. The work is in Latin, not Italian, and the names of the winds are the classical ones, but the author seems to be trying to construct a sixteen-wind system out of eight of the classical winds. Directional information is given in the first section of the work, which deals generally with the orientation of the coasts. Where the ports themselves are listed in the second part, only mileage is supplied. The author notes that his information comes not only from his own experience but also "a nautis et gradientes illorum" (from sailors and their guide-books, or portolans).[46] These words tell us that written sources of a sort already existed among the Mediterranean sailing community.

Gautier Dalché has diligently searched the accounts of the crusaders for similar specific sailing information, especially when the author notes that it came from sailors: "ut dicunt nautae" or "ad aestimationem nautarum." In the *Chronicle* (1202) of Roger of Hoveden and his *Gesta Regis Ricardi* (1169–92) are details which plainly come from his own experience traveling to the holy land on the occasion of the Third Crusade. Roger, a clergyman from Yorkshire, had been drafted into the administration of Henry II, and after the king's death went on crusade with his successor, Richard I. Gautier Dalché suggests that in addition to his own observations, Roger is consulting documents prepared for the royal expedition in order to plan their journey. On his return to England, Roger may have composed several works based on his experience. Surviving in two fifteenth-century German manuscripts, these have now been published in a critical edition by Gautier Dalché, who argues for Roger's authorship of all three works.[47] They include the *Expositio Mappe Mundi* (discussed in chapter 1), *Liber Nautarum* (The sailor's book), and *De Viis Maris* (On the pathways to the sea). The last work is an itinerary from the city of York in northern England to the holy land, passing through the Atlantic to Gibraltar and proceeding via Marseilles, Sicily, and Alexandria. The return journey, followed in 1191 by Philip Augustus, king of France, is recounted, from Acre to Otranto. In addition to a straightforward account of the sea journeys, Roger mentions alternative routes not actually traveled, as well as inland features not visible from the sea. In Marseilles, for example, he describes the wealth of holy relics found there, and in Sicily he gives details of the life of the monks at the Greek monastery of San Salvatore at Messina. The *Liber Nautarum* is based on the thirteenth and nineteenth books of Isidore's *Etymologies*, supplemented with the scholarly reading and practical experience of the twelfth-century author. For example, he supplements Isidore's description of the antique twelve-wind rota with the names of the eight winds used by southern Italian and Greek sailors. These three works together, argues Gautier Dalché, show a sophisticated geographical consciousness with no real conflict between the mappamundi (in *Expositio*) and the space consciousness of the portolans and sea charts.

Roger is not an isolated case—Gautier Dalché lists a number of crusaders' accounts from the late twelfth and early thirteenth centuries which contain similar information.[48] The presence of Northerners in the Mediterranean in the twelfth century may have had an impact on the development of a unified set of sailing directions and maps. Northern Europeans, most notably the Vikings, had their own history of navigational exploits.[49] In their elegantly designed long ships the Scandinavian adventurers crossed the north Atlantic, sailed down the rivers of Europe, and by the tenth century were established in Kiev, in southern

Russia, which they reached via the Baltic sea and a series of river routes, linked by portages. The best evidence of their navigational methods is in *The King's Mirror* (1250), in which a father gives advice to his son, who wants to become a seagoing merchant.[50] The narrator emphasizes observation of the sun and the polestar, as well as the prediction of the tides, a more significant factor in northern waters than in the almost tideless Mediterranean. Eight winds are described in the *Mirror*, which seem to have been derived independently of the classical tradition. From the sagas we note the importance of landmarks, the behavior of birds and sea mammals, and the formation of clouds. No source indicates use of the compass—in fact, there are numerous references to ships becoming lost in the fog or after a string of cloudy days waiting for the sun to appear so that they could reorient themselves. A recent discovery of a "solar compass," a sort of sundial, has largely been discredited.[51] There is no indication that the Vikings drew any maps—the first maps of Scandinavian origins date from the fifteenth century—and the oldest Scandinavian collection of written sailing directions is from the sixteenth century.

Mediterranean sailors learned their routes as well as their skills through long apprenticeship, but Englishmen, Scandinavians, and Flemings voyaging through the straits of Gibraltar in the twelfth century were in a world whose landmarks, currents, and winds were unfamiliar. They must have been eager for information of the type the portolans would have supplied. The participants in the First Crusade (1098) had mostly gone by land, struggling across the Balkan mountains and the deserts of central Turkey, but later crusaders frequently traveled by sea. The taking of Lisbon from the Moors in 1147 was aided by a collection of Northerners (Englishmen, Flemings, and "men of Cologne") en route to the holy land, although the besieged Muslims were deeply skeptical of their religious motives and accused them of being pirates.[52] Voyagers headed for the Third Crusade also came by water, sailing from England around the western coast of Spain and into the Mediterranean via Gibraltar, meeting the royal party, including Roger of Hovedon, at Marseilles. Others came on less savory expeditions, though it is not always easy to distinguish between pilgrimage, trade, and piracy, which were occasionally combined.[53] In *La Sfera*, a didactic poem written about 1400, Goro Dati described the use of the sea chart "by merchants and pirates who travel by sea, one for gain and the other for robbery."[54]

From the crusader accounts, one can imagine travelers from the traditional world of literacy and clerical scholarship, as well as those from court and castle, thrust into an environment completely new to them—a world dominated by sailors with technical expertise and practical experience. Gautier Dalché notes, "The crusaders discovered a world which doubtless was fascinating to them and which excited their curiosity, completely overturning their previously acquired ideas."[55] The cross-fertilization of ideas, exemplified in the mapmaker's conversation with the canon of Pisa, may have led to a systematic presentation of information in the form of the written portolan and the graphic representation of the map.

Navigation in the Middle Ages

Oar power was the primary means of locomotion in the Mediterranean Sea in the early Middle Ages. An oared galley carried a single fixed sail, whose usefulness was limited to sailing before the wind. In the thirteenth and fourteenth centuries, sails became more important as they changed their form (from square to triangular) and their number (from one to three), enabling ships to follow more complex courses under sail power and eventually to make headway in the Atlantic Ocean. Oarsmen were still important to Mediterranean navigation and, in Venice at least, were far from being "galley slaves." They were paid fairly well and were allowed to carry a certain amount of merchandise to trade on their own account. Oarsmen might rise through the ranks to reach as high a position as admiral.[56]

We know little about navigation techniques before the appearance of the first portolan in the mid-thirteenth century. Observations of the sun's position and of the stars went back to Odysseus's day, and these were used to find one's direction or position in the, alas very common, case of being blown off one's course. Aeolus's bag of winds was studied closely and wind names were used for directions. Winds could be identified by their characteristics. On the Hereford Cathedral map, the winds, drawn as puffing monsters, carry brief character descriptions. "Favonius [the west wind] gets its name because it fosters shoots and brings them to maturity. Also called Zephirus, it eases the hardness of winter; it produces flowers."[57] The north wind is dry and cold, the southwest wind brings thunderstorms, and so on. These descriptions are traditional, dating back at least to Isidore in the seventh century, and were doubtless elaborated by sailors. We find them in verse form and in circular diagrams in countless manuscripts.[58] Experienced sailors would have been skilled in recognizing the winds, but it is clear that wind direction could not have been refined to a single point. It is better to think of it as a segment of the circle, until use of the compass made more precise orientation possible. In the portolans directions were given in terms of winds ("per sirocco ver levante poco"—east southeast).[59] In the mid-thirteenth century Matthew Paris attempted to reconcile the eight- and twelve-wind systems. Instead of spacing the twelve winds equally around the circle, he grouped them in threes, each principal wind with its two collateral winds. This left him space to add four new winds at 45° from the cardinal winds to make a sixteen-wind circle. He labeled these new winds with vernacular English names, that is, northeast, southeast, southwest, northwest.[60] In 1400 Goro Dati attempted to rename the classical winds, making Chorus into Maistro, Aquilone into Tramontana, and so on.[61]

Estimates of distance were described in terms of days of sailing under the best conditions, that is, a twenty-four-hour period of sailing with a following wind, which worked out to about one hundred miles. As for other methods of measuring distance, the first written description of the use of the log and hourglass appears in a sixteenth-century English manual, but the practice was clearly older.[62] A floating object (a log or a piece of debris) was dropped into the sea at the bow of the ship, and the time it took to travel a fixed distance marked out on the ship's rail was calculated by use of a rhythmic chant. It was an approximate system that continued to be used well into the nineteenth century, the chant having been replaced first by the hourglass and later by the clock. The resulting mileage between port cities, even those quite far apart, was given in the portolans with fair accuracy.

Another tool of the medieval sailor was the lead and line or sounding lead. This was used both to measure depth and to bring up a sample of the sea bottom. The lead was hollowed out and filled with tallow so that material from the sea floor would adhere to it, and coastal charts were marked with characteristics—mud, sand, shells, and so on. This technique was especially useful in the British Isles and Baltic Sea, where the weather was often foggy and the sea was not so deep. Other observations, such as landmarks, currents, and the presence of certain birds or marine animals enabled experienced mariners to locate themselves. And there was always the method of asking for directions. The crusader accounts describe the bewildered denizens of a storm-tossed vessel inquiring about their location from Genoese or Saracen merchant ships.[63]

These techniques sufficed, more or less, for centuries of Mediterranean sailing—which is not to say that sailors did not get lost at sea. The trials of Odysseus, as recounted by Homer, are early evidence of the difficulties of sea travel. In the Book of Acts, chapter 27, we read the story of St. Paul's hazardous journey from Caesarea to Italy, which ended in shipwreck near Malta. Contrary winds and storms, the need to make observations by sun and stars, and the use of the sounding line are all mentioned. These basic practices were passed down through the sailing community by apprenticeship and experience. The saintly King Louis IX of France, on his way home from the Crusades in 1254, ran aground near Cyprus. His knights found him prostrate in front of the altar, thinking his last hour had come.[64]

In the later Middle Ages there appeared some new techniques for finding one's way at sea. The magnetic compass is first mentioned by Alexander Neckham (1157–1217), who describes it as a standard piece of sailing equipment: "Whoever wishes to outfit a ship has a needle placed upon an arrow; for it will turn and revolve, so that the point of the needle turns toward the north, and thus sailors will know which way they ought to go, when the Lesser Bear is hidden by stormy weather, even though it never sets on account of its small circuit (around the pole)."[65] After this, references to the compass abound, describing how it works and noting its usefulness at sea. None of these early sources indicate that nautical use of the compass is a novelty. By the fourteenth century, the compass was a familiar enough tool to be used by poets as a metaphor.[66] An illustration from an early fifteenth-century manuscript of Mandeville's travels shows a sailor intently studying a compass mounted in the stern of the ship (fig. 2.3).[67]

That the compass pointed to the magnetic north pole and not the geographic pole set up a basic conflict between two forms of orientation, by the compass and by the polestar. It is possible that some of the distortion on medieval charts is due to this angle of difference, which is from eight to ten degrees in the Mediterranean. As long as navigators stuck to charts designed by the compass, the discrepancy was not a great problem, but things changed when sailors began to venture into uncharted waters. By the fifteenth century the phenomenon of magnetic variation was noted, and it was observed that the angle almost disappeared at the longitude of the Azores. Some instrument makers constructed their compass cards to suit the variation apparent in their own region, but such an adjustment would only make trouble for the long-distance sailor.[68]

The relation between the compass, the portolan, and the marine chart seems obvious enough in the structure of the wind rose and the rhumb lines that fan out from it around the map. The marine chart was constructed so that the captain could plot his course between two points, using a particular compass direction. Naturally, a problem arose when the ship was unable to travel in a straight line, forced to tack from side to side or blown off course by contrary winds. Complex calculations were needed, and mariners had at their disposal a table known as the Toleta de Marteloio, first described by Ramon Llull of Majorca in his book *Arbor Scientiae* (1286). "How do sailors know how to measure their mileage at sea?" he asks. Llull, who had made numerous voyages in his quest to convert the Muslims to Christianity, answers:

les les tennent a qui que elles veullent qui a en compaigne a eulx. Et aussi à la terre est toute commune que li uns la tient un an et li autres un aut et prent la chose de quelque part quil veult. Et aussi tous les biens sont communes comme ble, et autres choses. car non est oidee nen est aultre, aussi prent chascun ce quil li plaist sans contredit. et aussi riche est bien comme lautre. mais ilz ont une mauvaise coustume. car ilz menguent plus voulentiers char homme que nulle autre char. Et si est le pays moult habondant deble, de poissons, etc. et taignent et des autres biens la vont les marchans et mainent leurs enfans pour vendre a ceulx du pays et il les a chatent. Et sil sont gras il les mengient tantost. et sil sont maigre il les sont engraisser. et dient que cest la meillour char et la plus douce du mon te. En celle terre ne en plusieurs autres partela on ne voit poit lestoille transmontaingue.

De lestoille de mer.

Cette estoille de mer qui ne se meut vers bise. mais on y voit une autre au contraire delle vers bise ainsi sont les mariniers deçà la par celle estoille trans mabi la quelle na pot point a nous et ceste trans bise na pot point a eulx. pourquoy on puet apperceuoir que la terre et la mer sont de ronde forme. Car les parties du firmament se voient un puis quil ne prent une en un autre. Et puet on bien ce peu

---

Figure 2.3. Using the compass at sea (from a manuscript of John Mandeville's *Travels* [1403]. Paris, Bibliothèque Nationale de France, MS fr. 2810, fol. 188v). This is one of the earliest illustrations of the use of the compass at sea, though it had probably been around for at least a century before this image was painted. This colorfully illuminated manuscript also contains the *Travels* of Marco Polo. Courtesy of the Bibliothèque Nationale de France.

Mariners must consider the four principal winds, that is to say the east, west, south and north winds; likewise they must consider the four winds which fall between these, that is to say, 'grecum, exalochum, lebeig and maestre' [here he uses the Catalan names for the northeast, southeast, southwest, and northwest winds]. And they look carefully at the center of the circle in which the winds meet at angles; when a ship travels by the east wind 100 miles from the center, how many miles would it make on the southeast wind? And for 200 miles, they double the number by multiplying and then they know how many miles there are from the end of each 100 miles in an easterly direction to the corresponding point in a southeasterly direction. And besides this instrument [meaning the table] they have a map, a portolan, a compass and the polestar.[69]

We find an almost identical description of the use of the *marteloio* in Andrea Bianco's atlas of 1436 in Venice and in sailors' handbooks from the fifteenth century.[70]

The compass was routinely used at sea by 1492, the time of Columbus's well-documented voyage.[71] In 1450 Fra Mauro had remarked it odd that native sailors in the Indian Ocean navigated "without a compass because they carry an astrologer (astronomer) on board who sits up in a high place and with an astrolabe in hand gives orders how to sail."[72]

What was the role of the compass in the construction of the early sea charts? We have no written evidence as to how the charts were made, though it is plain that the mapmaker needed correct directions in order to make an accurate chart. If the compass tended to shift too much at sea, particularly in its earliest version as a "wet" or floating compass, perhaps it was used to determine the direction to nearby ports while standing on land, sighting from point to point. It would have required hundreds of observations to compile a chart.

The author of the *Liber de Existencia Riveriarum* reported that he got his information from sailors and their guidebooks. Several of these sailors' handbooks from fifteenth-century Venice have been studied, and it is interesting what their authors found important to record.[73] Judging by the uniformity of their content, Piero Falchetta concludes that each author copied part of his material from a standard source circulating about the maritime community. Nearly every handbook contains a calendar, astronomical tables, rules for practicing astrology (such as determining the propitious time for a voyage), rules of arithmetic, medical recipes, regulations of the city of Venice, customs duties at various ports, information on ship construction, and even musical compositions. All of the handbooks contain portolans for stretches of frequently traveled coasts. The sailors' handbooks studied by Annalisa Conterio include instructions for the use of the *marteloio,* but they do not describe use of the compass or maps at sea, although one such handbook in the Cornaro Atlas of 1489 accompanies a collection of maps. However, the use of the *marteloio* definitely implies a compass and chart—how else was one to determine precisely the directions involved and plot the cumulative account of one's journey?

In contrast, the "Arte de Navegar" (1464–65), a more sophisticated manuscript studied by Claudio de Polo Saibanti, while it contains much of the same material (lunar tables, the movement of the sun through the zodiac, naval regulations, a portolan), does discuss the use of chart and compass at sea. The anonymous Venetian author describes himself as an experienced navigator and says that the sea chart "which he knows and uses" is constructed according to "winds, miles, and place-names" and must be used with a compass and a pair of dividers. The work is embellished with the trappings of humanism—classical quotations and references to the work of Ptolemy, who provided an overall view of the world. As a navigator as well as a humanist, the author stresses that the daily working experience of sailors must add to this theoretical picture by the drawing and perfecting of the nautical chart.[74]

The chart maker Francesco Beccari notes on his 1403 map that his information came from the "efficacious experience and most sure report of many, i.e. masters, ship owners, skippers, and pilots." He profited from these sources to correct the Atlantic coastal distances (too short) and to relocate the island of Sardinia (too close to Africa).[75] Later in the century Fra Mauro makes frequent references on his world map to information received from "experienced sailors" and "trustworthy men who have seen with their own eyes."[76] The sources could have been written as well as oral, drawing on the notebooks compiled by sailors.

Another technical marvel was the astrolabe, probably a Hellenistic invention reintroduced to Europe from the Arab world in the tenth century. By the late eleventh century there was quite a buzz among the intellectuals in Europe; Heloise and Abelard's celebrated son was even named Astrolabe in 1118. A complex and versatile instrument, the astrolabe is a model of the heavens projected onto a circular, flat surface. On one side is overlaid a rotating pierced disk, which shows the positions of the sun and the stars as they move through the sky. On the other side is a dial (alidade), which can be used to sight the altitude of celestial objects. Thus it can be used for telling time and so was indispensable to the Arab astrologer, who cast horoscopes with it. The underlying plate had to be redrawn for successive latitudes, and some astrolabes were made with numerous plates or tympans. It was too complicated to be used by everyone (despite that Chaucer thought that "little Louis," his son, should be able to master it at the tender age of ten years), and simplified versions were constructed for use at sea.[77] The so-called mariner's astrolabe was refined to a calibrated disk and pointer, the whole heavily weighted so as to hang steady in strong winds at sea. It could be used for sighting the sun or a star and was available in the thirteenth century. The quadrant, or ninety-degree instrument, was sufficient for locating objects between the horizon and the zenith and seems to have been in use by the thirteenth century as well.[78] How much these tools were used in practical navigation and how effectively is another question. An experienced mariner like Columbus had considerable difficulty with his quadrant and finally put it away, deciding it was "broken." A modern scholar thinks he was reading off the wrong scale.[79] Other tools—the Jacob's staff, the cross staff, and the *kamal* (imported from the Indian Ocean)—worked on essentially the same principle.

The astrolabe and its variants could be used in surveying and mapmaking, at least of a local variety, and were widely employed in this way in the sixteenth century. Medieval treatises on the astrolabe describe how to use it to discover the height of a tower or the distance between two places, but there is no evidence to show what role it played in the construction of early marine charts (fig. 2.4).[80]

Once Mediterranean sailors began to leave their home seas and venture to the west and far south, the navigational techniques that had stood them in good stead were no longer enough. The Canary Islands, rediscovered in the early fourteenth century, lay over eight hundred miles of open-sea sailing southwest of Spain. Later, in the fifteenth century, Portuguese navigators, moving purposefully south along the African coast, had to find some way to determine their position once they had "lost" the polestar. They were pioneers in developing navigation based on celestial observation, though early attempts were often in error. Eventually, in the sixteenth century, sea charts equipped with a latitude scale began to be produced, but the old-style rhumb-line charts continued to be made and used through the seventeenth century. Indeed, pilots resisted the change, preferring to rely on their charts, which they found more usable and reliable at sea.[81]

For the crucial period of the development of the sea charts, the thirteenth century, we are forced to conclude that our evidence of the use of technical resources is faulty. The compass and the astrolabe were both useful tools, if imperfect, but exactly how they were used in mapmaking is still uncertain.

Development of Charts, 1300–1500

The sea chart was not a static form but was continually being updated, as befits a practical tool designed for use and not for contemplation or for show. Early in the fourteenth century, the first atlas was produced by binding a set of maps

P R I M O.   45

Difegnifi dipoi con le fefte fopra un foglio ,, un cerchio grande a
modo noftro, fcompartendolo in 360. parti, o gradi, & il fuo centro
fia G, che raprefenti il punto della pofitura doue ftette nell'operare
lo Aftrolabio, quando fi prefono le diftantie delli alberi . Da que=
fto punto G, che haremo fatto ful foglio tirifi una linea diritta, lunga
a beneplacito noftro che fia G A, et quefta diuidafi in tante parti fra
loro uguali quante furono le braccia, che fi trouaron effere fra G &
A, quali prefupponemmo che erano 60.   Prefa dipoi la diftantia
de gradi, che noi trouammo effere nello Aftrolabio infra A & B.

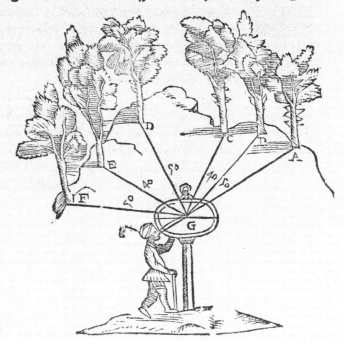

Figure 2.4. Surveying with an astrolabe (Cosimo di Bartoli, "Del modo di misurare" [Venice, 1564], p. 49). By the sixteenth century, when this surveyor's handbook was printed, astrolabes, mounted horizontally, were commonly used in surveying. We do not know how far back its usage extends. Courtesy of the Rare Books Division, Library of Congress.

together. Such a format was more portable, durable, and useable than a large, rolled, single-sheet chart. The format also made it possible to show more detail without extending a single sheet to unmanageable size.

Tony Campbell has done a thorough study of the place-names on the charts from this period and has shown that there was a continual transformation, with

new places being added and old ones dropped as political and other circumstances changed. He was able to use the accretion of names to date some previously undated charts.[82] Geographic forms were also modified, with the most
rapid development occurring in the early fourteenth century. Campbell demonstrates the increasing accuracy of the representation of the British Isles on the
series of charts by Pietro Vesconte. By that time the Genoese and Catalans were
regularly sailing to British ports, but we do not know exactly where Vesconte
got his information. However, the progressive development is clear in form and
toponymy.[83]

The charts extended their coverage as new routes became important. The
discovery of the islands of the Atlantic, beginning with the Canary Islands (the
Fortunate Islands of classical antiquity) in the early fourteenth century, expanded charts to the west. The Canaries appear for the first time on the 1339
chart of Angelino Dulcert.[84] The Azores, Madeira, and the Cape Verde Islands
became crucial way stations in later Atlantic voyages. As the Portuguese pains-
takingly explored the coast of west Africa, the charts recorded their progress,
showing new place-names and drawing the coast ever farther south. Despite the
alleged policy of secrecy, the data from these voyages were almost immediately
recorded. In Henricus Martellus Germanus's world map of 1490, we see the results of Bartholomew Dias's voyage round the tip of Africa in 1488.[85] Campbell
warns us, however, that "it must be stressed that many charts ignored new information, and it would be wrong to see the history of portolan charts purely in
terms of successive innovations and unrelenting progress" (fig. 2.5).[86]

Some sea charts continued to confine themselves to the bare outlines of the
coasts, but other chart makers, particularly those of the Catalan school, began
to give more definition to territories inland, depicting cities, mountain ranges,
and the courses of rivers. Dulcert's chart of 1339 included not only these features
but also an enthroned king in Africa, the Queen of Sheba in Arabia, and a set of
descriptive legends. He also extends the area covered to include more of Africa
to the south and Asia to the east as far as the Caspian Sea. The ornamental nature of this chart and others of this school indicate that the sea chart had gone
beyond utilitarian purposes to the decorative, and it is telling that some of these
were presentation copies for kings or popes. This format culminated in the lavishly decorated Catalan Atlas (1375), which was a gift from the king of Aragon
to the king of France, and has remained in the royal (now national) collection
ever since. (See chapter 3.)

The form of the sea chart was familiar enough to play with, as we can see
in the creations of the Italian Opicinus de Canistris (1296–1350). Taking a sea
chart, Opicinus transformed the geographical shapes into human figures. In one

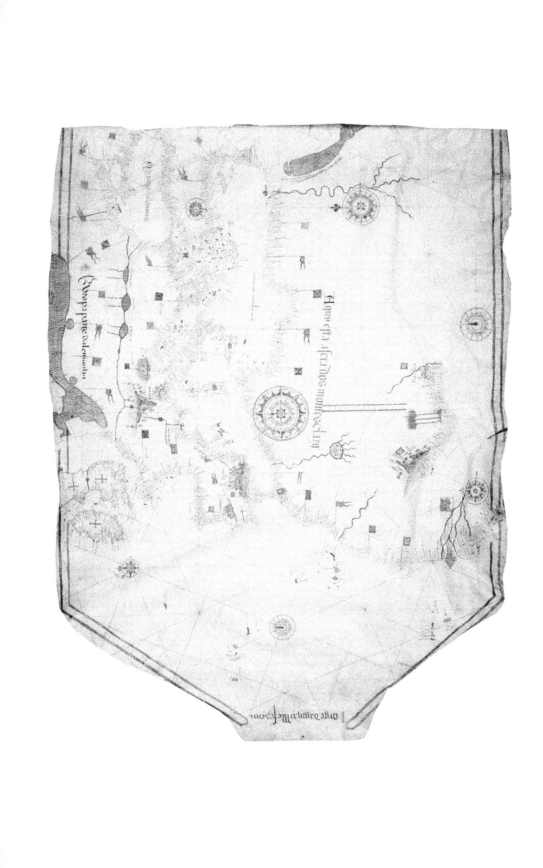

example, Europe is drawn as a crippled man, with Spain as his head, while Africa is a woman whispering in his ear. On some of his pictures, Opicinus noted allegorical and personal meanings. The female form of Africa, he says, represents literal knowledge "without spirit or virtue." In another drawing Europe and Africa are Adam and Eve, while the Mediterranean Sea is a devil, "mare diabolicum."[87]

We do not know the name of the maker of the Pisa Chart, but we do know many of his successors—forty-six names are recorded in Tony Campbell's catalogue of chart makers. The first man known to describe himself as a professional mapmaker is Pietro Vesconte of Genoa, who made his first map in 1311. He then moved to Venice where he made maps and atlases, a number of which survive. Some he signed, and one includes a picture of a mapmaker, presumably Vesconte himself, seated at a drafting table constructing a wind rose. In most cases, we know only the names of mapmakers and very little about how they worked or were paid. A recent study by Piero Falchetta surveys the identifiable mapmakers of Venice in the fourteenth and fifteenth centuries, searching Venetian records for additional information. He finds that most had extensive experience at sea and are listed in the city archives as shipowners, captains, and navigators, what he calls the "merchant-mariner class," with the famous exception of Fra Mauro, who was a monk.[88] However, it is important to recall that the co-author of Fra Mauro's great world map of 1450 was Andrea Bianco, a galley captain. Mapmaking seems to have been a sideline for these members of the merchant class, but perhaps this was not the case for all. Genoese records preserve the petition of cartographer Agostino Noli from 1438, describing himself as very poor and asking for a tax break, which he was granted.[89] Other now nameless chart makers may have worked in copy shops, where maps were not created but merely reproduced for sale. There is evidence that many surviving maps are not originals but copies, some more skillfully made than others. Fancy, made-to-order charts were beyond the budget of the average ship.

It has troubled scholars for years that most of the sea charts preserved in libraries give no evidence of use at sea. There are no holes pricked in them by

Figure 2.5. (*opposite*) Sea Chart of Jorge de Aguiar, 1492 (Yale University, Beinecke Library, MS *30cea/1492. 77 × 103 cm). This well-preserved sea chart is one of few Portuguese examples that survive. It is signed and dated by its maker on the neck end. Oriented to the south, it has a large decorative wind rose in Africa, as well as several smaller ones. In the north, the Baltic sea breaks through the frame. The depiction of Africa does not include the most recent Portuguese discoveries but shows the Cape Verde Islands, discovered in 1456. Courtesy of the Beinecke Rare Book and Manuscript Library.

the use of dividers, no lines drawn marking a course, no seawater stains. They seem to be like the First Lord of the Admiralty in *H.M.S. Pinafore*, who stuck close to his desk and never went to sea. Yet there are quite extensive reports of the use of charts at sea, beginning in the year 1270, when King Louis IX, setting out for his doomed crusade to Tunis, was in a storm at sea. After several days of being battered about, the king expressed anxiety to know exactly where he was. The Genoese sailors told him they were not sure, but they thought they were near land. Producing a mappamundi, they showed the king the harbor of Cagliari and said it was nearby. Fortunately, this proved to be the case, but theirs seems to have been a lucky guess, designed to placate the nervous king.[90] Campbell marshals a stream of evidence for maritime map use, which he calls "overwhelming," citing instances in which charts were found, used, or ordered to be used on ships.[91] In 1453 the work of Genoese chart maker Bartolomeo Pareto is described as "necessary for Genoese sailors."[92] Other authors have noted curious omissions. For example, Falchetta's study of the naval regulations of Venice, which were extensive and went so far as to condemn blasphemy on shipboard, never required the use of a map or chart.[93] He proposes that the charts, which were produced in large numbers in Venice, were not fully integrated into nautical practice before 1500. Of course, this circumstance may have been due to Venetian sailing routes, which followed well-known patterns in the Mediterranean and Black Seas.

The creation of the medieval sea chart and its corresponding text, the portolan, demonstrates an increasing rationalization of the practice of seafaring. What had been learned by practitioners, beginning at the rower's bench at an early age, had transcended the oral tradition of the trade and become documented and systematized. Eventually, the geographical information assembled here was transmitted to the world of scholarship and became incorporated into works of geography and onto the world map. The transmitters were most likely the merchant class, who possessed the literacy and numeracy necessary to keep records and carry on commercial correspondence. In the Italian city-states, education was not limited to the clergy. Frederic Lane records the will of Simone Valentini, a fifteenth-century Venetian merchant, who provided therein for the education of his children. They were to be tutored not only in reading, writing, arithmetic, and bookkeeping but were also to be acquainted with logic, philosophy, and the works of great authors. Yet he specified that they were to be merchants, not doctors or lawyers: "sed non fiant medici nec iuristi, sed solum mercatores."[94] This group of people moved between the rigidly defined classes of medieval society, not being exactly those who worked with their hands and certainly not those who fought or prayed. St. Thomas Aquinas had been very doubtful about whether trade could be considered as morally respectable work.[95] Research by Saibanti, Conterio, and Falchetta into the lives of seamen and map-makers in Venice, where to be a merchant was a source of pride, shows men rising dramatically through the ranks. Such men were in an ideal position to write down technical information from the world of practice and eventually to bring it into the realm of geographic theory.

Patrick Gautier Dalché suggests another avenue, the clerical class itself. Wrenched from their monasteries and bishops' sees by the siren song of the Crusade, these scholarly souls found themselves blinking in the bright Mediterranean sunlight on board ships headed for the holy land. This destination, romanticized and theologized into virtual unreality, lay at the end of a long, damp road, made up of risky harbors, contrary winds, and hazardous reefs. Some clerics, like Roger of Hoveden, were moved to translate their experience into writing. Gautier Dalché's discovery and study of the *Liber de Riveriarum*, by pushing the origin of chart and portolan back into the twelfth century, suggests that the crusading movement might have been the catalyst for these developments. Earlier scholars, such as John K. Wright, have puzzled over the lack of impact the Crusades had on geographical knowledge, but perhaps the right documents just weren't available.[96]

Eventually, the compilation of specific geographical material in written and visual form led to the modification of the traditional mappaemundi. The rectangular Mediterranean and the abstract land forms of Europe, Asia, and Africa no longer seemed quite so satisfying. Space traveled became a different kind of space. Beginning in the early fourteenth century, world maps incorporated a new kind of world picture based on measurement and the experience of motion through space.

CHAPTER THREE

Sea Chart and Mappamundi in the Fourteenth Century

The measured geography of the sea chart, with its practical uses, began to have an impact on world maps in the early fourteenth century. The great mappaemundi of the previous century had been sweeping statements of cosmology, theology, and history, as well as geography. How could they be combined with the mundane sea chart without losing some of their larger purposes? The first world map to incorporate the qualities of the sea chart was made through the alliance of a chart maker with a Venetian businessman who wanted to revive the Crusade.

Sanudo's Dream

Marino Sanudo was a man with a mission. Born about 1260 into a prominent Venetian family with extensive commercial interests, he had traveled all over the eastern Mediterranean world in the late thirteenth and early fourteenth centuries, living for a while in the 1280s in the city of Acre. The most important port and de facto capital of the surviving Christian enclave in the holy land, Acre housed a thriving Venetian colony, with three to four thousand permanent residents.[1] In 1291, however, this community was devastated when the city fell to Egyptian troops and was subsequently razed. Shortly thereafter, the remaining Christian possessions on the coast were taken over by the victorious Muslims, and refugees fled to Cyprus, the nearest Christian outpost.

Was this, then, the end of the Western crusading adventure? For almost two hundred years there had been a Christian kingdom in the holy land, always embattled but clinging to the dream of a physically realized spiritual homeland. The horrors, brutalities, and injustices in the establishment and maintenance of the European presence in the East did not go unremarked by the participants, but such atrocities were usually attributed to the all-too-human failings of in-dividuals and not to the dubious quality of the mission. The loss of Acre was blamed on the sins of the crusaders, and certainly greed, betrayal, and infight-ing had played prominent roles in its demise. But this unhappy event merely inspired many devout Christians to continue to dream of a kingdom in the holy land, a view expressed by Dante, among others.[2] One of the most persistent was Marino Sanudo.

Having received an excellent education in the palace of Doge Giovanni Dan-dolo, Sanudo began traveling on business on behalf of his immediate family, as well as acting as agent for a branch of the family which had established a dukedom on the island of Naxos in the wake of the Fourth Crusade. He moved in intellectual circles, too, for in 1304 he was in Palermo in the suite of Cardinal Ricardo Petroni of Siena, an advocate of the renewed Crusade. Sanudo's first lit-erary work was *Conditiones Terrae Sanctae* (1307), in which he put forward his ideas for the reconquest of the holy land. By 1320 *Conditiones* was expanded into the *Liber Secretorum Fidelium Crucis Super Terrae Sanctae Recuperatione et Conservatione* (The book of secrets for faithful crusaders on the recovery and re-tention of the holy land), an interesting mix of religious passion and hardheaded commercial sense.[3] The first step in the campaign, he asserted, was an economic blockade of the Egyptian empire, whose prosperity depended on Western trad-ers to bring spices from the East, slaves for the army, and wood for shipbuilding. The Pope had been decreeing embargos on trade with the infidel for years, but Sanudo put teeth into his proposal by planning to station a fleet of armed galleys off the Egyptian coast to waylay disobedient Christian merchants. After several years of blockade, he thought Egypt would be sufficiently weakened for a suc-cessful assault by a small, professional force of fifteen thousand foot soldiers and 300 horsemen. Once Egypt had been conquered, a more general crusade would be proclaimed, which would go on, ideally, to conquer the holy land and perhaps the Byzantine Empire as well.

Sanudo put his practical plan into an intellectual framework, including a complete history of the holy land, beginning with the division of the earth among the sons of Noah. He went on to describe the growth of Islam, the beliefs of the Muslims, and a history of the Crusades up through the fall of Acre. From this history, especially the modern phase, he drew important lessons for future crusaders, some technical (how to choose a battlefield and conduct a siege) and others moral (the importance of choosing a king who will follow the divine law). He also reviewed current events, describing the position and bellicosity of the Tartars and conditions in Cyprus. He followed with a geography of the holy land and a tour of the holy places.

Maps in the *Book of Secrets*

The more abstract sections of the book followed the first part, which was packed with specific details: lists of equipment and supplies, estimates of expenses, and, most important for our purposes, a set of maps. Four maps accompanied the first edition of the book, which was presented to Pope John XXII in Avignon in September 1321. In the introduction to the work Sanudo described these "mapas mundi" as depicting: (1) the Mediterranean Sea, (2) the land and sea, (3) the holy land, and (4) the land of Egypt. However, the manuscript, which is believed to be Pope John's presentation copy, now has six maps: (1) the Black Sea; (2) the southern half of the Balkan peninsula and the African coast; (3) the Eastern Mediterranean and Arabia; (4) Italy with part of Dalmatia, France, and North Africa; (5) Western Europe including Denmark and the British Isles; and (6) a world map.[4] The first five are sea charts from the workshop of Pietro Vesconte, and the last is a revolutionary amalgam of sea chart and mappamundi. Since these maps do not exactly fit the description of them in the prefatory letter, there is some thought that they were added later. Sanudo mentions giving the Pope two books; perhaps one was a separate atlas. The most likely candidate is an atlas signed by Pietro Vesconte and dated 1320, also in the Vatican Library. It includes five sea charts of the Mediterranean and Black Seas, a world map, a map of Palestine, and plans of Jerusalem and Acre.[5]

Over the years *Liber Secretorum* was copied numerous times at Sanudo's expense, and lavishly illustrated versions were sent to influential people throughout the West as a sort of lobbying effort.[6] The illustrations include decorated initial letters and scenes at the bottoms of the pages showing Crusaders and Saracens in battle, ships arriving, and so on. Thanks to all his energetic activity and also to the luxurious quality of the manuscripts, eleven texts survive, nine containing maps. A number of Sanudo's dedicatory letters also survive, as he was accustomed to recopying them and sending them along with later copies. The letters reflect changing political circumstances over two decades. The content of the map collection varies, with the frequent addition of city maps of Jerusalem and Acre and an unusual grid map of the holy land. In his letters, Sanudo stressed the value of the maps, urging his recipients to examine them with care. In 1332 he wrote to King Philip VI of France, "whosoever exercises the leadership of the crusade must wholeheartedly follow the directions as proposed in the *Book of Secrets*. The crusade leader should study and pay close attention to the map of the world, and pay very careful attention to the maps showing Egypt, the Mediterranean, and the holy land. If these precautions are followed, with the help of God, this venture will come to a victorious conclusion."[7] In his will, written in 1343, he is particularly anxious about the preservation of his book and his maps, specifying that they be deposited for safekeeping in a Venetian monastery until they can be sent on to the Pope in Rome.[8]

In line with Sanudo's practical sense and experience in overseas trade, the maps are derived from the tradition of nautical mapmaking described in chapter 2. In addition, he put into his text a portolan of the eastern Mediterranean coast, which is based on the same section of the *Compasso da Navegare* of 1250, translated into Latin. However, Sanudo adds details of the interior, noting distances to some inland sites and to political borders, as well as including the historical and religious importance of places.[9] The portolan follows his map of the area with great fidelity. Equipped with portolan and map, the would-be crusader would have no difficulty finding the way to go, even what sights to see. These are the first maps to survive from the Middle Ages which were designed for strategic military purposes.

The sea charts in the collection are the work of Vesconte, a Genoese resident in Venice in the early fourteenth century. His oldest surviving work, signed and dated 1313, is an atlas of five maps closely congruent with the sea charts attached to Sanudo's book.[10] Vesconte also signed the maps in his 1320 atlas, which may have been the maps originally presented to the Pope.[11] The choice to use sea charts in a book, in Latin, with some pretensions to intellectuality, shows an interesting correspondence between the practical world of the merchant-seaman and the higher world of papal and royal politics. What the relationship between Vesconte and Sanudo was is unknown; in all his vaunting of his maps, Sanudo never mentions his cartographer by name. It is one scholar's opinion that this silence reflects a cultural prejudice and the relatively low status of the mapmaker.[12]

The authorship of the other maps (the world, holy land, and city maps) has been the subject of considerable speculation. For a long time Sanudo was thought to have made them and was hailed as a pioneer of cartography, but once Vesconte was established as the maker of the sea charts, he became the leading candidate.[13] These maps share with the sea charts a dedication to physical accuracy over other qualities such as artistry or theological/historical interpretation.

The holy land map is distinguished by a grid that establishes the scale, each square equaling one league or two miles. Like the sea charts, it has its textual companion in the book, where the same places are described and the key given to its place on the grid.[14] For example, "In row twelve, square 67" is Massada, "an impregnable fortress built by Herod."[15] The map is not perfect—some sites are misplaced—but the author's intention of physical accuracy is plain. The mapmaker has taken the principle of the portolan—distance and direction—

and has applied it for the first time to an inland space. This map was to have an interesting afterlife. When "tabulae modernae" were added to printed atlases of Ptolemy's maps 150 years later, Marino Sanudo's map of the holy land was chosen to represent the biblical sites unknown to the classical world.[16]

The city map of Jerusalem draws on several sources. Earlier maps of Jerusalem tended to be schematic representations, considerably romanticized by the aura of so many sacred sites. The "Situs Hiersolymitae" series show a circular city with a few of the most important places greatly enlarged and with numerous pictures (pilgrims arriving by road, a battle between Saracens and Christians). The shape of the *Book of Secrets'* Jerusalem is irregular, not round, but otherwise it follows traditional representations. Nearly all the sites mentioned are biblical—the place where Judas betrayed Christ, the tomb of Isaiah, even the fig tree cursed by Christ. An interesting feature of this map is its emphasis on the city water supply, with ten references to water sources and storage places. The information comes from the pilgrimage account of Burchard of Mt. Sion (1280), and was of course very important to know in case the city were besieged.[17] Sanudo's map does not show any Muslim sites, although many of these had been restored after the Christians were driven out. He also shows the city completely surrounded by its wall, which had been partially destroyed a century before, in 1219.[18]

The map of Acre is more up to date, although it does show the city as it was before its fall in 1291, with its elaborately turreted fortifications, including the tower of the Templars on the coast, the last to be taken, and the Turris Maledicta, whose capture was thought to be the beginning of the end. The city is divided into walled sectors with the Venetians established near the outer harbor, flanked by the Genoese and Pisans. Also shown are churches, monasteries, houses of the military orders, the arsenal, and the palace of the patriarch or papal representative in the city. We do not know if Sanudo ever went to Jerusalem, but he lived for some time in Acre and may have contributed to the making of this map from his own experience.

Sanudo's World Map

The world map is particularly interesting for our purposes, as it is the first to attempt a fusion between the geographical forms of the medieval sea chart and the mappamundi (fig. 3.1). Oriented to the east, the map shows the three continents (Europe, Asia, Africa) surrounded by a narrow band of green Ocean, outside which there is a circle of white surrounded by a circle of red, symbolizing the spheres of air and fire which enclose the terrestrial and aqueous world.

A number of differences from the traditional mappamundi are immediately apparent. First, a network of rhumb lines covers the map, radiating from sixteen points on the rim. The names of the winds are based on the vernacular rather than the classical names, although an attempt has been made to Latinize them ("sirocus, magister"). The Mediterranean and Black Seas have taken their shapes from the sea charts. Africa curves to the south, with the Indian Ocean open to the east and dotted with islands, which seem to represent the sites of the Indian spice trade (the island of pepper, and sources of nutmeg, cassia, ginger). An indentation in the west coast of Africa suggests the Gulf of Guinea, which may be accidental, as it predates any other recorded evidence of European travelers there. The Caspian Sea is shown as landlocked, rather than as a gulf of the northern ocean, and there is an extra sea, labeled "Mare de Sara" between the Black and Caspian Seas. This is usually understood as another name for the Caspian Sea. The legends and place-names are a mix of classical and modern. Gog and Magog can be found in the far north, and the torrid and frigid zones are described as uninhabitable. However, there is no site marked for paradise, a staple of the mappaemundi.[19] The holy land is not enlarged, and so it has room for only a few places (Syria, Jerusalem). Modern place-names include Cathay, the home of the Great Khan, and many places in Europe, such as Cracovia, Estonia, Livonia, and Pomerania. On one manuscript, a reader, overcome perhaps by all this "novitas," has written "Archae Noe" in an appropriate location.[20] In Africa and Egypt the place-names are classical (Libia cirenensis, Pentapolis, Garamantia), possibly reflecting the author's reluctance to use Arabic names, which surely must have been known to him.

Surrounding the map is an anonymous geographical text of a conventional type, describing the division of the world into regions, delineating their borders, and giving the derivation of their names. The shape of the text, with Asia on top and twice the size of the other continents, Europe to the left and Africa to the right, mirrors the configuration of the map.[21] On a separate page is an account of "Minor Islands," mostly those in the Mediterranean for which there was no room on the map.[22] There is limited correspondence between the names on the map and those in the text, the map being more modern. The greatest similarity is in Africa, where, as has already been noted, many of the names are of classical origin. Conversely, there are significant differences between the Europe map and text. The map fills eastern Europe with current names, while the text populates it with tribes long since moved on and settled elsewhere: the Vandali, Rugi (conquered in 488), Eruli (Heruls, exterminated by the Lombards in 567), Turtlingi (Tervingi?), Vinuli (Vinnili, later became the Lombards), and so on. The text is not completely archaic, however. For one thing, it does not mention

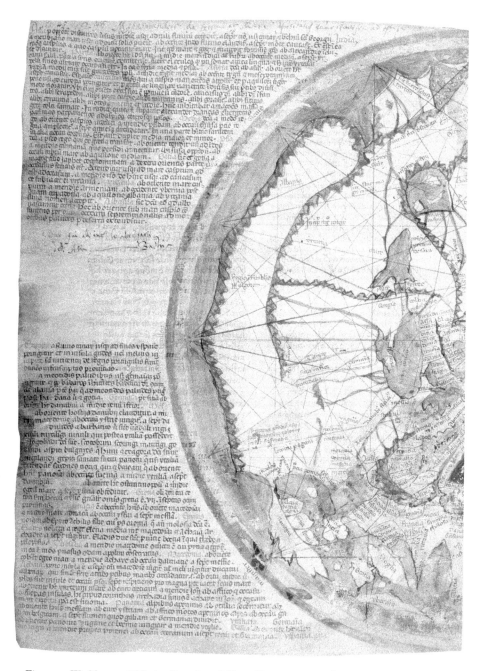

Figure 3.1. World map of Marino Sanudo and Pietro Vesconte, c. 1321 (London, British Library, Add. MS 27376, fols. 8v–9r 13″ diam.). One of a group of world maps attributed to Vesconte and made to accompany Marino Sanudo's *Liber Secretorum*, this was the first world map to employ the sea chart. Geographical forms in the Mediterranean, Black Sea, and Atlantic Ocean reflect those of the sea chart. East is at the top, but paradise is not shown. The Indian Ocean, open to the east, may reflect Arabic influence. Courtesy of the British Library.

paradise as a province of Asia. Speaking of Scythia and Hircania, it reads: "Moderns have divided and named [this area] differently, placing, where Scythia was, the kingdom of Cathay, which has the Ocean to the east; to the south, the islands of the Ocean; to the west, the kingdom of Tarsa [land of the Uighurs]; to the north, the desert of Belina."[23] The source of this text on Asia is the Armenian prince Hetoum's description of the east in his account of the Crusades, written just a few years before Sanudo's work.[24] The text and map were meant to be complementary, as there is a note at the bottom: "One must know that a mappamundi of this type is not drawn so that it shows everything, since this would be impossible."[25] The reader is further referred to the greater text, the *Book of Secrets* itself, for more information on the lands overseas. The world map is reproduced in a very similar form with almost identical nomenclature in all the surviving manuscript versions.[26] Some refinement of the Scandinavian peninsula appears on the British Library manuscript, where it has a more attenuated form and extends farther to the west.

Unlike the maps of the eastern Mediterranean and the holy land, the world map is not clearly connected with Sanudo's strategic purpose. He may have included it as a gesture toward the world of learning, as he wrote his book in Latin and included citations from classical authors. After all, his first audience was the papal court at Avignon, a center of Latinate learning. One function the world map could have, however, was to show the relatively small part of the world dominated by Roman Catholic Christianity after the defeat of the Crusader kingdom. In the *Book of Secrets*, Sanudo avers that the only Christian kingdom now in Asia is Armenia, and it is struggling for its existence. Even in Europe, Spain is partly controlled by Muslims, while eastern Europe is dominated by the schismatic Greeks. Looking at a world map, one can contemplate this sorry state of affairs.[27] Despite Sanudo's religious mission, his world map is a dramatic departure from the Christianized image of the world on the Hereford mappamundi, where Christ presides in judgment over the entire world and the apostles' missions reach out to all nations. The reality that most of the world is not Christian is the message of Sanudo's world map and a call for action.

Arab Origins?

Since the Sanudo world map is such a departure from the traditional form, scholars have wondered what its sources were. Joachim Lelewel was the first to suggest that Sanudo had seen an Arabic map, perhaps that of al-Idrīsī, which might account for the shape of Africa, the landlocked Caspian Sea, and the open Indian Ocean, but he was uncertain how Sanudo would have had an opportunity to see one of al-Idrīsī's maps. Tadeusz Lewicki has tried to carry this connection further.[28] He proposes that Sanudo in his travels in Sicily and throughout the eastern Mediterranean may have come across Arabic manuscripts and maps. Though he did not speak Arabic, it would not have been difficult for him to find a translator. Lewicki suggests that some of the names on Sanudo's map are taken from the Arabic. His examples are not very convincing, partly because of the difficulties inherent in transliteration, and partly because there are other sources from which Sanudo could have obtained these names. Most of the place-names Lewicki cites are along the Red Sea route to the Indian Ocean, such as Chus and Aden, which Sanudo had studied in connection with the spice trade. It is more likely Sanudo got his information from traders, Arabic or others, than from a twelfth century map. Al-Idrīsī's map does show an open Indian Ocean but is much less accurate in its delineation of the Mediterranean and Black Seas, and it does not include the British Isles. Also al-Idrīsī uses horizontal climate/latitude lines rather than the radiating rhumb lines of Vesconte's sea charts and world map.

In general, Arabic cartographic thought either followed the Greeks in being theoretical and scientific, or else maps emphasized itinerary rather than correct geographical forms. A newly discovered Arabic manuscript, *The Book of Curiosities of the Sciences and Marvels for the Eyes*, compiled in the late eleventh century, contains an interesting range of Arabic maps, ranging from a world map very like al-Idrīsī's to a series of local maps. It also includes maps of the Mediterranean Sea and Indian Ocean which are simplified to an oval shape sprinkled with regularly spaced circular islands. In the accompanying commentary, the anonymous author explains that these maps cannot be accurate for two reasons: one, that coastlines are continually changing, which would quickly render the maps out of date, and, two, that it would be difficult to show the locations and names of port cities without obscuring the shape of the coastline.[29]

Almost contemporary with the Sanudo/Vesconte maps is an interesting world map that appears in a manuscript of Brunetto Latini's *Li Livres dou Tresor* (Treasure-house of knowledge). Brunetto, a prominent political figure in Florence (1220–94), is perhaps best known for his appearance among the sodomites in Dante's *Inferno*. He spent some six years of his life in exile in France (1260–66), during which time he wrote his encyclopedic work in the French language. Along with a universal history, the book includes an account of the entire universe, a section on ethics taken from Aristotle, and several chapters on geography entitled "The Map of the World."[30] Only one manuscript out of many survivors of this popular work contains a world map and dates from 1310.[31] This map is in bad condition and without legends of any kind. Town symbols, rivers, and mountains seem to be arranged more or less at random, but the geographical forms, particularly around the Mediterranean, echo the sea charts of the day. There are, however, no rhumb lines. The Caspian Sea, described in the text as a gulf of the outer ocean, is not shown at all on the map. The configuration of southern Africa, bending around to the east, and the Indian Ocean, open to the east, are similar to Sanudo's world picture. The text is a little unclear but seems to imply that the Indian Ocean is also a gulf of the greater encircling world ocean.[32] The text is a mixture of contemporary information (the names of current bishoprics, a reference to the Saracens controlling part of Spain) and ancient tradition (the earthly paradise in the east, the double source of the Jordan River, the gold and silver islands of India, the cannibals in north Asia). Possible Arab influence is shown in the southern orientation of the map and the shape of Africa and the Indian Ocean. It is impossible to know what the relationship between the map in the Brunetto manuscript and the Sanudo/Vesconte work was, but their similarities suggest that both drew on some earlier work.

Paolino

When Sanudo presented his plan for a new crusade to the Pope in Avignon in September 1321, John XXII, in true bureaucratic style, immediately referred it to a commission for further study. Serving on this commission was Paolino Minorita, a Franciscan friar from Venice, who had already begun work on a universal history. A contemporary and fellow countryman, Paolino may have been acquainted with Sanudo before that meeting. They now became close friends, influenced each other's work, and remained in correspondence for the rest of their lives.[33] Paolino began his career in the service of the Venetian Senate. From 1319 to 1324, he was at the papal court in Avignon, and in 1324 he was named Bishop of Pozzuoli. He spent the rest of his life in and around Naples, primarily as Venetian envoy to the court of Robert of Anjou (reigned 1309–43), which was a lively center of the arts. His works include the *Epitoma* (1313), *Chronologica magna* (1321–34), and *Satyrica historia* (1334–39). These works were essentially collections of material drawn from other books, but they were marked by a strong even quirky originality in interpretation and arrangement. For example, Paolino constructed an immense chronological table in twenty-four columns in order to show the simultaneity of events throughout world history. Like Sanudo, he had his works copiously illustrated. He drew a detailed map of contemporary, boot-shaped Italy, which is his alone, and city maps of Rome and Venice, but he also included maps of the world, the eastern Mediterranean, and the holy land, which are virtually identical with those of Sanudo. Parts of his geographical text are also the same. Who was the creator and who the copier?

Paolino opens his *Chronologia* with a brief treatise, "De mapa mundi," which contains a geographical text similar to that surrounding Sanudo's world map. In the prologue he discusses the relationship between written geographical accounts and maps, explaining that both are necessary for a complete understanding of historical events.

> Here begins the prologue to the mapa mundi with its tripartite division of the world. Without the mapa mundi I would say it is difficult if not impossible to imagine or conceive in the mind those passages which speak of the sons or grandsons of Noah and of the four monarchies and of other kingdoms and provinces either in divine or secular writings. A double map is necessary, both in picture and in writing. You should consider neither one without the other to be sufficient, since a picture without writing shows the kingdoms and provinces confusedly, while writing without the aid of pictures does not mark the borders of the provinces which lie under the various parts of heaven in a fashion sufficiently clear, so that they can be seen as if by the eye. The images placed here have been composed from various maps copied from exemplars, which agree with the writings of illustrious authors, whom we imitate, that is to say: Isidore in his book *Etymologies*, Jerome in *On the Distance of Places* and *Hebraic Questions*, Hugh of St. Victor and Hugh of Fleury in his *Ecclesiastical History*, Orosius *De Ornesta Mundi*, Solinus *Of the Marvels of the World*, Gervasius *Of the Marvels of the Earth*, Pomponius Mela *Description of the World*, Honorius *Image of the World*, Eusebius, Bede, Justinus, Bishop Balderich of Dôle in *Overseas Journey* and of many other writers mostly on a description of the Holy Land and of the surrounding kingdoms of Syria and Egypt, which are needed for understanding many passages of Holy Scripture; whoever reads in these matters should understand the great work of the most studious doctor Jerome. Great caution must be applied so that the image is not spoiled by the painter.
>
> The world is first divided generally into Asia, Europe, and Africa. Though Asia is said by many to be a third part of the world, in fact its size is found to be half. End of Prologue.[34]

The geographical text follows. Although Paolino seems to have started with the same text as Sanudo, he edits it, particularly in the Paris manuscript. In the section on Asia, he adds a long digression on paradise, as one of its "provinces," noting that it touches the sphere of the moon and thus was raised above the waters of the flood. He describes the course of the four rivers of paradise, and mentions the preaching careers in Asia of Saint Thomas and Saint Matthew. He separates Syria and Egypt to deal with them in a separate treatise, which accompanies the map of the eastern Mediterranean.

In the prologue, Paolino echoes Sanudo's emphasis on the illuminating potential of maps, but where Sanudo advocates their usefulness for facilitating warfare and conquest, Paolino describes their importance to historical and biblical study. Like Sanudo, Paolino also backs up his maps with geographical text elsewhere in his book and uses the same description of the holy land. Sanudo's book was finished and supplemented with maps by 1321, while Paolino did not finish his first version until 1329. He did, however, put together some earlier work when he compiled it, and there is some discussion among scholars about an obliterated date, which may read 1320, on his map.[35] Paolino's world map in the Rome manuscript is less skillfully drawn than Sanudo's, which could mean either that it was a rough draft or a clumsy copy.[36] His Paris version is more finished looking, but both lack the rhumb lines, which may have been Vesconte's contribution, and there are some differences in place-names between the Rome and Paris versions. An interesting inscription near the Caspian Sea attempts to clear up the confusion between the Mare de Sara and the Caspian, which both mapmakers had presented as two seas. On the Rome map, Paolino labeled a gulf of the northern Ocean as the "mare Caspium," as well. On the Paris map, the gulf is unnamed, and he writes near the more westerly of the two Asian inland seas:

> This is called the Mare de Sara on account of the city where the emperor lives [Sarai], and it is also called the Caspian because it is near the Caspian Mountains and for the same reason it is called the Georgian Sea. In it there is a whirlpool where the water of the sea goes down, but on account of an earthquake it is stopped up. The sea rises a hand's width each year, and now many good cities have been destroyed. Eventually it seems that it will enter the Sea of Tana [the Black Sea], not without danger for many. It is 2,500 miles in circumference, and from Sara to Norgacium there are [illegible] miles. Around this sea for the most part is a sandy and inhospitable region.[37]

Norgacium is Novo Urgench (now simply Urgench), a city on the Amu Dar'ya or Oxus River just south of the Aral Sea. About 800 miles from the city of Sarai, it was on an important medieval trade route that linked the two cities. It was destroyed by the Mongols in Jochi's campaign of 1219–21.[38]

The other sea, farther east, is labeled "Caspium Mare" and is surmounted by the "Montes Caspii." The more polished appearance of the Paris map and its corrections lead to the inference that it was drawn after the Rome map, yet the two manuscripts are dated in reverse order. It looks as if the maps may have been drawn separately and incorporated into the manuscripts afterward.[39]

The confusion about the Caspian Sea was to continue, as traditional mappae-mundi and some classical authors (Pliny, Pomponius Mela, Isidore) had asserted that the Caspian was a gulf of the encircling ocean. William Rubruck was the first traveler to return from Asia to report on the Caspian as a lake rather than a gulf, and his words were emphatically echoed by Roger Bacon in his geographical text.[40] Paolino's text suggests that Marco Polo may have been his source, as Marco refers to the sea as the Mare de Sarai, after the Mongol capital, Sarai, on the Volga River, where the emperor resided.[41]

In the left margin of the page containing Paolino's world map in the Paris manuscript, there is a brief paragraph describing the provinces and borders of upper and lower Germany, giving some of the modern names (Saxony, Hungary, Austria, Russia) not in the geographical text. Parts of this text are identical with the description of Germany in Gervase of Tilbury's *Otia Imperialia*, written a century earlier.[42]

Those who have proposed that Paolino is the author of the maps point to his revolutionary map of Italy and his being more creative than Sanudo, the crusade-obsessed businessman, or Vesconte, limited to sea charts.[43] After 1321 it is clear that the Paolino and Sanudo collaborated, passing material back and forth. It seems, for example, that Sanudo added the portolan text of the eastern Mediterranean in the second edition of his work, which he expanded during his stay at Avignon. This revision may have been a response to Paolino's urging the necessity of map and text to give a complete picture. The collaboration may have begun even earlier, for there is evidence that Book III of Sanudo's work was originally written by Paolino. In two manuscripts, Vat.lat. 2972 (f. 71) and the one, now lost, used by Bongars in his seventeenth century printed edition, Paolino is noted as the author of this text.[44] It is almost impossible to determine the true prime mover in this close collaboration. Sanudo is certainly more likely to have had experience with the use of maps at sea and connections with the chart-making community in Venice. It is possible that it was he who converted Paolino to the appreciation of geographically precise cartography, even in a work as traditional as a universal history.

It is not surprising that Venice is the point of origin for the first world map to integrate the geographical forms of the sea chart. There the educated class was proudly mercantile, and the great families, who built their palaces on the Grand Canal, were well aware that their wealth came from the sea. Sanudo's upbringing as well as his purposes led him to disseminate an image of the world founded on the experience of commercial travel and not on the traditional picture purveyed in the mappaemundi.

The Catalan Atlas

The most ambitious and luxurious cartographic production of the fourteenth century was the Catalan Atlas, produced in Majorca in the 1370s and given as a gift to the king of France.[45] There are six double panels (sixty-five by fifty centimeters each) painted on vellum and mounted on wooden boards. The whole folds up like a screen into a relatively compact book form, which doubtless aided in its survival. It has been in the French national library since at least 1378, when it was first catalogued as part of the collection at the Louvre. The first two double panels are a compendium of geographical and astronomical texts and diagrams, while the map itself spreads over the last four panels, giving the entire sweep of the known, and not so well-known, world from the Canary Islands in the Atlantic to China and the island of Sumatra in the east. Painted in bright colors, with a generous application of gold leaf, and ornamented with pictures and text, it is truly a gift fit for a king, particularly a bibliophile like Charles V.

The author of the map has been assumed to have been Cresques Abraham, a Jewish chart maker from Palma in Majorca. This supposition is based on correspondence about the preparation of a similar map, designed as a gift from the crown prince of Aragon to the newly crowned French king, Charles VI. By 1381, when these letters were written, however, the map in question seems to have been already present in the French collection. The prince, summoning Cresques Abraham to the task, said that, if he could not be found, "it would be necessary to summon two sailors who would be capable of replacing him."[46] This statement implies the existence of a greater cartographic community closely connected with the seafaring enterprises of Majorca. If Cresques Abraham was not the author of the Catalan Atlas, it was someone very like him, and, as will be shown, probably a member of the Jewish community. The Cresques family was something of a dynasty; Abraham was succeeded by his son Jefuda, whom we find commissioned to make four maps as royal gifts in 1399. Cresques Abraham was described as a "master of maps of the world" and a compass maker. While mapmaking and the making of astronomical and nautical instruments were often associated in the same enterprise, "compass maker" may have referred to his role as the painter of the compass rose, the highly decorative directional indicator on nautical charts.[47]

At this time the Catalan commercial empire extended from their home ports in Barcelona and Valencia to the Balearic Islands, Sardinia, and Sicily—all under the rule of the Aragon/Catalan crown—and to the duchy of Athens, controlled by the independent Catalan Company. Catalan merchants were also entrenched in several North African port cities, while Catalan maritime law, written in the *Llibre del Consolat de Mar,* was known throughout the Mediterranean world. The earliest known Majorcan chart maker was Angelo Dulcert, who made a sea chart in 1339. Catalan charts developed in their own style. Although there are exceptions, they generally differed from charts of Italian origin in that they included more details of the interior and were more highly decorated, with por-traits of African and Asian kings, animals, and banners.

The Geographical/Astronomical Introduction

The map itself is preceded by two double panels of text and diagrams, similar to the material surrounding the Ebstorf wall map (fig. 3.2). There is a good deal of information on the calendar, including an essay and circular diagram of the days of the month, indicating days of good and bad luck, the so-called Egyptian days. Another diagram shows how to find the "Golden Number" important for calculating the date of Easter. Originally, a *volvelle,* or rotating figure, in this case a man, was in the center of this diagram. Positioned with his left hand pointing to the Golden Number of the year in question, his other hand pointed to Shrove Tuesday, the pommel of his sword to Easter, and the point of his cap to Whitsunday. A marginal note gives the dominical letters for the years 1375, 1376, and 1377, suggesting that this was the period when the map was made. In this section is also a "zodiac man," a drawing that shows the parts of the body governed by the position of the moon in the various signs of the zodiac. This correspondence had particular relevance for bloodletting, and an additional dia-gram aids in the calculation of good and bad times to undertake this procedure. Also on this page is a tidal diagram, enabling the reader to calculate the tides at fourteen locations in Brittany and England, reflecting the Catalan trade to these northern ports. This is the first appearance of the knowledge now known as "the establishment of the port," that is, differentiating between the theoretical time and the actual time of the turn of the tide in a specific location.[48]

On the other side of the first panel is a geographical text, which begins by defining the mappamundi as an image of the world, showing the "ages of the world," the regions, and the various peoples living upon it. What follows is largely derived from the *Imago Mundi* (1110) of Honorius Augustodunensis.[49] The world is round like a ball or an egg, giving the opportunity for an extended simile involving the yolk, the white, and the shell as the spheres around it. Next,

Figure 3.2. Calendar wheel, 1375 (Paris, Bibliothèque Nationale, MS Esp. 30, Panel 2. 65 × 50 cm / 26″ × 20″). The earth is in the center of this diagram, surrounded by the spheres of water, air, and fire. The other spheres include the planets, the signs of the zodiac, the cycles of the moon, and the calculation of the Golden Number. In the corners each of the four seasons holds a text, describing its characteristics. Courtesy of the Bibliothèque Nationale de France.

the four elements (earth, water, air and fire) and their qualities are described. Although the world is 180,000 stadia in circumference, we are told, "if a man looked down on it from high up," its irregularities would be smoothed out and it would look small. It is the center of the universe, and, as told in the Book of Job, it is "suspended on nothing" but kept up by the will of God. The text launches into a geographical description, beginning with the division into the three continents. For a map with such an interesting portrayal of Asia, that continent is hardly described, and the author skips that section of Honorius's work and goes directly to Italy, the first region described. Here he is more concerned with the various names it has had during its history and how they were derived than with its geographical features. The text rambles on to discuss Africa, Sicily, the Red Sea, Germany, and the world-encompassing ocean. Paradise, so often a staple of medieval geographical treatises, held a prominent place in Honorius's work but is not mentioned here. The last part of the geographical text, not from Honorius, is a rather technical analysis of the relationship between the tides and the moon, and the sun and the moon, describing lunar motions in terms of quarter-wind points, "according to the calculations of navigators." We also hear how to determine one's position when the sun and moon are not visible, a method that is a precursor of astronomical navigation, which was to become increasingly important in the next century.

On the second panel is an elaborate calendar wheel, showing the various cycles of time revolving around the earth.[50] An astronomer holding up a quadrant stands on earth, which is placed in the center and surrounded by the spheres of water, air, and fire in the style of many earlier medieval mappaemundi. Beyond these are circles that include the planets, with a description and personification of each, the signs of the zodiac in a gold ring, the position of the moon in relation to the fixed stars, the phases of the moon, and then six rings to show the complexities of the lunar motions. The outermost ring has a text on how to determine the Golden Number, while in the corners are four figures representing the seasons, each holding a descriptive scroll.

This astronomical/geographical presentation is the frame for the map that follows. Such material, which was common in earlier medieval computus manuscripts, also appears in nautical atlases and sailors' handbooks.[51] A modern observer may be puzzled at the sailor's pressing need to determine the date of Easter while at sea, as well as appalled by the idea of bloodletting while afloat, but clearly this information was considered important. In the Catalan Atlas, the opening panels are concerned with the role of the earth in the scheme of things—in relation to the entire universe or macrocosm, and in relation to the human being, or microcosm. The tides, so important to sailors venturing out into the Atlantic, were not only of practical significance but also demonstrated the interconnectedness of things (in this case the moon and the sea), just as "zodiac man" showed the intimate connection between man and the outer spheres of the universe.

The World Map

The world map in the Catalan Atlas has been described as "transitional," that is, a combination of sea chart and mappamundi.[52] The first two double panels (panels 3 and 4) are a sea chart of the Catalan type, with the names of ports written around the precisely drawn coasts of the Mediterranean, the Atlantic, and the Black Sea. In addition to the forms and place-names of the nautical charts, it is crisscrossed with rhumb lines and features a wind rose with the eight principal winds named in the vernacular. It is also in Catalan, instead of Latin. Over one thousand names of ports, capes, and other coastal features line the coasts, while interior characteristics, such as river courses, are somewhat abstract. (However, the city of Paris is shown divided by its river). Unlike the other continents, Europe has no pictorial elements, such as enthroned kings, but it does have flags, including an Arabic flag flying over Granada. The islands of the Mediterranean are gilded, while England is painted purple. The Danish peninsula is very large, and Norway is shown as a squared-off formation, surrounded by mountains. The island of "Stillanda" in the northwest ocean is perhaps meant to be Iceland. The legend says that the inhabitants speak the Norwegian language and are Christians.

The last two double panels (5 and 6) stretch from the Caspian Sea and Persian Gulf to the Ocean in the east, the literal ends of the earth, where a paraphrase of a prophecy from Isaiah is recorded: "I shall send those who are saved to the peoples of the sea, to Africa and Lydia . . . I will send to the isles afar off that have not heard my fame, neither have seen my glory; and they shall declare my glory to all people" (Isaiah 66:19). Like the traditional mappamundi, the Catalan map is placed in a larger context, in this case that of the universe as a whole, a unity of time and space. It is rectangular rather than round, but as David Woodward points out, the far eastern edge is curved as it would have been in a circular mappamundi.[53] It is lavishly decorated with pictures in Africa and Asia. Outside of Europe and the Mediterranean coasts it makes use of information from travelers rather than relying exclusively on antique sources; nevertheless, Alexander the Great is portrayed in northeast Asia, and the Red Sea is marked to show the passage of the twelve tribes of Israel.

Alessandro Scafi has defined the mappamundi by the presence of the earthly

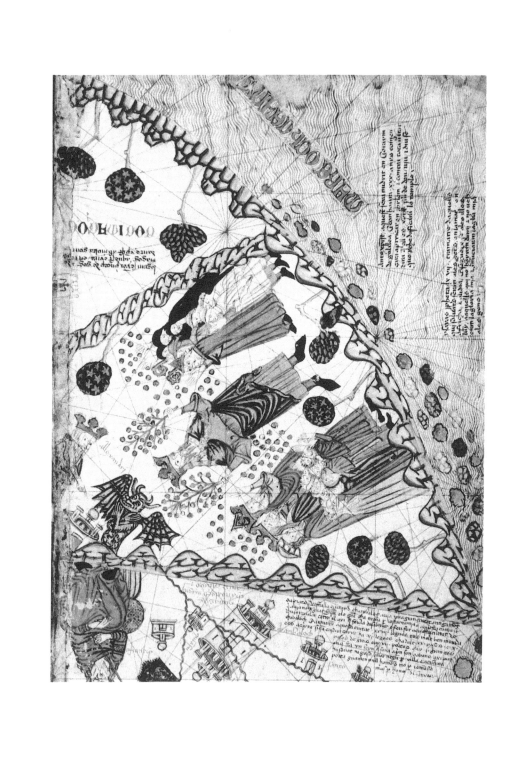

paradise. In these terms the Catalan Atlas is not a mappamundi, but it does show some curious scenes in northeast Asia (fig. 3.3). There is an enclosure surrounded by mountains and divided into two sections. To the west is a mounted king under a canopy followed by a small entourage wearing pointed hats and carrying a banner with the devil painted on it. The accompanying text reads that this great lord is the king of Gog and Magog, who will march out with many followers at the time of the Antichrist. To the east is an unidentified crowned figure with a tree in each hand. Two groups of people flank him—on the right are ecclesiastics, including monks, nuns, a cardinal, and men holding crosiers, and on the left are three crowned figures with a group of smaller figures in front of them. Above their heads and upside down is a crowned figure labeled "Alexander." He points to a black-winged devil who guards the gate to the enclosure. Outside are two bronze figures with trumpets, whom, we are informed, were made by Alexander to guard the gate. Alexander would have died in the Caspian Mountains, an inscription reads, had not Satan come to his aid and shut up the "Tartars, Gog and Magog," the tribes which eat raw flesh. Outside the enclosing mountains sits "Holubeim" (Khublai Khan), the greatest lord of the Tartars, implying that some of them have escaped already. A text in the sea nearby describes how the Antichrist will be born in Chorazim in Galilee. After thirty years he will preach in Jerusalem that he is the Christ, which he is not, and promise to rebuild the Temple. A recent study of this scene by Sandra Sáenz-López Pérez argues convincingly that it is the Antichrist, performing miracles to prove to his followers that he is the real Christ of the Second Coming. Here he is causing fruit to appear on dry branches.[54]

What are we to make of this? Although the crowned king, holding the trees, has been interpreted as Christ and his enclosure as paradise, there seems to be little justification for such a view.[55] The name Paradise does not appear, nor the four rivers, nor the figures of Adam and Eve who feature so prominently on earlier maps of paradise. Instead these scenes are accompanied by sinister texts of doom and savagery. And here we come to one of the most striking characteristics of the Catalan Atlas—it is almost devoid of Christian content. There are

Figure 3.3. (opposite) Alexander the Great (from the Catalan Atlas, c. 1375. Paris, Bibliothèque Nationale, MS Esp. 30, Panel 6). Northeast Asia on the Catalan Atlas is depicted with these vivid scenes, which have not been satisfactorily explained. "Alexander" is written next to the crowned figure talking to the winged devil, and a similar figure, possibly the Antichrist, holds up the two trees in the enclosure and is surrounded by admirers. The mounted figure in the adjacent enclosure (not shown) carries a scepter. This space is labeled with the names of Gog and Magog, and a text in the sea foretells the birth of the Antichrist. Courtesy of the Bibliothèque Nationale de France.

biblical sites (the Tower of Babel, Noah's Ark, Mt. Sinai, and the giving of the law, the Queen of Sheba), but these are references to the Old Testament, not the New. In the holy land itself are a series of mountains (Seir, Gilead, Nebo, Pisgah, Abarim) all mentioned in the book of Deuteronomy and important in the story of Moses. Another curious feature is the interlacing on the second panel of the map, which is made up of Hebrew letters.[56] A prophecy close to that of the Antichrist can be found in the Old Testament in the Book of Daniel, and the "mission statement" comes not from the Gospel of Matthew but from the Book of Isaiah, and refers to messianic Judaism rather than Christianity. In Isaiah's prophecy the scattered Jews will return home, and this quotation is put in interesting juxtaposition to Gog and Magog and the traditional story of the "enclosed Jews."[57]

The few Christian sites shown are the monastery of Saint Katherine, and the burial sites of Saints Matthew and Thomas. Jerusalem is represented by a church with a cross on the top and a cross on the altar shown within. The three wise men are shown in Tarsia (Persia) en route to Bethlehem. The text notes that they are buried in Cologne, "two days' journey from Bruges," as opposed to Sava, where Marco Polo said their bodies could be found. The author of the Atlas has made a startling, even subversive statement. While the sea charts themselves were generally without religious content and the mappaemundi were Christian, the Catalan Atlas is imbued with the ideology of Judaism.

Trade Routes in Africa

The sea charts, which formed the background of the Catalan Atlas, were made for merchants and sailors, and the atlas extends the commercial use of the map into the interior. In Africa prominence is given to the trade routes followed primarily by Catalan merchants, many of them Jews from Majorca. We see Taghaza and Tacost, the centers of salt mining in the Sahara, and the route the salt caravans followed to the coast via the oasis of Sigilmasa, shown surrounded by water, and through the pass in the Atlas Mountains, "Vall de Darcha." These caravans also brought from the south slaves who were sold in the markets of Majorca. A map fragment, now in Modena, shows a procession of slaves, chained by the neck, walking toward the coast.[58] Another interesting feature is a flag with a star of David flying over Brisch (modern Bissa), a coastal town in North Africa.[59] Many of these towns had large Jewish communities with close ties to Majorca—indeed the Jews who settled in Majorca after the Christian conquest in 1229 mostly came from North Africa. Majorcan Jews were also settled in the interior of Africa, particularly the town of Tlemcen and the area known as the Touat.[60] Here Jews had some advantages over Christians in obtaining trade privileges from local Muslim authorities, and they were able to form a network of way stations and trading partners throughout the region.[61] In identifying the names of towns in the interior, there are problems with the vagaries of transliteration from the Arabic. In regard to location, while the coastal charts had worked out coordinates of sorts for port cities, in the interior such precision was more difficult. Placing a town in the Sahara was as difficult as locating an island in the open sea.

Fourteenth-century merchants were intensely interested in the gold that came out of Africa. On the map is Musse Melli (Mansa Musa, reigned 1312–37), the king of Mali, holding a gold ball. He is positioned near the city of Tenbuch (Timbuktu) and a legend beside him describes his wealth, which was based on gold. He was a real, historical figure, though he was long dead by the time the Catalan Atlas was made. His pilgrimage to Mecca in 1324 was so spectacular that it remained in the folk memory for centuries, for he is said to have spread gold around Cairo with such lavishness that it caused a crisis in the currency.[62] The Africans kept the original sources of the gold supply secret in order to monopolize the trade, thereby piquing the curiosity of Western merchants who made a number of attempts to discover them. For instance, shown off the western coast of Africa is the small ship of Jacme Ferrer "who set sail on St. Lawrence's Day, August 10, 1346, bound for the Rio de Oro." Poor Ferrer was never heard from again, but his voyage was a precursor to the Portuguese expeditions of the succeeding century. On the other side of Africa, the Red Sea is noted not only for its color (here attributed to the sand beneath it) and its role in Hebrew history but for being the sea through which most of the spices from India were shipped to Alexandria. The important port of Chos (Chus) is also shown.

The Catalan Atlas is one of the first world maps to place the kingdom of Prester John in Africa rather than Asia. An inscription in the southeast near the city of Nubia informs us that "The king of Nubia is constantly at war with the Christians of Nubia, who are under the dominion of the emperor of Ethiopia and of the land of Prester John."[63] This interesting figure had been an obsession of the medieval European imagination since a mysterious letter, addressed to the Pope, arrived in Rome in 1165. The idea of a Christian ruler of stunning virtue and immense power and wealth in some remote land was intriguing, particularly if his aid might be enlisted in the fight against the infidel Muslims. Marco Polo located Prester John in central Asia and announced that he had been killed in battle by Genghis Khan. The Sanudo/Vesconte world map showed the elusive king on the eastern coast of Asia, in "India inferior." From the late fourteenth century on he was to appear in east Africa, and his story was to be marvelously confirmed by the eventual discovery of the Christian kingdom of Abyssinia. The original letter was, however, a forgery.[64]

Whereas Africa on the medieval mappamundi had been depicted with antique classical sites on the coast and monstrous races of humans and animals to the south, the Catalan Atlas gives quite a different picture. Africa is no serpent-infested wilderness but a productive region with towns and trade routes.[65] A legend on the atlas also notes its lack of centralized political control and its internecine struggles, particularly in the east, a point of some importance to merchants trying to transport valuable cargoes through the region. Cresques Abraham, or whoever our author was, drew on information from the Majorcan Jewish trading community to present a radically realistic view of the resources and commercial centers of Africa.

Islands of the Atlantic

In the southwest corner of the map are nine islands of the Canary group as well as the small Savage Islands to the north. The Canaries, discovered some time in the early fourteenth century, first appeared in part on the Dulcert chart of 1339. By the time of the Catalan Atlas they are integrated into the European world picture, and the Atlantic Ocean is shown not as a narrow band to the west of the known world but as a wide space speckled with islands, some real and some imaginary. The discovery of the Canaries is traditionally credited to a Genoese, Lanzarotto Malocello, whose name is preserved in that of the easternmost island of the group. The Genoese in the early fourteenth century were trying to outflank the Catalans' inland trade route by finding a sea passage to the lucrative sources of gold and slaves. It was a shock to the European Christians to find a pagan, neolithic population on the islands. Setting out to Christianize and exploit them, they eventually succeeding in exterminating them altogether.[66] The Canaries gained increasing importance in the two next centuries as a starting point for voyages to the New World.

Educated Europeans were quick to identify the Canary Islands with the Fortunate Islands of antiquity, most thoroughly (though vaguely) described by Pliny in the *Natural History*.[67] Martianus Capella, Solinus, and Isidore had purveyed information about them to the Middle Ages, and the islands appear on most medieval maps off the coast of Africa. The Hereford mappamundi has six islands, with names drawn from the classical tradition but also identified as the six islands of Saint Brendan. The Catalan Atlas places next to them a lengthy legend describing them as a land of milk and honey, with a mild climate, abundant garden produce, and trees 140 feet high. However, whatever pagans "of

India" might think, the author continues, this is not the earthly paradise. The legend cites Isidore as its source, even though Isidore's pagans were those of classical antiquity. There is also a representation of what appears to be the Madeira group, although we have no textual evidence of its discovery before the next century. The three islands, lying north of the Canaries, are labeled Porto Santo, Insula de Legname (or Wood Island = Madeira), and Insula Deserta.[68] The atlas also retains some of the mythical islands of the Atlantic, including two representations of the island of Brasil.

Mapping Asia

The Asiatic section of the map has been described as a traditional mappamundi, uneasily wedded to the western sea chart, but it is important to recognize that it has some definite mercantile features. Most prominent is the camel caravan making its way "from the Caspian Sea to the land of Cathay." The camels are loaded with boxes and bags and followed by one group on foot and another mounted on horseback. Lest we think this is a cakewalk, a lengthy legend details the horrors of the desert of Lop, where a traveler separated from his companions will hear spirits calling his name and leading him astray. The account is taken directly from Marco Polo.[69] Marco with his lively interest in other lands and peoples and his sharp eye for a profitable opportunity was well suited as a source for the maker of the Catalan Atlas. By the time it was made, some of his information was obsolete, as the fall of the Yuan dynasty and the breakup of the Mongol Empire had put an end to the overland traffic to Asia that thrived in the wake of the Polo expedition. The atlas's account of Asian matters is therefore dated, with "Holubeim" (Khublai Khan, d. 1294) still ruling in Chambalech (Beijing). A more modern note places Jambach (Janï-beg, d. 1357) in Sarra, north of the Caspian Sea. He had come dramatically to the attention of Western merchants when he laid siege to the ports of Caffa and Tana in the Black Sea in 1343.[70]

The maker of the atlas quoted freely from Marco Polo's *Travels* and took the names of twenty-nine Chinese cities from his book.[71] The cities are scattered about east Asia somewhat at random, which is not surprising as Marco had given little specific information on their location. The atlas also cites Marco's estimation of there existing 7,548 islands in the China Sea, all rich in treasures designed to make a merchant's mouth water: pepper, gold, silver, spices, and precious stones.[72] The entire southeast corner of the map is sprinkled with a brightly colored array of islands, further embellished by a double-tailed mermaid, a couple of naked fishers, and a Chinese junk with palm leaf sails (fig.

3.4). Only a few of the larger islands are named: Iana (Java), first introduced to the west by Marco Polo, and Trapobana (Taprobana) in the far southeast. Iana is ruled over by a queen, and her kingdom sounds like the land of the Amazons, here conflated with tales of the "Island of Women," found in *Arabian Nights* and in Marco Polo's book.[73] Taprobana was known to the Greeks and Romans and appeared on most medieval maps with a standard set of characteristics: two summers and two winters in each year, abundant fruit, elephants, and wild beasts. The atlas shows an elephant and a black king, and adds details about a race of fierce and stupid black giants, who eat up white men "whenever they can get them." Taprobana is usually translated as Ceylon or Sri Lanka, but as Europeans began to explore the Indian Ocean, it was pushed farther east and identified as Sumatra—Marco Polo does not mention it at all. The atlas introduces a new island, that of Saylam, corresponding to Marco Polo's Seilan and placed just east of Java.[74]

The Catalan Atlas is the first surviving world map to make extensive use of the work of Marco Polo. The legends on the map repeat stories from the text, such as tales of the spirits in the Gobi Desert, and copy descriptions such as those of Khublai Khan and the city of Chambalech. Numerous place-names derive from the book, including almost all the information on the kingdom of the Tartars. As we have already seen, the map charts the travels of other merchants, such as the Majorcan Jews in North Africa, and the way stations of the spice trade coming from the Indian Ocean. The travels of Ibn Baṭṭūṭah, which took place in the mid-fourteenth century, are often cited as a source. His work, which circulated in Arab lands, did not come into Europe until much later, but a Majorcan mapmaker could have had access to Arabic and African sources. We hear that Ibn Baṭṭūṭah returned home to Fez from Tunis on a Catalan ship about 1350, and we already know of extensive Majorcan contacts in Africa. One indication of the possible use of Ibn Baṭṭūṭah is the depiction of Tenbuch (Timbuktu) in the atlas. It is represented, unlike the other domed or turreted cities, by a building with a simple tile roof. This picture may be derived from Ibn Baṭṭūṭah's reference to the tomb of the poet, d'Al-Tu Waydjin, which he saw there.[75]

Albertin de Virga

Another example of a world map with features derived from the sea charts is that of Albertin de Virga of Venice. He made a sea chart in 1409 and a world map, dated 1411 or 1415, now, alas, lost.[76] In addition to the world map, an Easter table and zodiac circle appear on the sheet, with a "zodiac man" in the center. The map itself is round (sixteen inches in diameter) with the irregularly shaped

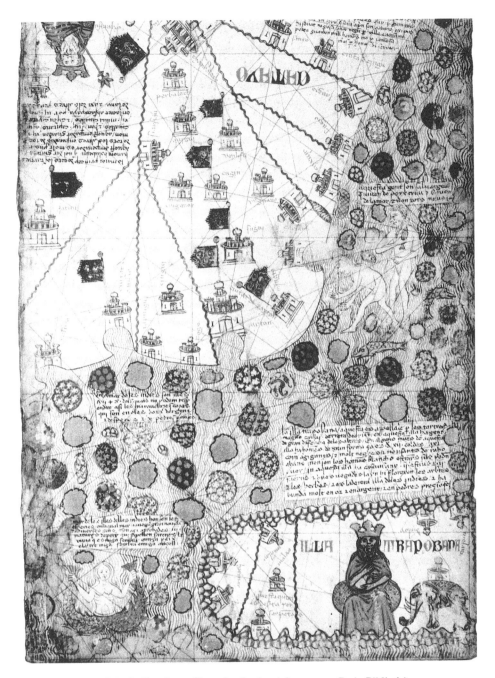

Figure 3.4. Islands of the Indian Ocean (from the Catalan Atlas, c. 1375. Paris, Bibliothèque Nationale, MS Esp. 30, Panel 6). Beautifully colored in the original, many islands are spread over the China Sea and Indian Ocean. Only a few are named, one being the island of "Trapobana," where an enthroned king poses next to a very small elephant. Cathay occupies the upper left portion. This section of the map reflects Marco Polo's report on the Far East. Courtesy of the Bibliothèque Nationale de France.

three continents surrounded by a greatly expanded ocean; in the Atlantic are found the Canaries and Azores. The coastlines of the usually traveled world show the influence of the marine charts, as does the presence of an eight-spoke wind rose. The Indian Ocean, plentifully sprinkled with islands, is open to the south and east, and the shape of Africa seems peculiarly modern. Two unusual formations appear in the northwest and southeast. In the northwest there is a large, triangular peninsula labeled *Norveca*, or Norway. While the southern part shows a good knowledge of the sea routes Venetian traders took to the Hanseatic ports, the rest is sheer invention. To the southeast a large island is labeled "Caparu sive Java Magna," which is thought to be a combination of Marco Polo's descriptions of Cipangu (Japan) and Java. Gunnar Thompson, an enthusiastic defender of the pre-Columbian discovery of America, has identified these two forms as "Greenland and North America" and "Brazil." His theory has not received much support.[77]

The mappaemundi of the late thirteenth century had striven for a timeless view of the world, drawing geographical information mostly from antique sources, classical and Christian, and putting the whole into a cosmological context. The marine charts had been of limited, practical use, and made no attempt to achieve a total world view—indeed they attempted to show only a small part of the world. In the fourteenth century a synthesis of these two map forms began to be made, starting with the Sanudo/Vesconte world maps.

The richness and large size of the Catalan Atlas catapults us into a new, contemporary world, crisscrossed with the trading ships and caravans of the enterprising denizens of the Mediterranean. Majorca, at the crossroads of maritime traffic in the western Mediterranean, was ideally placed to receive all types of news about the wider world. The school of Majorcan mapmakers, linked by their profession to the society of seagoing merchants, faithfully recorded the most up-to-date information. At the same time, the atlas is not simply a representation of geographical space. Historical and mythical figures and events continue to appear, such as the pygmies and the cranes, still fighting it out in the heart of central Asia as they had done since Homer's day, while the thousand-year-old history of Alexander the Great is also depicted.[78] It retains the cosmographic material that places the map in an astronomical, geographic, and scientific context but does not provide a Christian religious framework as its predecessors had done. We have no record of what the French king thought of this gift. Was he dazzled by its color and richness, or did he look in vain for the time-honored features of the mappamundi—paradise, the Last Judgment, the crucifixion?

In the fourteenth century, for the first time the sea chart and the world map were united into a single image, but the unity continued to be a troubled one for the next century. Albertin de Virga shows the limitations of a chart maker venturing into uncharted territory: once one has accepted the premise of making physically realistic maps, errors can no longer be tolerated. In addition, how can one show the historical and spiritual importance of places without altering physical reality? And what about places that are inaccessible even to the most indefatigable traveler, such as the earthly paradise? And how should the conflicts between the revered writers of classical antiquity and the sharp-eyed sailors of the present day be reconciled?

Traditional mappaemundi continued to be in demand and were made for several centuries more. One thinks immediately of the maps of Ranulf Higden in the mid-fourteenth century. (See chapter 7 for a fuller account of mappaemundi of the fifteenth century.) Consumers seem to have felt the need expressed by the thirteenth-century canon of Pisa, who, when confronted by a sea chart, complained that it did not show some of the most important aspects of the world. And since world maps were not much use to travelers, inevitably lacking the necessary detail, it is not surprising that not everyone desired the geographical realism of the sea chart. Even in the late sixteenth century, when Ortelius published his elaborate atlas, its purpose was more to educate and entertain its owners rather than supply a practical guide to finding one's way about the world. So, although chart and mappamundi were first integrated in the early fourteenth century, the two map forms continued their side-by-side existence, each serving its own purpose.

CHAPTER FOUR

Merchants, Missionaries, and Travel Writers

Long-distance travel was arduous and hazardous during the Middle Ages but never entirely ceased. Driven by spiritual or economic motives or by much-maligned *curiositas,* or inspired by a mixture of all three, medieval people took to the road or the seaways with surprising frequency. The Crusades impelled large numbers to leave home, while political circumstances caused Western monarchs to send embassies that attempted to form alliances against the Muslims. For those who remained at home, travel books were a less strenuous substitute for the real thing. Beginning in the late thirteenth century, eyewitness travel accounts began to have an impact on the Western concept of world geography and, subsequently, on maps.

For the Middle Ages the ultimate traveler was Alexander the Great, whose actual and fictional adventures encompassed the known world, traversing Asia until he was warned by an oracle to turn back. The popular Alexander romances included a trip beneath the sea in a submarine, a flight through the air in a balloon drawn by two griffins, as well as an attempt (unsuccessful) to enter paradise. Perhaps medieval people did not believe all these things really happened, but Alexander's purported letter to Aristotle, reporting on the exotic animals and peoples of Asia, was taken quite seriously. The story of the savage peoples, Gog and Magog, shut up in impregnable mountain fastnesses by Alexander appears on most medieval world maps, as do other representatives of the "monstrous races" he supposedly saw.[1] These tales were so compelling that travelers frequently reported seeing Alexandrian sights, such as the trees of the sun and the moon, the queen of the Amazons, and the barriers enclosing Gog and Magog.

Travel to the holy land was a long journey for western Christians, but pious pilgrims continued to attempt it, even though some died on the way or were forced to turn back.[2] Muslims made the obligatory trek to Mecca from their distant homes in Spain, Morocco, or central Asia. Some left written records, which included geographical material, though pilgrimage accounts tended to dwell on the spiritual benefits of the journey rather than such cartographically useful details as distance, direction, and precise location. In 1358 Petrarch wrote an itinerary to the holy land for a friend to keep him company, as Petrarch himself was not willing to make the long sea voyage. He was neither the first nor the last person to compose a travel guide without having traveled. Considering knowledge from books in this case superior to mundane experience, Petrarch confidently tells his friend what he will see.[3] Ignorant of the ultimate destination, Petrarch is quite knowledgeable about the first stage of the voyage from Genoa to Naples, and he describes the west coast of Italy in portolan-like detail. "After about twenty miles [from Genoa] you will find yourself before a promontory reaching out into the sea, which is called Capo del Monte, the gate of the dolphin, or, as the sailors say, Alfino, and hidden among the charming hills, Rapallo."[4] Once he progresses beyond familiar territory, he stresses the spiritual value of the rigors of travel, rather than the physical details: "Not far from where the Jordan flows into the Sea of the Sodomites is the place where one may see vestiges of the destroyed city and of the divine vengeance. Near here is the great expanse of the desert. The journey is hard but no road should seem difficult which is the way to salvation. Everywhere our enemy, the devil, will put many difficulties and annoyances, both of tiresome men and harsh places, in order to make you turn back or to impede you on your way, or if this is not possible, to render you less joyful on your holy pilgrimage."[5]

The way (or ways) to the holy land, though difficult and long, was at least well traveled and thoroughly written up. John Mandeville in his fourteenth-century *Travels* gives several alternative routes, weighing the advantages and disadvantages of each.[6] Beginning at the end of the eleventh century, the trickle of pious pilgrims was augmented by the masses of Crusaders bound on military missions to the east. Those who overcame their fear of the sea and traveled by water were aided by the Genoese, the Venetians, and other denizens of the Mediterranean seaports, who were thoroughly acquainted with the seaways of the east and were happy to take paying passengers. Colonies of Italian and Catalan merchants could be found all around the Mediterranean and Black Sea littoral, with their warehouses, moneychangers, and amenities such as bathhouses, churches, and marketplaces.

Travel to Asia

Travel beyond the coast into the heartland of Asia was rarer, and it was not until the mid-thirteenth century that we have the first records of European travelers crossing to the far East. Before this time merchandise exchange between East and West went through the hands of middlemen, peddlers, and caravaners, each going for relatively short stages of the journey.[7] The same process divided up the southern sea route into digestible chunks of space. Upheavals and incursions in central Asia, added to the great distance and harsh climate, made overland travel and trade hazardous before the establishment of the Mongol Empire, founded by Genghis Khan (c. 1162–1227). The sudden eruption of the Mongol cavalry into Russia and eastern Europe terrified the Christians, who surmised that these were the hordes of Gog and Magog, who had been locked up by Alexander but were now set loose to bring the prophesied reign of the Antichrist. "In this year [1240]," wrote Matthew Paris, "that human joys might not last long unmixed with lamentation, an immense horde of that detestable race of Satan, the Tartars, burst forth from their mountain-bound regions, and making their way through rocks apparently impenetrable, rushed forth, like demons loosed from Tartarus."[8] The death of Genghis Khan had given Europe a temporary respite, but by 1241 the Mongols were back, conquering southern Russia, Poland, Hungary, and Silesia, and were threatening Vienna. Refugees, fleeing to the west, brought tales of the invincibility and ferocity of the conquerors.[9] European Christians were heartened, however, to hear that the Mongols were not Muslim but pagan and, with the easy tolerance of such folk, expressed an interest in the practices, beliefs, and paraphernalia of the Christian religion. Perhaps they could be converted to Christianity and would form an alliance to defeat the Saracens, who were rapidly reclaiming the holy land. After the initial depredations of Genghis Khan, the Mongols settled down to build an empire, which, at its height, had an impressive communications systems, rather like the pony express. Privileged European travelers received golden passes that assured their safety and material support as they crossed the vast reaches of the Mongol domain.

The first traveler to write about his journey was a Franciscan friar, Giovanni di Pian di Carpini, dispatched by Pope Innocent IV, in 1245, bearing letters to the "Emperor of the Tartars."[10] On his way to the East, Giovanni was joined by "Benedict the Pole," an interpreter, and Ceslaus and Stephen of Bohemia. The papal letters explained the chief features of Christianity, upbraided the Tartars for their savage acts, and begged them to refrain from such in the future. The emperor was understandably unmoved by these communications and informed the Pope that it was God's will that other nations be subdued by him and that he expected the Pope to knuckle under as well.[11]

Other missions followed, including that of William Rubruck from 1253 to 1255, the European appetite having been whetted by the rumors of a Christian population in central Asia and the alleged conversion of some members of the Mongol royal family.[12] The Mongols in turn sent to the West their emissaries, charged with the task of forming an alliance against the Saracens in an attack on Syria. The idea was appealing, but the Europeans were too busy fighting among themselves to respond. The Mongols moved west, taking Baghdad in 1258 and Aleppo and Damascus in 1260. Another Mongol overture for an alliance in the 1280s came to an end with the Egyptian conquest of Acre and the other coastal cities of Palestine in 1291.

Travelers are not necessarily good sources of geographical information, either now or then. Not all the members of a tour group or a caravan know exactly where they have been. In addition, there is the human tendency to see what we expect to see, and late medieval travelers in Asia had to overcome their allegiance to a mass of knowledge inherited from the classical past before they could focus their eyes on what was in front of them. The purpose of the thirteenth-century missionaries was not the collection of geographical data, but several of them produced written accounts with valuable details of ethnography and geography which were to transform the European picture of Asia. Slogging across the steppes, miring down in swamps, scaling precipitous mountains, fording rushing rivers, parching in the interminable deserts, the European envoys eventually reached the Mongol court at Karakorum in Mongolia. The Mongols of that day, living in tents mounted on carts, moving from place to place, and guzzling fermented mare's milk, were a far cry from the hypercivilized court of Kublai Khan in Cambalech (Beijing) which Marco Polo found half a century later. Karakorum, William Rubruck tells us, was in 1254 not as big as the village of Saint Denis outside Paris.[13] The envoys, though welcome, were given skimpy rations and sent from prince to prince up the Mongol hierarchy, each many days of hard traveling apart. Seeking fellow Christians, they were dismayed to find Nestorian Christianity the dominant version of their faith in Asia. The Nestorian church had split from Rome in the fifth century, mostly for political reasons but also because of a different interpretation of the dual nature of Christ. Over the centuries, its rituals and customs had developed independently of the West—for example, priests could marry—and the Franciscans were shocked by what they saw as deviant practices.[14] Odoric denounced them as "vile and pestilent heretics."[15]

The most significant accounts of mid-thirteenth century Asia were those of Pian di Carpini (traveled 1245–47), composed for the Pope, and William of Rubruck, who was dispatched to the East by Louis IX of France. As Franciscans, their main interest was to get acquainted with the Mongol people and explore the possibility of converting them to Christianity. Pian di Carpini opened his account with a series of summary chapters on the location of the "Tartar" lands, the history of the people, their customs, and their mode of warfare. His story of the rise of Genghis Khan, with its many fanciful details, was apparently derived from Mongol folklore and is thus valuable in its own right. While only sixteen manuscript copies of the two versions of his report survive, we learn that he lectured at various monasteries on his way back to Rome and complained that some people copied parts of his report before it was finished. One survivor of this process is the "Tartar Relation," found in the manuscript with the controversial Vinland map.[16] It appears to have been put together from notes taken during a lecture by one of the traveling friars. So the word spread and reached, among others, Matthew Paris in his monastery at St. Albans, England, who recorded in his chronicle facts about the Tartars, interspersed with the sordid details of the endless quarrel between Pope and emperor, meteorological disasters, the effrontery of the new mendicant orders, and the oppressive taxes of the English king. In Paris, Roger Bacon interviewed William Rubruck and put some of his observations on Asia into the geographical part of his *Opus Majus*.[17] Vincent of Beauvais incorporated information from the report of Simon Saint-Quentin (not otherwise extant) and large parts of Giovanni di Pian di Carpini's *History of the Mongols* into his encyclopedia.[18] Completed in 1253, Vincent's encyclopedia was widely distributed and survives in over fifty manuscripts.

William Rubruck in his report to Louis IX gave more details about his traveling route than Pian di Carpini had. He discussed the source of the River Don, its course into the Sea of Azov, and the distance between the Don and the Volga. He also described the great size and configuration of the Volga River and the fact that it flowed into the Caspian Sea, not into the Black Sea as Pian di Carpini had surmised. Rubruck was the first to assert positively that the Caspian was not a gulf of the northern Ocean, as it appears on almost all medieval world maps up to this time, but was an inland sea, "four months' journey" in circumference.[19] He seldom estimated distances in any other terms than days' journeys, and his directions were frequently faulty—he obviously did not carry a compass.[20] Like Pian di Carpini, he gave the names of many peoples conquered by the Mongols and those as yet unconquered, some of which appear for the first time in the West, such as the Mordwins, the Uighurs, the Tibetans, and the Samoyeds.[21] Pian di Carpini retailed from Mongol folklore some stories of the monstrous races, which had appeared so colorfully on the mappaemundi. Rubruck, on the other hand, got a different response to his inquiry: "I asked (these same priests) about the monsters, or human monstrosities, of which Isidorus and Solinus speak. They told me they had never seen such, which astonished me greatly, if it be true."[22] Even Rubruck, however, had to admit the existence of the Cynocephales or Dogheads, and Friar Benedict of the Pian de Carpini mission said that he had seen one of their women at the Mongol court.[23]

The travelers named places that are now, even with the most heroic efforts of modern editors, unidentifiable.[24] The problems of dealing with unreliable interpreters and transcribing place-names into Latin script were formidable. In addition, the thorough devastation of central Asia by Tamerlane in the late fourteenth century, meant that some places ceased to exist or, when rebuilt, had another name.

The hardships of travel over land formed a constant theme—the harsh climate and extreme weather, the importunity of the impoverished people travelers came across, the scarcity of provisions, the incomprehension of language, and the lack of spiritual comfort in the form of fellow Christians. Going by sea was worse, however, in Rubruck's opinion. He mentioned the ruse of a Frankish traveler who was being sent back to the West with Mongol envoys who were thought to be spies. "He answered that he would put them to sea, so that they would not be able to know whence they came nor how they had come back."[25] In his concluding recommendations, Rubruck stressed the advantages of the land route to the East, where one was not "exposed to the dangers of the sea or to the mercies of the sailor men, and the price which would have to be given for a fleet would be enough for the expenses of the whole land journey."[26]

Missions on Maps

The great mappaemundi of the late thirteenth century were unaffected by the Tartar assault and the information of Western travelers. The depiction of Asia on the Hereford map is resolutely classical, based on the work of Solinus, Isidore, and Aethicus Ister, the most recent of these dating about 700 CE Although the map is more up to date in Europe, its conservatism in Asia may be noted in the closeness of its presentation to the text of *Expositio Mappae Mundi*, which was composed one hundred years earlier.[27] Russia is "Sithia," China is the land of the "Seres," and the Golden Fleece can be seen spread out in Colchis at the eastern end of the Black Sea. Most of the cities in Asia were those visited by Alexander the Great, if not by Jason, and the monstrous races are found in abundance. In north central Asia can be found the river Acheron, the entrance to hell.[28] The Ebstorf map, the date of which may be somewhat earlier than the Hereford, is equally traditional in its presentation of Asia.[29]

Matthew Paris was the first to recognize the existence of the Tartars in the map of the holy land which he made to illustrate his chronicle around 1253. In the northeast corner he shows the enclosure surrounding the peoples of Gog and Magog, and notes that from this place came the Tartars who "have conquered many territories and destroyed what is called India."[30] The first world map to take notice of these current events does not appear until the production of the Sanudo/Vesconte map corpus in 1320–21. On this world map Asia does not have much detail, partly due to its small size (between twenty-five and thirty centimeters in diameter) and partly because Sanudo's interests lay elsewhere, but we do find the Tartars in two inscriptions: "Hic fuerunt inclusi Tartari" and "Hic convenit multitudo Tartarorum." ("Here the Tartars were enclosed," and "Here gathered a great multitude of Tartars.") The last perhaps refers to the assembly held at the election of the Great Khan in 1246, recorded by Pian di Carpini. "Hic stat magnis canis" ("Here is the Great Khan") is placed vaguely in the area of Mongolia, and "Incipit regnum Catay" ("The kingdom of Cathay begins") marks the border with China. The map also makes note of "Carab, terra destructa," an unidentified city that could stand for many of those destroyed, particularly in the first wave of conquests. Thirteenth-century travelers commented frequently on the number of ruined cities and piles of bleaching bones they encountered.[31]

Other modern names appear in the geographical text that surrounds the world map in most manuscripts: Cathay, the Tartars, Georgia (instead of the classical Colchis), Turkestan, Corasmia, the city of Sarai, Russia (instead of the classical Scythia). The text also mentions a very rich province to the south of Cathay known as Sym, which seems to be Sin, or south China. We also find the Abcas or Abkhazes—Pian di Carpini uses the term *Obesi*—for an ethnic group among the Georgians,[32] which is described as now free from Tartar rule, regnum Tarsae ("Tarsi," or the Uighurs, found west of Cathay and east of Turkestan). Other features include the desert of Belina (north of Cathay), and Cumania, described as north of the Caspian Sea, though the map shows it north of the Black Sea.[33] These names appear to be interpolated into a more conventional geographical description as an updating and are mostly not on the world map, probably for lack of space. Sanudo almost certainly got this information from Hetoum of Armenia, *Flor des Estoires de la Terre d'Orient,* written in 1307. In Book I, Hetoum briefly describes the kingdoms of Asia, giving the same boundaries and directions Sanudo has here. Sanudo also repeats some unusual features mentioned by Hetoum, such as Mount Albors (modern Mt. Elborus or Elburz in Russia, 18,481 feet) and the hundred-day journey across the desert east of Corasmia. Sanudo refers to Hetoum's work several times in his main text.[34] The departure from the conventional picture of Asia presented by the mappaemundi is made clear by the fact that neither Sanudo nor Hetoum mention the location of paradise.

An Excursus on Prester John

Perhaps one of the oddest motives for the epic journeys of this era was the search for Prester John. No traveler was untouched by the legend of this Christian prince of astonishing wealth and equally astonishing virtue. Giovanni di Pian di Carpini, Marco Polo, Odoric of Pordenone, and others asked after Prester John and received varied answers. We now believe that Prester John was based on an actual historical figure, a Chinese prince named Yeh-la Ta-shih, who, driven from his kingdom, established himself in central Asia. Here, in alliance with the Christian Uighurs, he inflicted a crushing defeat on the Muslim Seljuk Turks in 1141. News of this battle filtered back to the hard-pressed inhabitants of the Latin kingdom of Jerusalem and eventually to the court of Rome, and a hope sprang up that the Asian prince might be a useful ally in the crusade against the forces of Islam.[35]

The excitement intensified in 1165 when a letter, purported to be from Prester John, arrived in Europe. The original of this letter was supposedly addressed to Manuel Comnenos, the Byzantine emperor, but no trace of this version has ever been found. Letters in Latin were sent to both the Pope and the emperor, Frederick Barbarossa. In the letter, Prester John described his kingdom, located in "the three Indias," "only three days' journey from Paradise." His land was a treasure house of gold and precious stones, inhabited by exotic animals and a full panoply of the monstrous races. Despite this diverse population, all lived together in utmost harmony, and there was no poverty, wrong-doing, or poisonous snakes. The king was so rich that his bedchamber was constructed of gold and gems with a roof of sapphire and his table, made of emerald, rested on amethyst pillars. Seventy-two kings paid homage to him, but he held no title other than priest (prester) because of its sacredness. It was his intention to visit the Holy Sepulcher in Jerusalem and to restore it to Christian possession.[36]

The letter appears to have been a forgery; in Umberto Eco's picaresque novel, *Baudolino,* he has a group of students at the University of Paris composing it while downing huge quantities of wine.[37] However, the Europeans took it quite seriously. The Pope solemnly replied and in 1177 sent his physician off in the general direction of the fantastic kingdom, but we do not hear that he was able to deliver the Pope's letter. The original letter from Prester John, copied and recopied and further elaborated, circulated throughout a society which was willing to be amazed. Over one hundred manuscripts survive.

At first everyone was certain that Prester John was to be found in Asia, and he appears there on the earliest world maps, those of Marino Sanudo and Paolino Minorita. But the term "three Indias" was vague enough to encompass Africa as well, and on Angelino Dulcert's map of 1339 Prester John appears in southern Africa west of the Nile.[38] Fifteenth-century maps generally locate his kingdom in Africa, but some, such as Andreas Walsperger (1448), continue to place it in Asia. Others, hedging their bets, put him on both continents. This is true of the Genoese world map of 1457 and the globe of Martin Behaim (1490). Of course there was a real Christian kingdom in Africa, but when Europeans finally arrived there, they were disappointed to find that it was not nearly so grand or perfect as legend would have it, and besides the Christians in question were heretical Monophysites.

Despite the fantastic qualities of Prester John's kingdom, the quest for it inspired travelers well into the age of exploration. In the fifteenth century we see the Portuguese king, Joao II, anxiously questioning travelers from Africa about the priest-king, and in 1487 he dispatched an expedition to cross Africa by land in order to find him. When Vasco da Gama arrived in Mozambique in 1498, he noted in his journal that the kingdom of Prester John could not be far off.[39]

Merchants

"There are only seven climes in the whole extent of the world," wrote Matthew Paris; "namely those of the Indians, Ethiopians or Moors, Egyptians, Jerusalemites, Greeks, Romans, and French, and there are none so remotely situated in the whole of the habitable part of the world, that merchants will not find their way amongst them."[40] Where missionaries could go, merchants could follow, the drive for profit being about as strong as the drive to make converts. The Mongols did not have much that was very attractive to a trader, except perhaps for the northern furs, but China, which lay beyond, was the producer of silk. To the south lay India and the islands that were the source of immensely valuable spices. When Giovanni di Pian di Carpini returned to the west, he reported the presence of a number of Western merchants in Kiev, including "Michael the Genoese, Manuel the Venetian, and Nicholas Pisani."[41]

The two Polo brothers of Venice, Niccolò and Maffeo, went to China almost by accident. The Polo family had an establishment in Soldaia in the Crimea, and, prompted by the possibility of trading their jewels, the brothers set off sometime around 1260 for the court of the khan of the Golden Horde near Sarai, on the Volga. No sooner had they arrived there, than the Greeks reconquered Constantinople from its Latin rulers and things became very hot for Venetians in the Black Sea. The brothers might have originally intended to travel south to return to the Mediterranean and home via Persia, but a war broke out between Berke Khan of Russia and Hulegu, the il-khan of Persia, making travel risky for merchants of any nationality. Heading north and east they eventually arrived at the court of Kublai Khan on the borders of Cathay.[42] Cordially welcomed by the Great Khan, the two stayed for some time at his court, returning charged with a mission to the Pope. The khan requested one hundred scholars, well versed in the Christian religion and able to hold their own in debate, as well as some holy oil from the lamp in Jerusalem. The Polos, along with their Mongol escort, were given one of the famous gold tablets, which worked as a pass throughout the Mongol kingdoms, and were sent on their way. After a lengthy journey via Ayas in Little Armenia and Acre, they arrived in Venice. A papal interregnum made it difficult for them to carry out their mission immediately, and they were already on their way back to the khan, accompanied now by Niccolò's son, young Marco, when they heard that a new pope, Gregory X, had finally been elected. Backtracking from Ayas to Acre, where the newly chosen pontiff was, they were given two Dominican friars (not one hundred) and the holy oil. But the two friars were less hardy than the Venetian merchants and dropped out early on, frightened by the prospect of war between the Mongols and the Mamluks of Egypt in Armenia. The date was now 1271.

The Polos retraced their steps to Cathay, meeting up with the court at its summer retreat at Shangdu. They were to remain in China for the next two decades. Marco's book was written in 1298, while he was enjoying some enforced leisure as a prisoner of the Genoese, having been captured in a sea battle. The book, ghostwritten or "as told to" Rustichello, a writer of contemporary pulp fiction, was at first greeted with some suspicion. The richness and size of China, a country barely heard of by Europeans and on the other side of the world, was surely exaggerated. Marco's description "millions of people," "millions of houses," and "millions in revenue" led to his nickname "Il Milione."[43] We know now, however, that his information was founded on fact and that the kingdom of Cathay was as grand and wealthy as he had portrayed it. His book, partly based on his own travels and partly on Mongol sources, was full of information of particular interest to merchants.[44] The size of the markets, the rich goods traded there, and the profits that could be made were all faithfully recorded, culminating in the epic account of Quinsay (Hangchow) with its ten huge market squares visited by forty to fifty thousand shoppers three days a week. After listing the many commodities for sale, Marco pinned down his argument by stating the amount of pepper, a luxury item in the West, consumed in Quinsay each day: forty-three cartloads of 223 pounds apiece, for a total of nearly five tons.[45]

The Polos returned from China via the Indian Ocean in 1292, having been entrusted by the khan with the transportation of a Mongol princess, Kokachin, intended as a bride for Arghun, the il-khan of Persia. Marco's description of the China Sea and the Indian Ocean islands was a revelation to Europeans; even the learned men of classical antiquity had little knowledge of India beyond its western coast. There are 7,448 islands in the China Sea, he announced, and 12,700 in the Indian Ocean, many of them rich in desirable trade goods, such as camphor, sandalwood, nutmeg, cloves, and assorted varieties of pepper. Java was the biggest island in the world, while Sumatra and Ceylon were not much inferior. Although there were hazards at sea and some rather terrifying phenomena ashore, such as cannibals, enormous griffins, and "idolaters," there were also tempting commercial opportunities. Of the islands in the China Sea, Marco reported, "there is no tree that does not give off a powerful and agreeable fragrance and serve some useful purpose—quite as much as aloe wood, if not more so. There are, in addition, many precious spices of various sorts. The islands also produce pepper as white as snow and in great abundance, besides black pepper. Marvelous indeed is the value of the gold and other rarities to be found in these islands . . . When ships from Zaiton or Kinsai come to these islands, they reap a great profit and a rich return."[46]

The Polos continued, being marooned in Sumatra, probably waiting for the monsoon, for five months. They went on to explore the west coast of India, which Marco called "the best of all the Indies," thanks to its profusion of spices.[47] They sailed into the western Indian Ocean, and the book contains descriptions of Zanzibar, Madagascar, and Abyssinia, though the Polos almost certainly did not visit them. Delivering the princess to Persia, they discovered that Arghun had died in 1291, but the young lady was married off to Ghazan, his son, who succeeded to the throne in due course. The Polos traveled on to Trebizond, and through the Black Sea to Greece and then to Venice, where (the story goes) they were at first not recognized and then treated as though they had come back from the dead.[48]

Mixed with fantasy, some of it from traditional Mongol lore and some embellished by the professional writer of romances, *The Travels* provided a treasure trove of new geographical material. John Larner says, "Never before or since has one man given such an immense body of new geographical knowledge to the West."[49] The chapter on Japan was the first notice of that island's existence, although Marco himself did not go there and most of the information is bogus.[50] China, southeast Asia, and Indonesia were all news, as were the wealth and populousness of this previously unknown part of the globe. The names of cities, provinces, rivers, and peoples, as well as indications of distance, gave a shape and content to what had formerly been at best misty, or perhaps empty, territory. It is probably not surprising that this huge indigestible heap of information did not immediately get transferred onto maps.

One problem, of course, is that Marco appears to have been completely innocent of any knowledge of classical geography. The Scythians and the Seres do not stand in the way of his description of Russians, Mongols, and Chinese. The absence of the traditional framework made it difficult for the educated classes to absorb what he was telling them and, even more, to believe it. His successors were to agonize over Ceylon, Sumatra, and Java, trying to decide which was the classical Taprobana. Marco either did not know about Taprobana or did not care. When the information he provided finally found its way onto a map, the Catalan Atlas of 1375–80, it was placed in slightly uneasy juxtaposition to more ancient geographical lore. (See chapter 3.)

Despite the appalling rigors of the journey, the Polos had successors. We hear of merchants, as well as missionaries, well established in major Chinese and central Asian cities in the early fourteenth century. At that time, Francesco Balducci Pegolotti, a Florentine merchant employed by the Bardi bank, compiled a handbook for trade. His book *La Pratica della Mercatura* gave information on trade routes, customs duties, weights and measures, and the availability of various products for trade. Pegolotti himself probably went no farther east than Cyprus, where he compiled most of his data, but he included several pages on trade with China, based on the report of Genoese merchants.[51] He described the route departing from the port of Tana in the Crimea and going by way of Astrakhan, Sarai (on the Volga), Urgench, Otrar, and Almalyk to Quinsay, traveling by camel cart, river boat, horseback, and donkey. The major problems were the need for reliable interpreters and guards. Though Pegolotti said, "The route from Tana to Cathay is very safe both by day and by night," he recommended hiring an armed escort, as well as traveling with a caravan. His sources explained how paper money, the currency of Cathay, operated, and gave opinions on the best goods to bring along: fine linens (for they do not care for the coarser kind) and silver. The product they were buying was principally silk. Although the quality of Chinese silk was not particularly good, it was inexpensive, even including the cost of transport. The first record we have of Chinese silk in the west was at the fairs of Champagne in 1257, and from then on it was in demand.

The evidence for the presence of Western merchants in Asia is limited by the notorious secretiveness of their trade. Merchants were not likely to proclaim where they were obtaining profitable merchandise, and, in the records of the formation of commercial partnerships, citizens of Genoa are very sly about their intended destinations. For example, the Vivaldi brothers, who set sail around

Africa in 1291, listed their objective as "Romania" or the Byzantine empire, which lay in the exact opposite direction from that in which they planned to go.[52] Marco Polo is most unusual in his fulsome description of the East in which he had sojourned, perhaps because he was thinking not as a merchant but as an administrator, in which capacity he was employed by Kublai Khan. But even he inclined to secrecy about mercantile activities. In Campichu we learn that Marco and his uncle Maffeo spent a year "on business of theirs not worth mentioning."[53] We hear of Genoese merchants employed as envoys going back and forth from the khans to the west, useful for their knowledge of the languages and customs on both ends of the journey.[54]

The Travels of Friar Odoric

Other references to the mercantile presence appear in works such as the travel book of Odoric of Pordenone (c. 1318–30). Odoric, an Italian and a Franciscan, set out for the East in order to "win some harvest of souls."[55] According to his book, he left Venice for Constantinople and traveled across the Black Sea to the port of Trebizond. From there he went by land to Tabriz and Sultaniyeh, and then to the port of Hormuz, from which he sailed to India, where he collected the relics of the four martyred Franciscans whose story dominates the center of his book. His route took him around India to Ceylon, through Indonesia to Vietnam and thus to China. In Zaiton he deposited his precious relics at a Franciscan monastery.[56] His return route is not clear, and it seems unlikely that he saw, as he indicated, Tibet, the kingdom of Prester John, the haunt of the Old Man of the Mountain, and hell itself. He did not tell of making any conversions during his trip, but tradition attributes the harvest of twenty thousand souls to him, and he was nominated for sainthood shortly after his death, eventually receiving beatification in the eighteenth century. Despite his holy mission, Odoric was quick to notice the commercial possibilities of every place he visited: salt is so plentiful in Tabriz that one can get it free and without tax, pepper is harvested in Malabar, gold and tin are abundant in Sumatra, and spices in Java. He also exclaimed frequently about the low prices, for example, in the teeming food markets of Censcalan (Canton), where one could buy three hundred pounds of ginger for less than one groat.[57] He said he would not even begin to tell of the greatness of Cansay (Quinsay, Hangzhou), except that so many people from Venice had already been there and perhaps would vouch for him.[58]

A number of Odoric's tales bear a suspicious similarity to Marco Polo's text, such as the statistics on Quinsay (one hundred miles in circumference and 12,000 stone bridges) and the tale of the Old Man of the Mountain.[59] Perhaps he was familiar with *The Travels* or had heard some of Marco's tales in Venice. Odoric's simple, straightforward style and good stories made the book an instant success, and it circulated through the network of the Franciscans. At least a hundred manuscripts survive until this day, and—the surest sign of flattery—his work was copied by others, most notably, John Mandeville.[60] Odoric's book was also incorporated in the great travel compendium, the *Livre des Merveilles,* assembled by Jean Le Long of Ypres in 1351. By the end of the century not only Marco Polo but John Mandeville himself had joined the collection.

Odoric found his own church established in several parts of the East, with Franciscan houses in Zaiton, Hangzhou, and Yangzhou in China. In Cambalech (Beijing) he reports that "we Minor Friars have a place assigned to us at court."[61] The Dominicans also had establishments in Asia, and bishops were installed in Tabriz, Sultaniyeh, Samarkand, Urgenj, and Beijing, though their constituencies came mostly from the foreign population. All was not smooth sailing, as is witnessed by the martyrdom in India of the Franciscans, whose bones Odoric went to collect. Within a few years other Christians in Asia were to meet the same fate, as the Mongols, now converted to Islam, became less tolerant of religious diversity.

Ibn Baṭṭūṭah

Of all the late medieval travelers, the one who seemed to be most strongly motivated by curiosity alone was Ibn Baṭṭūṭah (1304–68).[62] Although his journeying began with a pilgrimage to Mecca in 1325, he did not return for thirty years, visiting most of the countries of the Islamic world from southern Africa to eastern Asia. Born in Tangier, he had been educated in the law and was able to put his skills to use as a judge in some of the far-flung places he landed. The universal use of Arabic in the lands of Islam was an additional advantage to a traveler. Ibn Baṭṭūṭah used the respect accorded to an educated man to demand special privileges and was quite annoyed when his reception was not as honorable as he felt to be his due. He traveled through the islands of the Indian Ocean, to China, and into the dreaded Sluggish Sea, from which he was lucky to escape. His book is really an autobiography, in contrast to Marco Polo's more impersonal account. We hear about his minor and major illnesses, his love affairs, his triumphs and setbacks. The book, written with the aid of a professional writer, was popular in the Islamic world, and Ibn Baṭṭūṭah today is remembered fondly in Tangier, where the airport, a street, and a hotel are named after him. Whether his book had any impact on the European West is another question. It is possible that the Catalan Atlas records some of his findings, for the barrier between the Islamic and Christian worlds was most porous in the western Mediterranean. (See chapter 3.)

On his way home through Syria in 1348, Ibn Baṭṭūṭah was a horrified witness of the ravages of the Black Death. By the time he reached Damascus the death toll was two thousand per day, and the entire population of Jews, Christians and Muslims had joined in praying for mercy to their respective gods.[63] The epidemic was one reason why the century of lively communication between the West and Asia was to come to an end. The dramatic drop in population—Cairo is thought to have lost almost half its people—meant a smaller labor force and a concentration on basic needs, such as food production, rather than manufacture for the luxury trade.[64] At the same time the Mongols were fighting among themselves, and in the 1330s the il-khanate of Persia broke up into hostile factions. An attack on the Christian population of Almalyk in 1340 was an additional discouragement, and Christian churchmen in their small enclaves in Asia now more often found martyrdom instead of converts.[65] Commercial crises in Italy, such as the collapse of the Bardi bank, were unrelated to either plague or Mongol warfare but contributed to the decline of long-distance trade. Finally, in 1368 the Yuan (Mongol) dynasty was overthrown by the native Chinese Ming, which was much less interested in international outreach, expelling all foreign Christians in 1369.[66] None of these factors can be considered decisive, but we can see the fragility of an extended system of trade like that which thrived from 1250 to 1350. A few breaks in the chain and merchants were no longer willing to venture themselves or their capital on a too-risky proposition. After the mid-fourteenth century, there were no more eyewitness travel accounts of the Far East. There were travel books, but the authors were not travelers themselves but writers and compilers of the works of others.

Travel Writers

Travel literature is still a popular genre, but it is also treated with suspicion; a recent anthology of excerpts from travel classics was entitled *Travelers and Travel Liars*.[67] Giovanni di Pian di Carpini gave as a guarantee of the authenticity of his information that he either saw it all with his own eyes or heard it from "Christians worthy of our trust."[68] He admitted the possibility that some of his stories might not be believed due to their exotic nature but thought it cruel that a man who only wished to do good should be defamed by others. Here we look at two fourteenth-century travel books, both of which have a special relationship to cartography: *The Travels of Sir John Mandeville* and *El Libro del Conoscimiento de Todos los Reinos* (The book of knowledge of all kingdoms).

The most popular travel book of the late Middle Ages was the work of Sir John Mandeville, a self-described knight of St. Alban's in England, who tells us that he set out to travel the world in 1322, beginning with a journey to the holy land. Before he came home again some thirty years later, he had been to Armenia, India, Indonesia, Mongolia, Cathay, Tibet, the kingdom of Prester John, and even within earshot of the rivers of paradise. Who Mandeville actually was and where he actually went have been the subject of scholarly disputes. His book is in part a pastiche of excerpts from other books, drawing heavily on the work of William Boldensele for the holy land, and Hetoum of Armenia and Friar Odoric for Asia, though apparently not Marco Polo.[69] What we would call plagiarism was looked upon quite differently in the Middle Ages than it is today. Some medieval writers hastened to assure their readers that they were not responsible for a single word of their books but instead copied from the very best authorities. We hear no such humble protestation from Mandeville, who instead makes a positive virtue of his first-person experience, whether that experience was actually his or another's, and everywhere inserts his personality into his book. He does not just copy his sources but rearranges them, expands them, and adds his own observations and conclusions.

Mandeville's world view, as he begins his book, appears to be that of the mappamundi with Jerusalem in the center, the earthly paradise in the Far East, and the monstrous races on the edge of the world. Like the mappamundi, his account of the holy land takes up almost half the book, emphasizing its importance to his Christian readers, whose desire to see it in person was thwarted by its present inaccessibility.[70] He occasionally steps back to give a cartographic view of space, for example, when he describes how to get to the holy land from western Europe by way of Constantinople.[71] After a tour of the holy places, he returns to the question and gives a number of alternative routes to Jerusalem, both by sea, via Cyprus, which he says is the quickest way, and by land, for those who fear to travel by sea.[72] Again, when he leaves the holy land, he gives an overview of the world that lies to the east:

There are many diverse kingdoms, countries and isles in the eastern part of the world, where live different kinds of men and animals, and many other marvellous things. Those countries are divided by the four rivers that flow out of the Terrestrial Paradise. Mesopotamia and the kingdom of Chaldea and Arabia are between the two rivers, that is to say the Tigris and the Euphrates; the kingdoms of Media and Persia are between the Tigris and the Nile; and the kingdoms of Syria, of Palestine, and Phoenicia are between the Euphrates and the Mediterranean Sea, which stretches from the city of Morocco on the Spanish Sea to the

Black Sea. So it stretches 3,040 Lombardy miles beyond Constantinople. Towards the sea which is called Ocean is the kingdom of Scythia, which is surrounded by hills. Below Scythia from the Sea of Caspian to the River Don is the land of Amazonia, which is the Land of Women, where women live by themselves with no man among them. Then there is the realm of Albania, a great land; it is called Albania because the people of that land are whiter than the people of the lands round about . . . Then there is the land of Hircany (Hircania), of Bactrice (Bactria) and many others. Between the Red Sea and the Great Sea Ocean to the south is Ethiopia and upper Libya. For Libya the Lower begins at the Spanish Sea, where are the Pillars of Hercules, and reaches to Egypt and Ethiopia.[73]

The layout of Asia and Africa and the place-names he uses are very close to those on a thirteenth-century mappamundi. It is interesting that, in the pages that follow, he does not repeat these names but instead substitutes modern ones. Instead of Scythia, Bactria, Hircania, Albania, and Amazonia, we hear of Tartary, Turkestan, Khorasan, Manzi (south China), and Cathay. As fourteenth-century mapmakers began cautiously to blend the traditional geographical lore of the mappamundi with the modern experience of travelers, Mandeville does the same. But rather than denying the old construct, he usually just piles the new one on top. For example, he describes the numerous islands in the Indian Ocean which had been explored by Marco Polo and Friar Odoric, but he retains the gold and silver islands of Chryse and Argire, which date back to Pliny's *Natural History* and can be found on most medieval world maps.[74] Christiane Deluz creates an interesting pair of maps in her book, one based on the text of Brunetto Latini's encyclopedia and the other on Mandeville's *Travels*. While Brunetto's is a conventional circular mappamundi, Mandeville's is a double-hemisphere map, closer to Behaim's globe than to the Hereford map.[75]

If Mandeville were looking at a mappamundi, he needed to account for the monstrous races, which in the Hereford map are arranged mostly along the southern rim in Africa. He never went to south Africa, despite the grandiose claims in his opening passage, so he puts the monsters on suitably remote islands in the Indian Ocean.[76] This location is not so different as it might seem to us, for in the fourteenth century in Europe (and earlier in the Arabic world) world maps were beginning to show an Africa that curved around to the east, as a southern boundary to the Indian Ocean. Contemporary travelers, such as Marco Polo and Odoric, had been unsuccessful in finding these paragons of human diversity, and Mandeville jumbles them all together in a page or two, without giving much detail. They had to be included as part of the traditional picture of the world, but they were becoming something of an embarrassment. His illustrators, however, were more anxious to feature them, and the monstrous races tend to dominate the pictures that accompany various editions of the *Travels*.[77]

What was original in Mandeville's work? Deluz is willing to concede that he did go to Constantinople and the holy land, where he comments disparagingly on other writers and adds specific details of his own, such as precise measurements of distance and dimensions of buildings. He also uses words derived from Arabic and includes some anecdotes not found elsewhere, which he must have picked up locally.[78] Deluz's list of Mandeville's sources for the holy land include Jacques de Vitry, William of Tripoli, Hetoum (Hayton) of Armenia, Vincent of Beauvais, and Brunetto Latini, as well as his major source, William of Boldensele. William, a Dominican from Minden in Germany, went to the east in 1333 and wrote his book in 1336 upon his return.[79]

Lest one imagine Mandeville as some sort of fourteenth-century graduate student, grubbing in the library, it is important to note that most of his sources could be found in one place, the *Livre des Merveilles*, compiled by Jean Le Long in 1351. This compendium included William of Boldensele, Odoric of Pordenone, and Hetoum of Armenia, among other shorter works. Mandeville's dependence on this source is revealed by the fact that he usually follows Le Long's French translation instead of the Latin of the originals.[80] The problem with diligently tracking sources is that much of the material Mandeville reports was circulating not only in books but in the oral tradition as well. To take some obvious examples, when he says that Jaffa is the oldest city in the world, founded by Japhet, or that Muslims are forbidden to drink wine, he could be getting this information from his own experience rather than finding it in a book. It is the rare traveler who sees what absolutely no one else has seen. In addition, some of the most interesting passages of the book are those that seem to be unique to Mandeville. After a brief survey of the beliefs of Islam (mostly accurate, though very much from a Christian point of view), he describes a dialogue with the sultan of Egypt. The sultan's trenchant criticism of Christendom, in its neglect of its own principles, is the forerunner to Gulliver's dialogue with the King of Brobdingnag. After hearing the sultan's harsh judgment, Mandeville is forced to concede the moral superiority of the Muslims, saying, "It seemed to me then a cause for great shame that Saracens, who have neither a correct faith nor a perfect law, should in this way reprove us for our failings, keeping their false law better than we do that of Jesus Christ."[81] In contrast, in the midst of his account of Egypt, William of Boldensele launches an attack on the falseness of Islam and the perfidy of Mohammed, who "first preached to savage and ignorant men in the Arabian desert and imposed his diabolical law upon them."[82]

Mandeville frequently makes comments, draws conclusions, or expands upon his sources. Odoric, for example, is almost staccato in style. Although he occasionally recounts a personal observation or an anecdote, he often just lists sights or cultural characteristics and not infrequently concludes, "And there be many other marvellous and beastly customs which 'tis just as well not to write."[83] Mandeville is never deterred by the marvelous and beastly, but regales us with gusto. He is also much less judgmental than some of his sources. As we have already seen, he credits the Muslims with a strong moral sense, and, as he travels around the world, he views the beliefs of others with sympathy or at least an attempt at understanding. Marco Polo had dismissed the population of most cities in China as "idolaters," while Odoric shakes his head over "abominable superstition" and "detestable custom" in the places he visits.[84] Mandeville, after making note of the worship of idols in India, reflects at some length on the difference between simulacra, made in the likeness of some natural being, and idols, which are not natural. The worship of simulacra he sees as honoring some famous person or being made by God. He may have taken this distinction from Isidore, but certainly not his comparison of simulacra to crucifixes and statues of the saints and the Virgin Mary.[85] As for ox worship, he explains that the ox is the most useful being on earth and also the holiest, "for it does much good and no evil." He gives a sympathetic description of a Tibetan funeral, where the chief mourner boils the flesh off his father's skull and offers bits to his friends, preserving the skull as a drinking cup.[86] He is even kind to schismatic Christians, held in special abhorrence by the traveling friars, and in his conclusion he announces: "Know that in all those lands, realms and nations, except for those inhabited by men lacking reason, there is no people which does not hold some of the articles of our faith. Even if they are of divers beliefs and creeds, they have some good points of our truth. And generally they believe in God who made the world."[87] The exception to his tolerance is the Jews. Mandeville blames them for the torture of Christ and accuses them of a failed plot to poison the world. He also says that the fierce tribes walled up with Gog and Magog are Jews, and, when they finally burst forth, they will join with the Jews of the West in destroying the world.[88]

Some of his reflections are scientific, and these probably extended his popularity into the age of discovery. Odoric describes how the polestar begins to disappear below the horizon on the island of Lamory, perhaps a part of Sumatra. Mandeville adds that there is an Antarctic polestar (which there is not) to guide people in the southern hemisphere, thus proving that the earth is round, "according to what I have seen."[89] He describes sightings with an astrolabe in the northern and southern hemispheres, followed by some mathematical calculations that lead him to aver that he has traveled over three-quarters of the world in terms of latitude. In fact, he says—and this observation was to inspire generations of explorers—"So I say truly that a man could go all round the world, above and below, and return to his own country, provided he had his health, good company, and a ship."[90] He gives an estimate of the circumference of the earth (20,425 miles) and tells a fascinating anecdote, which he says he heard in his youth, of a "worthy man" who inadvertently circumnavigated the world until he came to a place where he heard his own language spoken.

Mandeville's book was so popular that over three hundred manuscripts still survive, as well as thirty-five editions in six languages printed before 1500. Its popularity was not based solely on its readability but also on its reputation for credibility. Mapmakers used Mandeville as a resource, perhaps beginning with the maker of the Catalan Atlas. The atlas contains a number of features mentioned in his book—burial customs in Asia, the configuration of the Caspian Sea, and the existence of several towns in China not named by Marco Polo—but these may have been derived from another, similar source, such as Odoric. Mandeville's work continued to be consulted throughout the fifteenth century, as he was quoted at length on Martin Behaim's globe of 1492. Even in the sixteenth century he was treated with respect, and we hear that Abraham Ortelius made a special pilgrimage to his supposed grave at Liège in 1584 and wrote a letter about it to his fellow cartographer Gerard Mercator.[91] In his world map of 1569, Mercator, more circumspect, had credited Mandeville with providing accurate information about Java and the southern hemisphere but described him as "an author unbelievable in other respects."[92]

Of course, some of this esteem for Mandeville was due to his use of reliable sources, such as Odoric of Pordenone. Another relevant point is that land travel across central Asia slowed dramatically in the mid-fourteenth century, and there were few travelers to dispute his information until the Portuguese began to penetrate the Indian Ocean at the end of the fifteenth century. Also, like other medieval books, editions of Mandeville vary a great deal, and readers could pick up a version or section that appealed to them and leave the rest.[93] It is fascinating to reflect, however, that Mandeville's book was in the hands of many of the great explorers of the Age of Discovery. In 1576, Sir Martin Frobisher took it to the northwest of America as a useful source in case he got through to China, and Sir Walter Raleigh quoted it respectfully in his *Discovery of Guiana* (1596).

Libro del Conoscimiento de Todos los Reinos

Another travel book that had an impact on maps and explorers is the *Libro del Conoscimiento de Todos los Reinos* (The book of knowledge of all kingdoms). The anonymous narrator described his extensive and somewhat improbable travels all over the known world, and some of the unknown, beginning in Spain.[94] The style is dry and the itinerary summary: "I departed Fez, which they formerly called Cotamanfez, and went to Miquynez and to Ribate, and on to Tanjar on the seacoast, and from there to Arzila, and I went along the coast to Raxy, and from there to Cale, a city on the coast of the Western Sea."[95] It sounds as though the author is simply listing place-names from a map, rather than making a journey through actual space. Occasionally, the author mentions fellow travelers; he "entered the boat of some Germans," and "after this I departed Persia with Christian merchants who were coming from Catayo."[96] Mostly, however, he simply lists places, only rarely breaking into a brief historical excursus or elaboration, all of which are so concise that they could easily fit upon a map as legends. Scholars have long been impressed with the similarity of the *Libro* to fourteenth-century maps of the Catalan school, in place-names and legends as well as general structure. In style and content it is a sea chart cum mappamundi presented in prose.

In general, the nomenclature is modern. When the author uses a classical place-name it is nearly always in a historical context, telling what a place used to be called. He does not completely omit traditional features of the medieval world map. For example, he describes Alexander's confinement of Gog and Magog in North Asia but adds a touch of verisimilitude by telling us: "And in this castle of Magot I lived for a time because every day I saw and heard marvelous things."[97] More contemporary is his description of the Indian Ocean as open to the Eastern Sea and with a large continent to the south, comprising one tenth of the earth's surface. He places the earthly paradise near the Antarctic pole, instead of in the east, and says that the noise of the four rivers rushing forth has deafened the population in the immediate vicinity, a detail also recounted by Mandeville.[98] What he calls Sçim, or Sym, is clearly China. He also puts "Trapovana" west of Java and distinguishes it from Ceylon, which was its classical identity. This is the beginning of a long dispute over the whereabouts of this island. He describes the Caspian Sea and gives a sampling of its numerous names: the Sea of Sara, of Bacu, or Jorgania, of Quillon, as well as the Caspian Sea.[99]

The book dates from the third quarter of the fourteenth century and was early on attributed to a "Spanish Franciscan."[100] Of course, the Franciscans were great travelers, but there is no internal evidence to suggest that this particular author was a member of the order, and in two of the surviving manuscripts there is an author portrait of a man in definitely secular costume.[101]

The most arresting part of the author's travels is a purported trip along the west African coast, rounding Cape Bojador (here Cape Buyter) and going south to the River of Gold.[102] The cape was purported to be a great navigational hazard, and the Pillars of Hercules were moved from Gibraltar to this point in order to indicate that sailors should go no further.[103] No other fourteenth-century source, book or map, shows the African coast south of the cape, and in 1434, when a Portuguese ship dispatched by Prince Henry the Navigator sailed past the said cape, it was heralded as the first to do so. On this particular point, the veracity of the *Libro*, described as a "geographical novel" by its latest editor, has been considered particularly shaky.[104] However, we know that the cape had been rounded at least as early as 1401, and in general our knowledge of the chronology of the exploration of the Atlantic is sketchy. For example, the exact dates of the discovery of the various island groups are in dispute. The Azores, colonized by the Portuguese in 1439, appear on the Catalan Atlas (c. 1380) and are certainly here in the *Libro* several decades earlier.[105] It is even possible that Henry's sponsorship of navigation along the African coast and into the Atlantic was inspired by the *Libro* itself. He certainly was familiar with the epic *Le Canarien* (1402), which described an expedition by a group of French adventurers, themselves using the *Libro* as a guide. The Navigator would surely have been intrigued to hear that the merchants trading along the coast "made a great profit."[106]

Another feature of the *Libro* is its author's inclusion of coats of arms for all the kingdoms he purportedly visited. Over one hundred shields, some fanciful and some strictly accurate, are illustrated in the manuscripts. For example, he shows the shield of England: four quarters, two with gold lilies to indicate the king's kinship with the French royal house and two with three gold leopards. This flag was first used by Edward III in 1340 and ceased to be used in 1406, bracketing the period when the book was written. For countries more remote from Europe, our author probably invented a few flags, such as the green palm tree flanked by two gold keys for the legendary kingdom of Organa in Africa. The book's obsession with heraldry has caused its latest editor to speculate that the author might have been a herald himself. Similar shields, though not so many, appear on sea charts of the fourteenth century.[107] So our traveler may have been purely or partly a traveler on a map rather than one in reality, yet, like Mandeville, he was to have his believers. Several later maps, such as the Catalan world map at the Biblioteca Estense in Modena, show evidence of the use of his book.[108]

Other fourteenth-century travel writers did not even pretend to have made the journeys they described but based their writings on books. The French writer Christine de Pisan wrote *Le Chemin de Long Estude* in 1401, drawing on Mandeville, among others. Since her journey is frankly imaginary, she has no hesitation in approaching the earthly paradise, which even Mandeville feared to enter, with the Sybil as her guide.[109] Another example is the lengthy verse introduction to Ptolemy's *Geography* composed by Francesco Berlinghieri and printed in 1482. Here Ptolemy acts as his guide, and Berlinghieri takes on the daunting proposition of turning his dry prose, including the technical instructions, into lively verse. Dubbing his poem "The Seven Days of Geography," Berlinghieri occasionally inserts his own material, adding modern toponymy in Europe and occasionally putting in a few historical, mythical, or ethnographic tales. In addition to Ptolemy, Berlinghieri apparently had a sea chart, for placenames on the Mediterranean coasts reflect this use.[110]

Conclusion

A world-wide economic boom from the mid-thirteenth through the mid-fourteenth century created possibilities for trade and travel which had not existed since the days of the Roman and Han Empires. Goods for export, an increasing demand for exotic imports, and merchants in eager pursuit of profits fuelled, if briefly, a global economy.[111] Textiles, easily transported and high in value, were one of the primary goods shipped back and forth—linens and woolens from the West and silks from the East. At the same time, the political situation changed. The Pax Mongolica made travel along the great central Asian caravan routes safer than it had been, and the caravan towns such as Urgench, Almalyk, and Samarkand, enjoyed a burst of prosperity. In Europe, the Crusades had encouraged travel to the eastern Mediterranean, and, once the Latin kingdom began to fall apart in the late twelfth century, Christian rulers were interested in finding an ally against the great Muslim empire centered in Egypt. Rumors of a potential Christian ally in the East caused ambassadors to join the steady stream of merchants, missionaries, and adventurers traveling the long roads between East and West.

At home readers had always had an appetite for travel books, but now they were intrigued by those of their own time. The early crusaders had written exciting accounts of the conquests of 1099. Not all of them were writers, but others returned with tales of adventure, heroism, and hardship to tell to those at home. As more Christians settled in the East, those at home began to be aware of a wider world beyond the holy land, learning by such messengers of events from farther Asia. Writers such as Jacques de Vitry and William of Tyre, and his continuators, added to their books information about other lands and peoples formerly unknown to the west. The appeal of the eyewitness challenged the classical lore, possibly appealing to a more popular audience, and readers were thrilled by the line, "and I was there and saw it with my own eyes." It was this desire that inspired Giovanni di Pian di Carpini to add to the eight analytical chapters of his *Historia Mongolorum* a ninth retailing his itinerary and his sufferings on his epic journey. The first-person adventure book became so popular that it even inspired those who had not been away to compose travel fictions, based on the works of other, real travelers.

The travelers of the thirteenth and fourteenth centuries presented a challenge to the traditional world map format. Those who went to Africa south of the Sahara or to China or to the Indian Ocean saw not only places formerly unknown in Europe but also an arrangement of geographical space different from that conveyed by the mappaemundi. If we think of the mappamundi as a paradigm of the conception of the organization of the world in the High Middle Ages, we see that the data brought back by explorers, coupled with the challenge of the new way of mapping seen in the sea charts, began to break it apart. Mapmakers took the challenge of eyewitness reports seriously and tried to find room on their maps for such places as Tartary, Cathay, Java, and Ghana. The enclosed Caspian Sea and the open Indian Ocean, with its wealth of islands, had to be incorporated. In some early attempts, such as the Sanudo/Vesconte world map and the Catalan Atlas, the mapmakers attempted to blend the new data with the old picture. The tension between these two world visions became even greater in the fifteenth century.

CHAPTER FIVE

The Recovery of Ptolemy's Geography

The excitement in intellectual circles in Florence was palpable in 1397 when Manuel Chrysoloras arrived from Constantinople bringing a copy of Ptolemy's *Geographia*. The wealthy book collector Palla Strozzi immediately took Chrysoloras under his wing, and in his will, made in 1462, he bequeathed the precious volume to his sons and nephews.[1] The *Geography* Chrysoloras brought was, naturally, in Greek, and he began at once to make a translation into Latin. The work was completed by Jacopo Angeli da Scarperia (1360–1410/11) around 1409 and dedicated to Pope Alexander V (1409–10).[2] A mania of copying ensued, as many intellectuals were eager to possess or at least peruse a copy.

Ptolemy, the great Alexandrian scientist of the second century CE, was already well known in the West for his other works, especially the *Almagest*, his magnum opus on astronomy, which had been translated into Latin in the twelfth century. Although the *Geography* had disappeared from the West in the early Middle Ages, it had never been entirely forgotten. It is mentioned in a scattering of medieval works and apparently existed in some form in the Arab world from the ninth century on, but the entire work that now appeared was a revelation, offering a new way of seeing and charting the world.[3]

The copy of the *Geography* that Chrysoloras brought to Italy was accompanied by a world map and twenty-six regional maps.[4] These maps had been constructed, according to Ptolemy's instructions, in Byzantium in the fourteenth century some time after 1295, when a copy of the text (without maps) had been resurrected by Maximus Planudes.[5] As far as we know, no Ptolemaic maps had survived from antiquity; there is even some discussion about whether he in fact had made them. The first Latin translation of the *Geography* did not include the maps. They were more complicated to produce than the text, but copyists were at work on them by 1415. The oldest surviving Latin manuscript with maps is at the city library of Nancy in France. The text was copied in 1418 at the order of Cardinal Fillastre of Reims, and the maps were added in 1427. The date of another early version with maps is uncertain but it is from about the same time.[6]

Early Humanist Geography

The humanist interest in geography dates back to Francesco Petrarch (1304–74), who had sought out ancient works on the subject, successfully unearthing copies of Pliny's *Natural History* and Pomponius Mela's *Cosmographia.*[7] He filled the margins of these books with annotations and also wrote some geographical works himself, for example, *Itinerarium ad Sepulchrum Domini* and *Africa.* Petrarch's main concern was the correct identification of classical place names. He complained that it was difficult to understand a work such as Virgil's *Aeneid* if one was not sure whether the author was referring to a mountain, a river, or a headland. In his reading of antique works, Petrarch consulted contemporary maps, as well as later medieval works, such as that of Guido of Pisa (early twelfth century). Thus began the "geographical philology" of the humanists, a heroic attempt to understand and correct the works of the past, which had suffered from repeated copying. The first Pliny to arrive in Florence in the early 1400s, for example, was a tenth-century manuscript, many times removed from its original.[8] In addition, place-names had changed, river courses had meandered, coastlines had altered, and cities and towns had dwindled in importance or been wiped out completely, while new cities had grown up and prospered.

Petrarch's interest in geography was taken up by his contemporary Boccaccio (1313–75), who copied into his notebook lengthy geographical excerpts, interspersed with his own comments and corrections. For example, he wrote out a good deal of the geographical part of Paolino Minorita's *Satyrica historia* but added a string of insults: "imbractator" (gilder), "bergolus" (Venetian!), "smemoratus" (forgetful, negligent), and "bestia" (ignoramus).[9] He also copied out Hetoum of Armenia's description of Asia and a letter from a Florentine merchant in Seville reporting the discovery and early exploration of the Canary Islands (1336–41). In addition, Boccaccio had recourse to modern sea charts in order to clarify points of geography which were unclear in the texts he was consulting.[10] He began to write in 1359 his own geographical work, *De Montibus, Sylvis, Fontibus, Lacubus, Fluminibus, Stagnis seu Paludibus, de Nominibus Maris* (Of mountains, forests, springs, lakes, rivers, swamps or marshes, of the names of the sea).[11] In the preface he explains the need for a work of this type to aid the reader of classical texts, who is often bewildered by the profusion of unfamiliar names. By the time he wrote *De Montibus*, he had gained access to some useful classical works of geography, particularly those of Pliny and Pomponius Mela.

The *Geography* of Ptolemy

Neither Mela nor Pliny came equipped with maps—their geographies were purely descriptive. What Ptolemy supplied, which was new, was a scientific basis for mapping, the culmination of geographical thought in the ancient world. First, he analyzed the problem of map projection, that is, projecting a spherical body onto a plane surface. European scientists were acquainted with the spherical projection of the heavens which appeared on the back of the astrolabe but had never applied this method to terrestrial mapping.[12] Ptolemy admitted that a globe gives a superior representation of the whole earth, but globes are difficult to make and unwieldy to consult. He proposed three solutions to the problem. The first was a conical projection, in which the latitude lines are drawn as concentric circles and the longitude lines converge at a single point somewhat beyond the pole. The meridians (longitude lines) are also deflected inward at the equator. The second method, called a pseudo-conical projection, shows the longitude lines curved like the latitude lines (fig. 5.1). The third projection, which was seldom employed, was based on the armillary sphere with its rings representing the circles of the celestial sphere. One should imagine an observer standing out in space, seeing the rings projected onto the globe. The resulting image was very close to that of Ptolemy's second projection.[13] The cylindrical projection, used by his predecessor Marinos of Tyre and harshly criticized by Ptolemy, was useful for regional maps on which there was not the same need to account for the earth's curvature. In this projection, latitude and longitude are drawn as straight lines. Modern readers will be familiar with this as the Mercator projection, whose shortcoming is its distortions of distance and geographical form as one moves away from the equator.[14]

Medieval mappaemundi had shown the inhabited landmass as circular, surrounded by a thin band of ocean. There was an alternative world picture throughout the era, however, the zone map, which showed half the globe, with the *ecumene* in the northern temperate zone. These two images were seldom reconciled, although Matthew Paris had grumbled about the shape of the inhabited lands, which he said should really be like a chlamys, or cloak.[15] Lambert of Saint Omer made a world map (c. 1120) that combined the two and showed a large continent in the southern half of the eastern hemisphere, empty of geographical forms but full of writing instead, a text drawn from Martianus Ca-

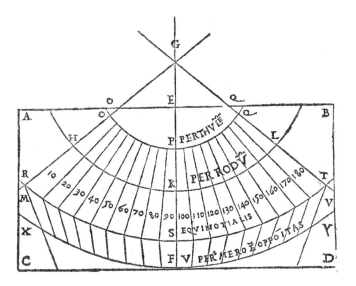

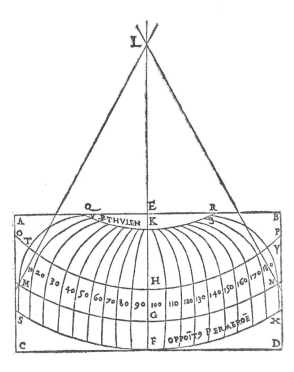

Figure 5.1. Ptolemy's Projections (Ptolemy, *Geographia* [Rome, 1490], pp. 45 and 47. Library of Congress, Rare Books, Incun. 1490. P8). The opening chapters of Ptolemy's *Geography* discussed the problem of projecting a spherical object (the earth) on a flat surface. The first, a simple conical projection, uses straight lines for the meridians or longitude lines, while the latitude lines are curved. Ptolemy shows the world from the latitude of Thule, 63° north to "anti-Meroë," 16 5/12° south. The second, a pseudo-conical projection, better preserves the impression of the curved surface of the earth. It covers the same area as the first. Courtesy of the Rare Books Division of the Library of Congress.

pella discussing the Antipodes.[16] The shape of Ptolemy's world was longer from east to west than it was wide (from north to south) and occupied only part of the eastern hemisphere, indeed more like a chlamys than a disk.

The sea charts of the late Middle Charts were "projectionless," that is, the chart makers did not consciously take into account the earth's curve, though plotting courses by the compass naturally led meridians to converge at the magnetic pole. While such a method was good enough for smaller areas, it had already begun to result in inaccuracies in a body the length of the Mediterranean Sea. Once the Europeans began to move south along the African coast, it became clear that a new method was required, particularly if a world map was to be constructed using a single system.[17] The chart makers' interest in "constant-course navigation" led to results similar to the Mercator projection of the sixteenth century.[18]

The other method Ptolemy employed was the construction of a map based on a grid of astronomically determined coordinates of latitude and longitude. This concept was known to Western scientists, who had been using the astrolabe to find coordinates and compile tables during the Middle Ages in the Christian and Islamic worlds, but they had been used primarily for casting horoscopes. Roger Bacon had proposed the idea of making a map based on a latitude/longitude grid and may have drawn a prototype, but it does not survive and his suggestion was not followed up by anyone else. Gautier Dalché suggests that Bacon may have gotten the idea from his indirect knowledge of the content of Ptolemy's *Geography*. Bacon himself says he used the *Almagest* and the Alphonsine Tables.[19] First, a massive amount of data collection was required. What was truly astounding about Ptolemy's *Geography* was the completeness with which he had carried out his concept.

Of course, even Ptolemy did not really have the necessary information. His list of eight thousand places and their coordinates included many approximations and errors. Most of his figures were based on travel time rather than astronomical observations. For example, to calculate the distance across Asia from the Black Sea to the land of the Seres, he used travelers' estimates of the length of their journey and then reduced them by a certain percentage in order to account for changes of direction made necessary by geographical obstacles. His error was enormous, overestimating the length of the *ecumene* from Spain to China as 180° instead of 150°. Even in a smaller and better known area such as the Mediterranean Sea, the error was 20°—60° instead of 40° in its east-west extent.[20]

Although Roger Bacon's map does not survive, several maps constructed on a grid have come down to us. The map of Palestine which accompanied Marino Sanudo's book (1320) was covered with a network in which each square was one league or two miles. Sanudo provided a guide, telling in which square various places could be found. This method did provide a measure of proportion, even though it was not entirely accurate, but it was not based on coordinates of latitude and longitude. The oldest sea chart, the Pisa Chart (c. 1275), has a grid drawn upon it, but it looks as if it were used to aid the mapmaker in replicating a preexisting chart. In the early fifteenth century, a group of scholars working at the monastery of Klosterneuberg near Vienna seems to have been experimenting with the use of an astronomical grid, supplemented with measurements from itineraries.[21] The maps we have are reconstructions, and the dates of some are not clearly established. They could have been influenced by Ptolemy's *Geography*, news of which spread rapidly throughout intellectual circles of Europe, particularly after all the cardinals and bishops returned from the Council of Constance in 1418. A copy of the *Geography* was in Klosterneuberg by 1437.

Ptolemy's *Geography* offered mapmakers a systematic and measured vision of the ordering of space, based on the abstract principles of Euclidean geometry. While nautical chart makers had presented excellent detail, the charts were difficult to assemble into a coherent whole, particularly with the distortions inherent in mapping larger areas without taking the curvature of the earth into consideration. Unlike the mappaemundi, Ptolemy's maps were just maps, not histories or theologies, and were devoted to depicting *space*, not the idiosyncrasies of *place*. Legends on his maps were place-names only, without descriptive tags. For the next several centuries, mapmakers made up for this "defect" by rimming the maps with puffing wind heads, an array of the monstrous races, and the sons of Noah, but the map in the center was resolutely focused on space.[22] The design of the Ptolemaic world map also meant that there was no privileged "center." Jerusalem, which had dominated the later mappaemundi, was now just one more location on the grid.[23] The unknown lands were affected as well, for all fell under the network of the grid, whether those blank spaces around the fringes of the *ecumene* or even places on the other side of the world. Ricardo Padrón calls this "positive emptiness," emphasizing the discoverability and mappability of places as yet unvisited.[24]

The enthusiastic and instant acceptance of Ptolemy in the West has also been linked to developments in painting.[25] During the fourteenth century, northern artists had become fascinated by "topographical realism." The Limbourg brothers included detailed portraits of the castle of the Duke of Berry in their books of hours, and also drew a map of Rome with accurate images of the most important structures. Jan van Eyck is said to have designed a globe, which does not survive.[26] In Italy the approach to representing reality was more systematic and was founded upon geometry. Imagining a "pyramid of sight" with its base on the picture plane and its apex at the viewer's eye, the artist was able to create a convincing illusion of three-dimensional depth. The work of Leon Battista Alberti (1404–72) in *Dalla Pittura* was the first to explain this technique, which took Italy by storm. Its relation to surveying was evident—indeed some of the same tools were used. The realistic depiction of three-dimensional space on the picture plane has an obvious connection to Ptolemy's projections used for map construction. For the fifteenth century artist there was no agonizing about the nature of reality versus appearance. The goal was to reproduce, as convincingly as possible, what was seen by the human eye. Appearance, in other words, was reality. This attitude can be seen in the way people now began to look at maps.[27]

The Council of Constance

In 1414 a major Church council was convened by the emperor Sigismund at the German city of Constance. Its chief purposes were to resolve the ongoing schism in the papacy and to deal with the growth of the Hussite movement in Bohemia. As churchmen assembled from all corners of the Christian world, they concerned themselves, outside the formal meetings, with an intensive exchange of books and ideas. It was, according to Jean-Patrice Boudet, "a great fair of manuscripts."[28] One of the especially hot items was Ptolemy's *Geography*, and another was the *Chorography* of Pomponius Mela, a descriptive geographical work of the first century CE, which had also been out of circulation for most of the medieval period (fig. 5.2).[29]

Among the French cardinals at the council, one of those most deeply interested in geography was Guillaume Fillastre of Reims (c. 1348–1428). He had heard of the reappearance of Ptolemy's work as early as 1407, when he was traveling on church business in Rome, and was eager to obtain a copy for himself as well as for the library he was establishing at Reims. During and immediately after the Council, he had more than fifty manuscripts copied, many of which remain today in the municipal library there.[30] He ordered two copies of the Latin version of Ptolemy's *Geography* and one of Pomponius Mela, the originals of which he obtained from his Italian colleagues. He appended writing of his own to these works to aid his French readers. To Ptolemy he added notes on the regional maps, updating some of the place-names, and to Mela, a set of glosses and an introductory essay, which not only outlined the contents of the book but also included his own reflections.[31]

Looking at Fillastre's commentary, we can see that the initial reception of classical geography among the humanists was enthusiastic but not without a critical sense. For one thing, the two ancient authors did not agree. While Ptolemy showed a landlocked Indian Ocean, Mela argued that it was a gulf of the world ocean and that one could enter it by sailing around Africa. Similarly, Mela asserted that the Caspian Sea was a bay of the northern ocean, while in Ptolemy's description, it was landlocked. Fillastre used the Bible as the deciding authority, for Genesis 1:9–10 says that all the waters of the earth were congregated into one. He felt this proved that all large bodies of water were connected, and so both the Caspian Sea and the Indian Ocean had to be gulfs of the great world ocean.[32]

Ptolemy and Mela also disagreed about the habitability of the torrid and frigid zones of the earth. Mela repeated the conventional wisdom, but, as Fillastre pointed out, he contradicted himself by describing people living in these allegedly unlivable places. He also noted that Ptolemy referred to the Ethiopians living in the torrid zone and extended the habitable world to 16° south of the equator. Again he turned to divine wisdom to solve the problem. Would God, who created the earth for mankind, have made large parts of it uninhabitable? He also made the sun for human benefit, not to give more heat than humans can endure. Fillastre agreed that some areas of the world are uninhabitable due to the sterile nature of the soil or because of beasts hostile to mankind, but he commented that, "God has made these things to show the marvels of his power."[33] Fillastre's references to Christian theology illuminate the nature of French humanism in the early fifteenth century, which was still firmly established on a Christian base. One should also recall that his potential audience consisted of the canons of Reims cathedral, who might be disturbed by a geography completely divorced from the theologically based geography that they knew.[34]

These two points—the continuity of the seas and the habitability of all zones of the earth—were especially important to the development of practical geography, that is, travel and exploration. The ocean could now be seen not as an obstacle but as a highway, open in all directions to human passage and habitation. After discussing the likelihood of the existence of the Antipodes, Fillastre concluded: "from east to west the passage, so far as concerns the shape of the earth, is simple and easy, if the division and malice of human beings does not prevent it."[35] Fillastre stressed that Mela's evidence was not purely theoretical but based on experience, both in his own time and in the era of Alexander. Fillastre added the voyages of King Solomon's fleet, and, more recently, the travels of the Portuguese "who live in the farthest reaches of the west." These intrepid adventurers crossed the Mediterranean to Syria and Mesopotamia and went on to the Euphrates and Tigris rivers. From here they could sail down to India and into the eastern ocean as far as Taprobana.[36] The habitability of lands in the ocean was proved by another contemporary event—the settlement of the

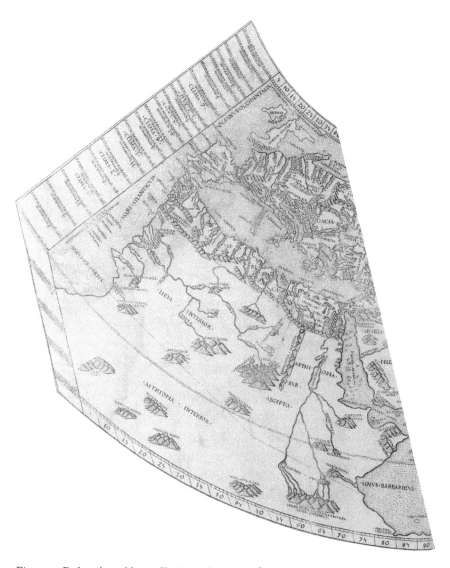

Figure 5.2. Ptolemy's world map (Ptolemy, *Geographia* [Rome, 1490], pp. 99–100. Library of Congress, Rare Books, Incun. 1490.P8. 54 cm / 22″ long). This map shows the essential features of Ptolemy's world. Oriented to the north, it extends 180° from east to west and from 63° north to 16° south of the equator. The mythical island of Thule breaks into the degree scale at 35° top left. The Indian Ocean is enclosed, and southern Africa and eastern Asia are vaguely drawn. The map is at its best within the confines of the old Roman Empire. Courtesy of the Rare Books Division of the Library of Congress.

Canary Islands in the early fifteenth century—not to mention the inhabited condition of the British Isles, which lay not only in the midst of the ocean but also in the supposedly unlivable frigid zone. "And the same may be true of the eastern ocean, although we are ignorant of it," he added.[37] Fillastre observed that there were many lands still unknown, but that did not mean that they were uninhabitable.

In the Reims manuscript Fillastre bade his readers to regard with caution the small circular mappamundi that appeared in the opening initial "O" of the text, as well as the large wall map in the Reims library, now no longer extant but apparently of similar design (fig. 5.3). He warned that the earth does not really look like this but extends much farther east to west than from north to

Figure 5.3. T-O map, Reims, 1417 (Drawing from Konrad Miller, *Mappaemundi*, 3:139. Reims, Bibliothèque Municipale, MS 1321, fol. 13r 10 cm / 4″ diam.). This decorative little map fills the initial *O* of "Orbis," the opening word of Pomponius Mela's *Description of the World.* An introduction by Cardinal Guillaume Fillastre, who gave the manuscript to the Reims library, warns the reader to beware of such maps, which did not adequately represent the world. The east and India (named three times) are at the top, Seres (China) at the upper left, and Prester John in southern Africa.

south. He added that the ocean is not circular and that the small size of the map gives the erroneous impression that every one lives near it, when actually some peoples are landlocked. In the Rome manuscript of Mela's *Chorography*, which also includes Fillastre's introduction, there is a more suitable map, one of the earliest Ptolemy-style world maps to be made in the west (1415).[38] We know nothing about Pirrus da Noha, the mapmaker, except that he worked as a copyist for Cardinal Giordano Orsini, but we may assume the map was copied from one of the manuscripts of Ptolemy in circulation at the council (fig. 5.4).

In addition to the text of the *Geography*, which was copied during the council, Fillastre commissioned the copying of the maps and added the first of the *tabulae novae*, a map of northern Europe by Claudius Clavus, a Dane.[39] It was clear to thoughtful fifteenth-century readers of Ptolemy that his information was faulty when it came to northern Europe. Scandinavia and even Iceland, unknown to Ptolemy, had been appearing on mappaemundi since the tenth century. Adam of Bremen in his geographical account of 1075 had added the names of Greenland and Wineland, rather vaguely described as islands "in the northern ocean."[40] On mappaemundi they floated around in the ocean, usually to the northeast, as the circular form of the mappamundi did not leave much space for additions in the northwest. Sea charts from the fourteenth century included a more finely detailed drawing of the Baltic coast than Ptolemy was able to show, as well as the Scandinavian peninsula with its characteristic forests and some of its ports.

In his commentary on Ptolemy's eighth map of northern Europe (fig 5.5), Fillastre wrote: "Beyond that which Ptolemy put here, there are Norway, Sweden, Russia and the Baltic Sea dividing Germany from Norway and Sweden. This same sea, further to the north, is frozen for a third part of the year. Beyond this sea is Greenland and the island of Thule, more to the east. And this fills all the northern region as far as the unknown lands. Ptolemy made no mention of these places, and it is believed that he had no knowledge of them. So that this eighth map might be more complete, a certain Claudius Cymbricus outlined the northern regions and made a map of them which is joined to the other maps of Europe, and thus there are eleven maps (instead of ten)."[41]

Clavus's map was doubtless based on other maps, probably sea charts, as well as verbal descriptions that included the northern countries, though he did add the coordinates of latitude and longitude in the new style to fit in with a Ptolemy atlas. Fillastre's own information about northeastern Europe was enriched by his participation in 1421 at a trial at the papal court between the Teutonic knights and the king of Poland. The Teutonic knights had been commissioned as crusaders to fight pagans in eastern Europe back in the early thirteenth century, but now, with the conversion of the Lithuanians, the last holdouts, in 1387, they were unwelcome interlopers. In the end, the knights lost their case. Among the evidence was a map especially prepared by Clavus to show that the territories seized by the Teutonic order were within the borders of Poland and cut that kingdom off from the sea.[42] This particular map is lost, but its use as evidence in a court case shows the growing practical use of maps and recalls the purpose of Sanudo's maps in planning a crusade a century earlier.[43]

The map of northern Europe which does survive can be found in Cardinal

Fillastre's personal copy of Ptolemy's *Geography* now at the Nancy municipal library.[44] Where Ptolemy had shown the Baltic Sea empty of land north of the Danish peninsula and Germany, save for the small island of "Scandia," Clavus inserted a large Scandinavian peninsula filled with about twenty place-names, including Oslo, Stavanger, Bergen, Stockholm, Halsingborg, Lund, and the island of Gotland. His map stretches over twenty degrees of latitude, from the fifty-fourth to the seventy-fourth parallel. At 66° he notes that the length of the longest day is twenty-four hours. He also shows Greenland as a huge northern territory attached to Scandinavia and running along the northern and western edges of the map. In the farthest north, in those "terrae incognitae," he records the presence of the pagan Carelians, unipeds, maritime pygmies, and griffins.[45]

Fillastre's little world map in the Reims manuscript does not include Scandinavia. In the 1415 manuscript of Pomponius Mela in Rome, the small map by Pirrus da Noha is the first Ptolemy-type world map to incorporate this new material. It shows a very large Scandinavian peninsula with "Norvegia" and its forests, possibly adapted from Claudius Clavus or from the sea charts that had shown this feature.[46] It is ironic that this map actually contradicts Mela's text, showing a closed Indian Ocean and Caspian Sea.

Fillastre had the most to say about northern Europe, but in his comments on the other maps, he added more modern place names. For the three classical regions of Spain, he substituted the kingdoms and territories of the fifteenth century: Castile, Leon, Portugal, Algarve, Granada, Aragon, Catalonia, and Navarre. In Africa he mentioned the kingdom of Prester John and said that two ambassadors came from there to the court of Aragon in 1427.[47] In Asia he added Tartary and Armenia Major (where there used to be Amazons) and observed that all of Asia Minor was now ruled by the Turkish emperor.

Pierre d'Ailly

Also in attendance at the Council of Constance was Fillastre's fellow cardinal and countryman, Pierre d'Ailly (Petrus Alliacus, 1350–1420), bishop of Cambrai. In 1410 he had completed his influential cosmographic work *Imago Mundi*, which is a handy summing up of the accumulated knowledge of the West. Rely-

Figure 5.4. (opposite) World map of Pirrus da Noha, 1414 (Vatican City, Biblioteca Apostolica Vaticana, MS. Arch di San Pietro H.31, fol. 8r, 20 × 28 cm / 8″ × 13″). Also in a copy of Pomponius Mela's work, this map is a striking contrast to Figure 5.3. It is one of the very earliest Ptolemy-style world maps, but it adds a Norwegian peninsula to the northwest, a feature not present in the *Geography.* Courtesy of the Biblioteca Apostolica Vaticana.

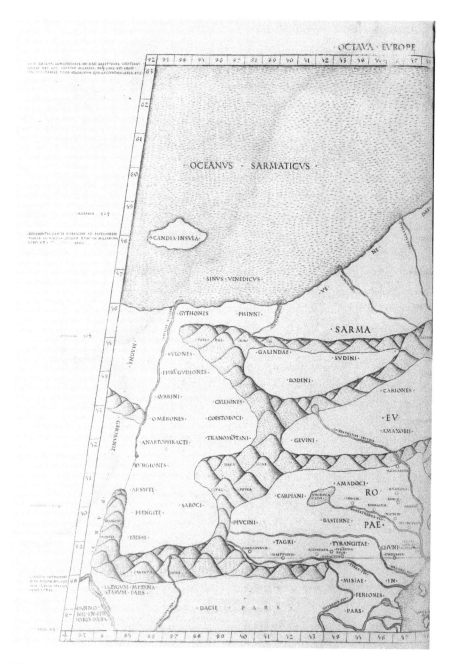

Figure 5.5. Northern Europe (from Ptolemy's *Geographia* [Rome, 1490], 8th map of Europe. Library of Congress, Rare Books, Incun. 1490.P8). One of the first reactions in Europe to the rediscovered work of Ptolemy was that his depiction of northern Europe was sadly deficient. Despite this judgment, maps of the region continued to be made throughout the century according to his directions. North of Germany is the Oceanus Sarmaticus with the island of

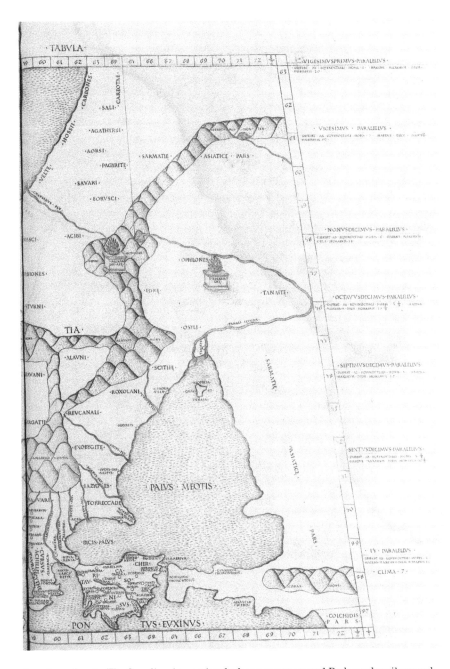

Scandia or Scania. The Scandinavian peninsula does not appear, and Ptolemy describes much of this part of the world as "unknown." Courtesy of the Rare Books Division of the Library of Congress.

ing on such venerable sources as Isidore, Pliny, and Orosius, he also consulted more modern ones—Sacrobosco, Roger Bacon, and Nicholas Oresme. He illustrated his work with several diagrams, including a schematic climate map.[48] Oriented to the north, the circular map is divided by horizontal bands into climates in the northern hemisphere (fig. 5.6). In the southern hemisphere are lines for the tropic of Capricorn and the Antarctic Circle. Geographical names are scattered throughout the northern hemisphere, more or less in relation to their locations, but no geographical forms are drawn. Most of the names are traditional and can be found on any medieval map. One oddity is the appearance of the mythical city of Arym in the center of the map on the equator. This place-name comes from Hindu cosmology, via the Arabic world, though its origins are somewhat mysterious. It was introduced to the West by Petrus Alfonsus, a Spanish Jew converted to Christianity, who put it on a climate map in 1110.[49] Arym is also mentioned by Roger Bacon, who identifies it with Syene but observes there must then be "two Syenes," one on the tropic (as the original Syene in Egypt was) and one on the equator. D'Ailly repeats this curious information.[50] The importance of Arym as a center point is that it was astronomically determined, as equidistant between the extreme east and west of the habitable world, unlike the center point of Jerusalem, which was theologically determined.

Like many other unfortunate authors, d'Ailly became aware of the reappearance of Ptolemy's *Geography* soon after his book was completed. In a manuscript of the *Imago Mundi* at Cambrai, his anonymous "research assistant" added a note near the end that Ptolemy in the *Geography* contradicts his view, expressed in the *Almagest*, that the climate at the equator was temperate. Instead he now says that the Ethiopians living near the equator were burnt black by the force of the sun's rays.[51] The heady intellectual exchanges at the council caused d'Ailly to produce another book *Compendium Cosmographiae*, a summary of the *Geography*, of which he now seemed to possess a copy.[52] D'Ailly did not accept Ptolemy unconditionally. He noticed that there are conflicts between Ptolemy and other ancient authorities. For example, Ptolemy said the extreme north of the habitable world is 66°, but Pliny said that the Hyperboreans live beyond this point. He also reorganized Ptolemy's data according to climate rather than continent and proposed to make a more useful map, using straight lines for latitudes instead of curved ones. No such map survives.

It was d'Ailly's *Imago Mundi*, however, that enjoyed lasting popularity, appearing in an edition printed in Louvain in 1483. Christopher Columbus owned a copy of *Imago Mundi* and carefully annotated it. The bit of information that most appealed to Columbus was d'Ailly's remark in chapter 11 that, "according to Aristotle and Averroes," the sea is narrow which separates the extreme west

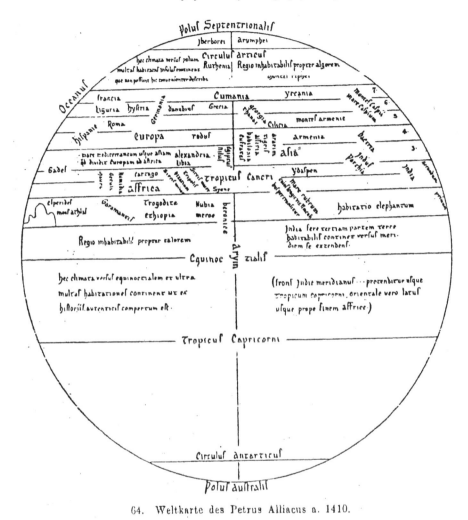

64. Weltkarte des Petrus Alliacus n. 1410.

Figure 5.6. World map of Pierre d'Ailly (from *Ymago Mundi* [Louvain, 1483], from Konrad Miller, *Mappaemundi*, 3:128. Original, 1410. 20.3 cm / 8″ diam.). The world map of Pierre d'Ailly is rather abstract, as befits the content of his theoretical work on the universe. It is a zone map, and the upper temperate zone is divided horizontally into seven climates as well. North is at the top. The mapmaker notes that the northern polar and central torrid zones are uninhabitable. Place-names are located more or less correctly, but there are no geographical forms.

of the habitable world from its eastern edge, a passage which Columbus copied into the margin of his book. Of this intellectual exchange a modern scholar notes, "The lesson to take away seems rather that a major leap forward in the history of humanity can originate in part not from progress but from scientific backwardness."[53] D'Ailly's work was primarily theoretical—he generally did not incorporate material from travelers or sea charts, and he used hardly any modern geographical names. In the case of the Caspian Sea, he did note the controversy: Was it one sea or two? Did it flow into the northern ocean or was it landlocked? On this issue he concluded, "Although there are diverse opinions, it is not for me to decide among them, for, on the one hand, I do not dare to doubt the witness of the Ancients and, on the other hand, it is impossible to contest the affirmations of modern eye-witnesses."[54] John Larner dryly notes, "His is a work which enjoys all that clarity which comes from being unburdened by too much knowledge, and, as such, was to prove a useful textbook for Christopher Columbus."[55]

The geographical work of Fillastre and d'Ailly illuminates the state of cartographic studies in the early days of the reappearance of Ptolemy's *Geography*. Rather than solving all the problems of Western geography and mapmaking, it is apparent that this event began a debate about the form and features of the earth which would have far-reaching consequences. Fillastre was not, like Petrarch, mainly interested in the correct rendering of classical place-names from literature. Though he was concerned with the theology inherent in an image of the world, he was equally intent on the world of his own day. Pierre d'Ailly, who can probably be absolved of much interest in the contemporary world, was trying to catalogue traditional knowledge, to reconcile theology and "astrology."[56] It is ironic that his remark about the distance from Spain to India was to have such an impact on subsequent events. Two recent events—the 1402 expedition by two Frenchmen to the Canary Islands, sponsored by the crown of Castile, and the Portuguese conquest of Ceuta in 1415, which was the first step in the African venture—were harbingers of the era of exploration that was to follow. The ocean, indeed, though not so narrow as d'Ailly proposed, had become a highway rather than a barrier.

Geography in Florence

Between the close of the Council of Constance and the opening of another church council in Florence in 1439, there was a kind of ongoing symposium on geographical matters in the intellectual circles of the city. Humanists, including such luminaries as Leonardo Bruni, Paolo Toscanelli, Niccolò Niccoli, Giorgio Antonio Vespucci (the uncle and mentor of Amerigo), and Palla Strozzi, gathered almost daily at the cell of Ambrogio Traversari in the monastery of Santa Maria degli Angeli.[57] Copies of Ptolemy were being made at a great rate, and there was an intense interest in other geographical works. The translator of Ptolemy, Jacopo Angeli, commented in his dedication to the work that the city of Florence was "sparkling with genius" as it enjoyed the revival of the liberal arts after their long slumber.[58] Poggio Bracciolini described coming into the library of Niccolò Niccoli and finding "as was usual" Cosimo de' Medici and Carlo Marsupini poring over a copy of Ptolemy's *Geography*.[59]

The interest in far-flung lands was not merely theoretical. Many of the leading luminaries of humanist Florence were merchants or from mercantile families, and they were definitely interested in markets abroad and potential profits. Gregorio Dati, in his history of the city from 1380 to 1405, commented that Florentines "know all the entrances and exits that are in the world" and that they "have spread their wings over the world and have news and information from all corners."[60] The fortunes that merchants had made helped to fund the copying of expensive works like the *Geography* as well as the purchase of books. The profits of the Medici bank were used by Cosimo himself to found three separate libraries in his home city. After Niccolò Niccoli's death in 1437, Cosimo paid off his debts in return for his library, which he then established at the monastery of San Marco.

The Council of Florence was convened partly to shore up the power of the Pope against the obstreperous and apparently eternal Council of Basel, but one of its higher missions was an attempt to reconcile the long-sundered Greek Orthodox and Roman Catholic churches. The Greeks were interested because they needed military aid from the West to fend off the ever-encroaching Turks. An agreement was signed at the council but was roundly repudiated at home, and, of course, the Byzantine Empire came to a complete end at the fall of Constantinople in 1453. In the meantime, however, a fascinating group of international scholars had assembled in Florence. The Greeks were represented by the Byzantine emperor himself, Ioannis Palaiologos, as well as Ioannis Bessarion, bishop of Nicaea, and Isidore, bishop of Kiev. Also in the Greek delegation was Gemistos Plethon, who enthusiastically entered into the discussions of geography. He thought the Florentines were too obsessed with Ptolemy and introduced them to the geography of Strabo.[61] Isidore provided information about lands, lakes, and mountains of Russia which had been unknown to Ptolemy.

The council also attracted churchmen from even more remote locations. Among them were delegates from Ethiopia, seeking to resolve theological issues that divided the Coptic Church from the Roman. Peter, their deacon, announced in an address to the council that he belonged to an "extraterrestrial people," as the Ethiopians lived outside the supposed borders of the habitable world. Flavio Biondo was on a commission that interviewed the Ethiopian delegation at some length, asking questions not only about doctrinal matters but also about geographical matters such as the stars visible at that latitude, the climate, the variation of length of days and nights, and whether the sea at the equator was boiling. Poggio Bracciolini also talked to the Ethiopians, inquiring particularly about the source of the Nile. When he found that they came from the region in which the river rose, he was tremendously stirred to think that he might now learn something that was completely unknown to Ptolemy and the other ancients.[62]

Travelers

It was during the Council of Florence that Nicolò de' Conti, a Venetian traveler, showed up in Florence after twenty-five years of wandering in the East. He was seeking absolution from the Pope for his forced conversion to Islam while he was in Egypt. His penance was to write an account of his travels, which had taken him to India, Ceylon, Thailand, Burma, Sumatra, Borneo, and Vietnam. His essay came into the hands of Poggio, who adapted it as part four of his book, *De Varietate Fortunae*, completed in 1448.[63] Poggio saw Nicolò's adventures as a perfect illustration of the vicissitudes of fortune, as well as giving information about a part of the world where no one from the west had gone since an ambassador from Tiberius Caesar. (Marco Polo seems to have been forgotten here.) Nicolò was a merchant working in Damascus, who first set out with a caravan to Baghdad. Like Marco Polo he had a merchant's interest in markets and produce. Among other things, he reported that Venetian ducats were circulating in abundance in India.[64] Nicolò's journey was extracted from Poggio's work, translated into the vernacular and handed around as a separate piece. The material in it almost immediately found its way onto maps. The Genoese world map of 1457 set out lengthy legends written on decorative scrolls, incorporating some of the more arresting details of South Asian life, such as the practice of tattooing, the eating of snakes, and, in Sumatra, the use of human skulls for money (see fig. 7.7).[65]

To Florence also came the news of the Portuguese voyages down the West African coast, undertaken under the sponsorship of Prince Henry in the 1430s. In 1434 the first ship passed Cape Bojador, and by Henry's death in 1460 the Portuguese had gone as far as the present site of Monrovia in Liberia, 6° 19" north of the equator. Prince Henry the Navigator no longer receives the acclaim he once enjoyed as a disinterested pursuer of knowledge.[66] His famed academy at Sagres apparently was a fabrication, and he seems to have been more interested in the quest for gold and profits from the slave trade than in the accumulation of geographical information for its own sake. However, it was under his encouragement that the voyages began, first to the Atlantic islands, which were settled under his sponsorship, and eventually to the exploration of the west coast of Africa. Genuinely interested in cartography, he ordered his navigators to map the coastlines they sailed along. After his death the expeditions continued un-til 1487, when Bartholomew Dias finally rounded the Cape of Good Hope and found himself in the Indian Ocean.

The Portuguese captains were energetically involved in trade, regularly re-turning home with cargoes of slaves and ivory, but they also carried out their mapping responsibilities. Despite the alleged policy of secrecy, the details of the coast rapidly made their appearance on Italian and Catalan sea charts and world maps. Part of the reason for the swift transmission of knowledge was that not all the sea captains were Portuguese. Genoese, Venetians, and Flemings were all involved, and one of these, a Venetian seafarer named Alvise Cadamosto, wrote an account of his two voyages of 1455 and 1456, as well as that of Pedro de Sintra, who sailed a few years later. Returning in 1464 to Venice, where he composed his reminiscences, Cadamosto must have communicated his information to the mapmaker Graziosa Benincasa who, in 1468, showed the African coast explored by Sintra and Cadamosto on a sea chart as far south as "Cape Mesurado," at 6° 19" above the equator. Here forty new toponyms appear, all in Portuguese and not all found in Cadamosto's text, which implies that either he gave more infor-mation orally or Benincasa had an additional source.[67]

Ptolemaic atlases were less responsive to innovation than the sea charts. De-spite the early inclusion of a modern map in Fillastre's 1427 edition and the acute awareness of Ptolemy's shortcomings, the next atlas to add modern maps did not appear until 1468. This was the second recension of the atlas made by Nicholas Germanus and included separate maps of Spain, northern Europe, and Italy.[68] The maps of Spain and Italy were not much changed from Ptolemy's version, except that modern names were added. The map of northern Europe, however, where Ptolemy had put only the island of Scandia in the Baltic north of Germany, was based on the Clavus map, with added nomenclature on Green-land and Iceland from another map made by Clavus. The second map is lost, but Clavus's description survives in two manuscripts in Vienna.[69] The island of Thule, which broke the northern frame in the Ptolemy map, was displaced from its "ultimate" position and was now shown south of Norway. The new configu-ration of northern Europe also appeared on the world map, but to fit it in, and to place Greenland far enough north, the mapmaker had to break through the upper frame. The peninsula of Greenland, attached to Russia, arcs through the polar region and comes down west of Norway.[70] In the third recension, Nicholas Germanus altered the map of northern Europe once more, moving Greenland (Engronelant) to the north of Scandinavia, with Iceland the farthest to the west (fig. 5.7; see also fig. 5.5). Nicholas also added France and a map of the holy land based on the Sanudo/Vesconte version of 1320. Of course, Ptolemy's picture of

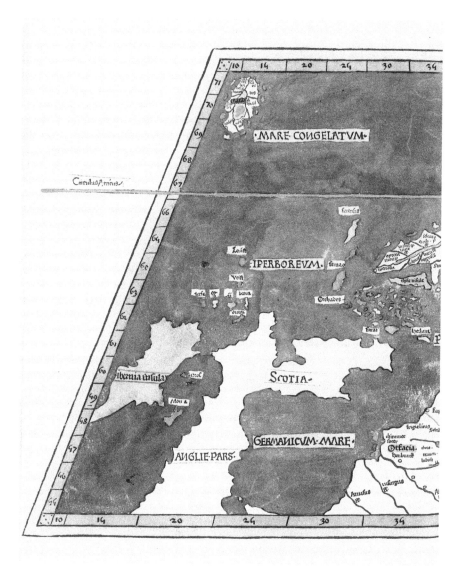

Figure 5.7. Northern Europe, Nicholas Germanus, 3rd recension, 1482 (from Ptolemy, *Cosmographia* [Ulm 1482], 4th map of Europe, "Tabula Moderna." Washington, D.C., Library of Congress, Rare Books, Incun. 1482.P8). In 1468, Nicholas Germanus drew a new map of northern Europe, showing a large Scandinavia and the peninsula of Greenland to the north and west. It is believed that he was influenced by the information on the Clavus map. This is one of the "tabulae modernae" that were added to Ptolemaic atlases in the last half of the fifteenth century, attempting to bring the classic work up to date. A few years later he drew this version, moving Greenland from the far west to north of Scandinavia. Iceland is in the upper left-hand corner in the Mare Congelatum (Frozen Sea). Courtesy of the Rare Books Division of the Library of Congress.

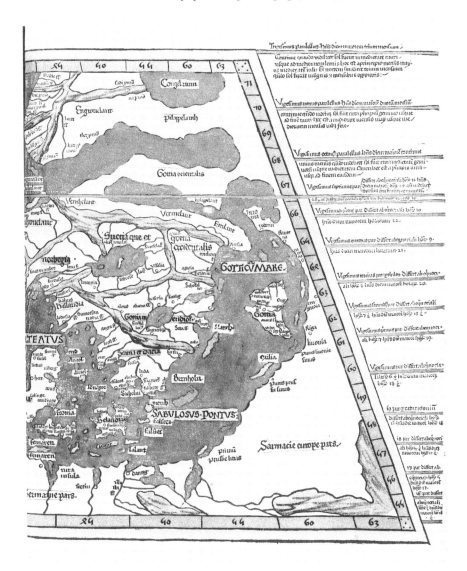

this region had been hopelessly pagan, and many of the sites dear to the hearts of late medieval Christians had been omitted.

By this time printing presses had begun to operate. The first printed map was not a Ptolemaic one but a small T-O type map in an edition of Isidore's *Etymologies,* published in Augsburg in 1472. The first printed version of Ptolemy's *Geography* came out in 1475 at Vicenza, though without the maps. The maps followed in the next edition, printed at Bologna in 1477. Printing made a uniform product possible—though of course this could include uniformity of errors—and a greater number of copies. Five hundred copies were made of the Bologna edition. Still, lavish manuscript copies of Ptolemy's work continued to be made for important patrons: Lorenzo de' Medici; Federico Montefeltro, Duke of Urbino; and the Duke of Calabria, later to be King Alfonso III of Naples.[71]

Revising Ptolemy

Criticism of Ptolemy, as we have seen, surfaced almost at once, but major structural changes were slow to come. A Ptolemy atlas, like any other ancient work, tended to preserve its integrity. New regional maps, when added, were placed alongside old ones but did not replace them. As for the world map, with the exception of Pirrus da Noha's map of around 1415, which included the Scandinavian peninsula, no early versions showed any features that were not pure Ptolemy.[72] Any sea chart could demonstrate a better outline of the northern coast of Africa, a more correct length of the Mediterranean, a better orientation of the Italian peninsula (shown running almost due east-west in Ptolemy's version), as well as additional features in northern Europe. As for the closed Indian Ocean, doubts had arisen almost immediately, partly from the contradictory testimony of other antique authors and increasingly from the experience of modern navigators. Arabic world maps had depicted an Indian Ocean open to the east since the ninth century. In 1490 Henricus Martellus Germanus made a world map that showed a circumnavigable Africa and an open Indian Ocean, documenting the voyage of Bartholomew Dias, completed just two years before (see fig. 8.2). The first world map in a printed edition of a Ptolemy atlas to show an open Indian Ocean was not produced until 1504. A vestige of its eastern rim, the "dragon's tail" in the far east, was to have an even longer history. (See chapter 8.)

Europeans also continued to show a fondness for the geographical information they had inherited from the Middle Ages. The fact was that Ptolemy was difficult to read and boring. The main part of the book was a bunch of lists, the opening chapters were highly technical, and considerable space was devoted to the denunciation of Marinos of Tyre's work, otherwise unknown. Francisco's Berlinghieri, in his 1482 atlas of Ptolemaic maps hardly used Ptolemy's text but substituted a work of his own, composed in Italian in terza rima, which he called "Septe Giornate della Geographia" (Seven days of geography). He fleshed out Ptolemy's stark geographical data with historical, mythological, and contemporary digressions, drawing material from other ancient authorities such as Pliny, Strabo, Pomponius Mela, as well as some modern authors, including Flavio Biondo and Cristoforo Buondelmonti.[73] Berlinghieri also apparently used a contemporary sea chart in an effort to match up ancient and modern names, point-ing out that it was tedious for readers to try to decipher them on their own. His atlas contained thirty-one maps, four of them modern maps of Spain, France, Italy and the holy land, all derived from those in Nicholas Germanus's third recension of the atlas.[74] In 1482 Berlinghieri's atlas was printed in Florence, in the same year that the atlas of Nicholas Germanus came off the presses at Ulm. These atlases contain the first modern maps to appear in print. (Although sev-eral editions had already been published, none had included any other but the standard array of Ptolemaic maps.) Berlinghieri's atlas was also the first to ap-pear in a modern language.

Another early printed edition of Ptolemy, made by Johann Reger in Ulm in 1486, helpfully provided the reader with not only an annotated index ("Regis-trum alphabeticum") but a treatise entitled "De locis ac mirabilibus mundi."[75] This text, drawing on Isidore of Seville, Vincent of Beauvais, and a Latin trans-lation of Jean Germain's "La mappemonde spirituelle" (1450), informed the reader of the whereabouts of paradise and about fabulous beasts, monstrous hu-mans, and foreign customs, as well as other traditional and entertaining lore. The Registrum was fully cross-referenced to enable the reader to move between Ptolemy's place-names and maps and the expanded information in "De locis." The editor's confidence that this material would not be deemed old-fashioned shows that the revolutionary quality of Ptolemy was not fully appreciated as such in the late fifteenth century. Ptolemy brought a new system of mapping but did not devalue what had been known before. In fact, his purely math-ematical geographical content was felt to be lacking in what we would now call human geography, the kind of ethnographic as well as botanical and zoological details that made a bare list of places come to life.

While Ptolemaic atlases were appearing in print, sea charts continued to be made by hand and were quickly responsive to new discoveries of interest to sea-farers. Ptolemy's maps would have been completely useless for their purpose, with his antique names and twelve winds, as opposed to the eight-wind system used by sailors. It was not until 1511 that a Ptolemaic atlas was published which claimed to have corrected the maps by using a sea chart. The author, Bernar-dus Sylvanus, was not especially successful, since he seemed to have changed latitude and longitude figures somewhat arbitrarily, but he did produce a world map that broke dramatically with the standard Ptolemaic model.[76]

Did the rediscovery of Ptolemy's *Geography* lead to a dramatic revolution in European thought, transforming cartography from a theologically based en-terprise, largely indifferent to physical reality, to a scientific one, founded on measurement and objective observation? Or, on the contrary, did Ptolemy's cartography lay a heavy hand, weighted with the authority of antiquity, on the vital contemporary development of mapping linked to the nautical chart? Clearly, neither of these extreme statements is completely true. Ptolemy's work was greeted with cautious enthusiasm. Its shortcomings were immediately apparent, and, if manuscripts and the first printed editions were produced with scrupulous fidelity to the master's instructions, sea charts continued to be made and used, not only by navigators but by geographers and the makers of world maps, who incorporated the fifteenth-century voyages of discovery into their world picture.

A third theory has been recently championed, primarily by the French scholar, Patrick Gautier Dalché, that Ptolemaic cartography was not really new but had been thoroughly anticipated by late medieval mathematicians and geographers.[77] The method Ptolemy introduced, especially that of establishing location by astronomical observation of latitude and longitude, was known theoretically in the West. But the appearance of this massive ancient work, which reinforced what was already known, was an inspiration. Sailing based on astronomical observation was cautiously introduced only in the late fifteenth century but did not completely replace traditional methods for some time.[78] Columbus, for example, was primarily a compass-and-chart sailor who had constant struggles with his quadrant. Ptolemy's work was a model of how a universal map based on precise measurement could be made, and he lent his venerable authority to such a project.

Sea charts, mappaemundi, and Ptolemaic atlases coexisted throughout the fifteenth century, each having its own role to play in the graphic presentation of geographical information. It was a Venetian, Fra Mauro, who made a heroic attempt to integrate these three disparate traditions into a single map.

CHAPTER SIX

Fra Mauro

The Debate on the Map

One of the great monuments of the late Middle Ages reposes in splendor at the Biblioteca Marciana in Venice. Nearly six feet in diameter, painted on parchment glued to wood panels, it is the world map of Fra Mauro. The circular map is mounted in a heavy, square frame hanging in its own room and is usually covered by a curtain to protect it from the light. When the curtain is drawn aside, one sees a rich abundance of color and detail: ships sailing on seas rippled with blue and white, innumerable tiny castles mounted on hills, rivers winding throughout the land, irregularly shaped islands scattered in the sea. In addition to thousands of place-names, the map is covered with two hundred descriptive texts, sometimes mounted on scrolls that are pasted onto the surface. These not only give further geographical information but also present the debate about geography which was raging in fifteenth-century Europe (fig. 6.1).[1]

The usual theory is that the original map was commissioned by the king of Portugal shortly before 1450. Last seen in the royal monastery of Alcobaça in the seventeenth century, it is now lost, perhaps a victim to the great earthquake of 1755. The map in Venice has been considered as a copy made for "questa illustrissima signoria," the Republic of Venice, and completed in 1459. Recent discovery of relevant documents in Venetian archives suggests that these two events should be reversed, the first map being made for Venice about 1448–53, and a copy produced for Portugal between 1457 and 1459.[2] The map we have today is written in the Venetian dialect of Italian; we presume the Portuguese version was translated into Latin. The maps were produced at Fra Mauro's monastery of San Michele di Murano, a house of the Camaldolensian order. Its establishment in Florence was Santa Maria degli Angeli, which the reader may remember was the site of almost daily gatherings of humanists in the mid-fifteenth century, often for the purpose of studying geography.[3] We do not know if Fra Mauro ever traveled to Florence, but he did spend some time at another monastery, San Michele al Leme in Istria in the northeast Adriatic, where he made a map of the monastic estates. This map survives in a seventeenth-century copy. In 1444 he was on a commission charged with changing the course of the Brenta River, which flowed into the Venetian lagoon. Both of these projects show a practical approach to the use of maps. Another map attributed to Fra Mauro is an extended sea chart, covering the world most familiar to Europeans, from the Caspian Sea in the east to the Atlantic islands in the west, and as far south as the northern part of the Persian Gulf and Red Sea. This map has a number of legends identical to those used on Fra Mauro's world map, and Almagià has proposed that it was a copy Fra Mauro made of a Portuguese chart supplied by the royal government and used as a source for his world map.[4]

Fra Mauro was dead by October 1459, when he was honored with a medallion bearing his profile and the inscription "Frater Maurus S. Michaelis Moramensis de Venetiis Ordinis Camaldulensis Chosmographus Incomparabilis."[5] The idea of a cloistered monk becoming one of history's greatest cartographers recently caught the fancy of a novelist with more imagination than historical sense, who wrote a fantasy called *A Mapmaker's Dream: The Meditations of Fra Mauro, Cartographer to the Court of Venice.* He puts Fra Mauro in the sixteenth rather than the fifteenth century, thereby skewing almost everything important about the cartographer's work, but his image of the monk interviewing travelers and seafarers about the world outside (based on a fictional journal) has a certain appeal.[6] In actuality, Fra Mauro had a well-traveled collaborator on his project, Andrea Bianco, who had sailed a good part of the Western world with the Venetian fleet and already had several important maps to his credit. (See the introduction.) The third collaborator, Francesco da Cherso, was apparently a painter.

A Traditional Mappamundi?

The map itself is circular, showing the three traditional continents surrounded by the ocean. Compared to the classic mappamundi, their shapes are more irregular, the coastlines are indented, and the seas have opened up, particularly in the southeastern part, to show an Indian Ocean open to navigation and filled with many islands. As we have already seen, the mappamundi form began to alter in the early fourteenth century, when Pietro Vesconte included elements from the marine charts and altered the outline of the world with an open Indian Ocean flowing into the thin band of ocean which formerly surrounded the inhabited world. Like Vesconte, Fra Mauro has incorporated the geographical

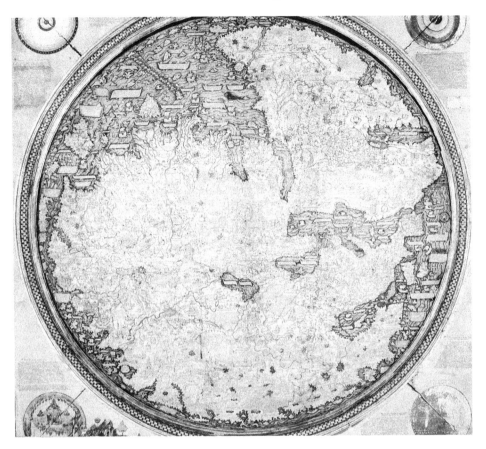

Figure 6.1. World map of Fra Mauro / Andrea Bianco, c. 1450 (Venice, Biblioteca Nazionale Marciana. 1.96 × 1.93 m / 6′ 6″ × 6′ 5″. The world map of Fra Mauro is oriented with south at the top and features an open Indian Ocean and a circumnavigable Africa. The Mediterranean, Black Sea, and Atlantic coasts are influenced by the marine chart. The Scandinavian peninsula is at the lower right, but there is no Greenland. Courtesy of the Biblioteca Nazionale Marciana.

forms of the sea chart in Europe, western Asia, and North Africa. There are no rhumb lines drawn on the map but around the perimeter are medallions for the eight winds of the sea charts (*griego, maistro, auster,* etc.). There is also a wind rose, unlabelled and half buried beneath the land, in the eastern Mediterranean. The pictures and legends evoke the encyclopedic tradition of the great world maps of the thirteenth century.

Outside the frame of the map in each corner is a circular diagram surrounded by a discussion of related cosmographical subjects, a characteristic of the mappaemundi, which liked to put the terrestrial realm into the larger context of the whole universe. In the upper left-hand corner, there is a depiction of the spheres that surround the centrally placed earth, as well as the four elements (air, fire, water, earth). The text presents the views of various authorities on the total number of heavens beyond the obvious eight (sun, moon, five planets, stars) and tries to reconcile these with the system of the four spheres of the elements. It concludes that, beginning with the sphere of the moon, there are ten spheres, counting the empyrean, "chome apar qui ne la presente pictura" (as the picture here shows). Nearby are given the diameters of the planets and the distances from one sphere to another (e.g., 73 million miles from the surface of the earth to the eighth sphere, that of the fixed stars.) In the top right-hand corner is a diagram of the moon's orbit around the earth, with an accompanying text describing the lunar influence on the tides. Another paragraph explains how, by the providence of God, part of the earth is raised above the water, so that air-breathing creatures can live on it. This was an important issue in the later Middle Ages as scientists struggled with the Aristotelian concept that the sphere of the earth was surrounded by the sphere of water. If so, how could any part of the earth be above water and thus habitable? Fra Mauro argues that dry land is lighter and more porous than the land beneath the sea and so can rise above it. This question was important for exploration and to reconcile new discoveries of land with cosmic theory. This section concludes with a discussion of the force of gravity, as the "natural appetite" of earthly things. In the lower right-hand corner is a depiction of the five zones of the earth. The texts above and to the side discuss the question of the habitability of the nontemperate zones and the relative quantities of the four elements in the sublunar realm, using the techniques of Euclidean geometry. To support these small dissertations, the author calls upon "natural reason."

Finally, in the lower left is a little painting of Adam and Eve being instructed by God within the walled garden of Eden.[7] The Garden of Eden, located in the east, had been a standard feature of the mappaemundi, but its location had become an increasing problem for fifteenth-century mapmakers, as further exploration had rendered the East less vague. On the Genoese map of 1457, the mapmaker noted that some had put paradise in Africa, following the theory that the climate around the equator was the most "equable" and therefore suitable for a garden of delights.[8] On Fra Mauro's map it has been displaced from the map proper to the northeast corner, outside the frame. At the gate stands the angel with the flaming sword, about to bar humanity from ever entering the garden again, and at his feet flow the four rivers. The text, "Del sito del paradiso terrestro," tells us that paradise is not only a spiritual place but also, according to Saint Augustine's authoritative views on the subject, a real physical place on earth. It is in the east but remote from human habitation and knowledge. The four rivers that flow from it form the principal hydrography of the world, being the source of four of the greatest rivers of all. Fra Mauro's text is thoroughly orthodox, but it is contradicted, or at least challenged, by the picture he presents. Eden is in the east to be sure but outside the map entirely. The four rivers flow out into a small landscape nearby and disappear, with no apparent connection to the great rivers of the world. On the map itself the source of three of these rivers (the Ganges, Euphrates, and Tigris) is described as being in the mountain range that traverses Asia, while the source of the Nile is pondered in a number of different texts.[9] What appears to be a border of water surrounds this vignette, so perhaps it is meant to be an island.

Alessandro Scafi argues that the placement of the Garden of Eden here is "thoroughly orthodox," a brilliant solution to the problem of separation and connection with the world, balancing "revealed truth" with observed reality.[10] Angelo Cattaneo suggests that Fra Mauro shows the garden as part of the "cosmographic" space but not the "chorographic" space of the map, that is, the known and inhabited regions of the earth.[11] It is still hard to ignore its position outside the frame of the map. This placement, despite the conventional text, is a radical change, a precursor to the boxed Garden of Eden attached to Bible maps from the sixteenth century on. As the adventurous Europeans extended their knowledge of the world in the mid-fifteenth century, the idea of a place on earth not accessible to human travelers was less acceptable than it had been in the past.

Looking at the map itself, we find other traditional features such as Jerusalem, Gog and Magog, Noah's Ark, the pyramids, and the adventures of Alexander the Great. With many pictures, long legends, beautiful colors, and lavish use of gold paint, this map seems securely located in the world of the mappamundi. The texts themselves, however, destroy this sense of security. The centrality of Jerusalem, for example, a staple of the later mappaemundi, causes a problem for our Venetian cartographer. The great expanse of Asia has pushed the holy city slightly to the west, and Fra Mauro explains this in a bit of tortured prose: "Jerusalem is in the middle of the inhabited world according to the latitude of the inhabited world, although according to longitude it is too far west. But because the western part, Europe, is more heavily populated, it is still in the middle according to longitude, not considering the physical space of the earth but the number of its inhabitants."[12] The physical center of the circular map is east of Babylon in Persia and appears to have no symbolic significance. In the conflict between symbolic meaning and the accurate representation of space, the latter has won out.

Gog and Magog, those enclosed peoples in the northeast who will burst forth to assist the Antichrist at the end of days, are the subject of several texts. Fra Mauro puts together the various conflicting traditions about these folk, who were said to live in various places, one being the Caspian Mountains, or Iron Gate, east of the Black Sea. Fra Mauro thinks this cannot be true, for in his day the area was frequented by Western travelers and no trace of the mysterious enclosed people could be discovered. "In truth," he says disgustedly, "this error is put forth by those who like to draw on the Holy Scripture to support their own sentiments."[13] Moving to northeast Asia, in the land of Tenduch, he echoes Marco Polo, who was informed that the correct names of these tribes was "Ung and Mongul," and that they live "at the ends of the earth, between the northeast and north winds, and are surrounded by very rugged mountains." As to whether they were shut up by Alexander the Great, Fra Mauro doubts that Alexander got that far and seems to think the enclosure was a more natural phenomenon, being that of "rugged mountains and the ocean sea." After having read all this, however, the observer cannot fail to note a tidily locked gate at the southern end of the rock-walled enclosure.[14] Here the picture is conventional, while the text is not (fig. 6.2).

As on the mappaemundi, the three continents of the known world are shown, their forms somewhat altered by the mapping of the sea charts, and Europe is distinctly reduced in size in relation to the other two. More surprising than these physical modifications are the comments the mapmaker posts about the divisions between the continents. "The River Don," he says, "rises in Russia and not in the Riphaean Mountains, but at a great distance from them . . . And anyone who wishes to deny this should know that I have information from trustworthy persons who have seen this with their own eyes. One can in fact say that this river does not make a good boundary between Europe and Asia, first because it traverses a great part of Europe, second because of its form like a Roman V, third because its source is not in the place where the written sources have put it."[15] He thinks the river Edil or Volga, long confused with the Don, is a better candidate for the boundary, "because its course is more direct and has a better form and rises in the place which is the origin of that division."[16]

As for the division between Asia and Africa, he writes, "Because I didn't have space in Europe to give Ptolemy's true opinion about the division of Africa from Asia, I say here that he seems to make two divisions. The first begins from the horn of Africa [la ponta de Ethiopia] and goes along the coast of the Arabian gulf. Then he says, in order not to divide Egypt, he places it in Africa, and I agree."[17]

His scattered comments on the topic are put in more compact form on the Borgia map in the Vatican:

About the division of the earth, that is to say, Asia, Africa and Europe, I have found among cosmographers and historians opinions more diverse than I can well say. As it is a rather interesting subject, I will review my opinions briefly here. The Ancients, among them Messala the orator, [Marcus Valerius Messalla Corvinus (64 BCE–13 CE); his works are lost] who wrote a genealogy of the family of Augustus, and Pomponius Mela and those who borrowed from him, affirm that the Nile separates Asia and Africa, and the Tanais, Europe and Asia. Among them Ptolemy affirms that Africa is separated from Asia by the mountain chains of Arabia, which are found beside Nubia and extend across Abyssinia as far as southern Ethiopia. Others express modern opinions, affirming that the delimitation of Africa by the Nile or by these mountains makes it too small, and say that Africa is best bounded by the Red Sea or by the Gulf of Arabia. Likewise, considering that the Edil (Volga), which flows into the Caspian Sea, flows more directly from the north than the Tanais, they judge that that river is a better boundary between Europe and Asia. This latter opinion seems more clear and convincing and requires much less tracing of imaginary lines than with those authors proposing the previous division.[18]

It is interesting that he should here refer to the division of the continents as the "tracing of imaginary lines." He does not, however, do away with the concept of continents altogether but, by shifting the borders, makes them potentially unstable.

One dramatic change Fra Mauro makes to the mappamundi format is the orientation of his map to the south. Nearly all medieval world maps had been oriented to the east, with the garden of paradise at the top, even, in the case of the Hereford map, surmounted by Christ in glory. Fra Mauro has not only displaced Eden from its dominant position but has reoriented the map completely. This is generally thought to be the result of Arab influence, as many Arabic world maps were south oriented in the Middle Ages. Orientation seems not to have been such an important issue for Arabic culture, as we have found no real justification for southward orientation. For Europeans, however, it was important, and abandoning the east-on-top tradition with its important significations was a decisive and radical step.[19]

Another staple of the mappamundi is the engaging display of monsters, both beast and human, which exhibited the great variety of creation. On the Hereford mappamundi, they appear throughout Asia and Africa but are most concentrated on the southern rim of Africa. The Fra Mauro map is strangely devoid of animal or human figures, outside the Garden of Eden. One has to read the text to find references to a few imaginary creatures, such as the phoenix and the

Figure 6.2. Detail, northeast Asia. "Chataio" (Cathay) appears in the lower left above an elaborate monument labeled "the imperial sepulcher," or tomb of the Grand Khan. At the bottom (center) is a gate enclosing a mountainous region where Gog and Magog were contained, "according to popular belief." The Caspian Sea is to the right of center and a part of the Black Sea appears at the far right. Courtesy of the Biblioteca Nazionale Marciana.

dragon, as well as real but exotic ones, such as parrots, elephants, apes, and polar bears. Fra Mauro speculates in several places about the monsters traditionally described in maps and geographies and where they might be found:

> Since there are many cosmographers and very learned men who write that in Africa, and especially in Mauritania, there are many monstrous men and animals, it is necessary to note my own opinion, not that I wish to contradict the authority of so many, but in order to say I have inquired with diligence into all the new information I could find in Africa for many years, beginning from Libya, Barbaria, and all of Mauritania from the river of Gold and from the seven mountains across from the country of the blacks outside the first climate and afterwards beginning from Binimagra, Marocho, Fessa, Siçilmensa and along the coast of the mountains and toward the west through Garamantia, Saramantia, Almaona, Benichileb, Cetoschamar and Dolcarmin and more toward the east through the kingdom of Goçam and toward the south and in Abassia and in their kingdoms, which are Barara, Saba, Hamara and further toward Nuba through the kingdom of Organa and the island of Meroe and through all these lands of Negroes, I have found not one person who could give me knowledge of what I have found written; wherefore not knowing any more that I can confirm, I leave to those who are curious to seek to understand such novelty.[20]

He is not opposed to monsters on principle—just skeptical of rumors that are not substantiated by reliable reports. He refers to the secrets of nature, saying "the many things we know are but a small part of those which we do not know, and those which we know are little esteemed due to their familiarity, and those which are unfamiliar are not believed, and this happens because Nature exceeds the intellect." He concludes that "therefore those who wish to understand must first believe."[21] Thus he uncritically records the existence of a seven-headed serpent in India, a dragon in Parthia with a miraculous healing stone embedded in its forehead, and an island in the Indian Ocean where metal objects are turned to gold.[22] Sailors' stories—the dark sea in the southeast from which no ship ever returned, the fish in the northern ocean that can bite a boat in half—are also recounted.[23] But perhaps he has not included enough marvels. As for "those that desire miraculous things and other monstrosities, let them read Julius Solinus Polyhistor, Pomponius Mela, Saint Augustine, Albertus Magnus, Saint Thomas Aquinas in his book against curiosity. Also the 'Metaure' [Meteorology] of Aristotle and Pliny on the marvels of the world, and they will see that of a thousand things I have said scarcely one."[24]

Although Fra Mauro tries to work within the conventional format of the mappamundi, his words explode its certainties and leave questions to be resolved by future explorers and cartographers. The earth is reoriented to the south, paradise removed from the top of the map, and Jerusalem displaced from the center; the monster population is depleted, and the tripartite structure undermined. What remains of the traditional world picture? In the opinion of George Kish, "The world has not yet changed structurally, but a revolution (a de-centering) is already at work in the interior of the ancient structure."[25]

Sources of the Map

Among Fra Mauro's sources we find the usual classical authors (Solinus, Macrobius, Martianus Capella), augmented by then recently recovered works by writers such as Pliny, Pomponius Mela, and, above all, Claudius Ptolemy. The Bible, with its rather puzzling observations on the structure and layout of the physical world, is helpfully interpreted by generations of scholastic philosophers from Augustine and Jerome to Isidore, Bede, Saint Thomas Aquinas, and Albertus Magnus, whose works are cited, particularly in the corner texts. Then there was the testimony of travelers, especially the seagoing Venetians and the Portuguese, who provided some of the material for the map. These folk Fra Mauro referred to with respect as "nautici," "homini degne di fede" (trustworthy men), or those who have "veduto ad occhio" (seen with their own eyes). Visual sources included traditional world maps and diagrams, nautical charts (a number of which were made in Venice), and the maps of Ptolemy's *Geography*. Andrea Bianco's solution had been to place these three map types side by side, but now, working with Fra Mauro, the daunting project was to incorporate them into a single image of the world.

Although Christ does not appear embracing the world, as he does on the Ebstorf map, the Christian religion is by no means banished from Fra Mauro's picture. The sites of the Old Testament, an elaborate shrine of the holy sepulcher in Jerusalem, and the places where the saints preached and/or were martyred can all be found, but there is a difference from how these sites had been represented in the past. First of all, the size of the holy land is greatly reduced, putting its small area in proportion with the scale of the rest of the world. Some medieval mapmakers had stretched the size of the holy land in order to accommodate all the places of interest, and Fra Mauro feels he has to apologize for omitting them: "Those who are knowledgeable would put here in Idumea, Palestine and Galilee things which I have not shown, such as the river Jordan, the sea of Tiberias, the Dead Sea and other places, because there was not enough room."[26] It is true that the place-names in "Palestina" are few and mostly modern, such as the cities along the coast, while we seek in vain for Bethlehem, Cana, Nazareth,

Hebron, and Mount Quarantana, where Jesus fasted. It is worth noting that the long explanatory inscription for Jerusalem takes up considerable space in the holy land where place-names could have been put. Beside the Red Sea, there is a brief notice of the crossing of the children of Israel, but, most unusual in a medieval map, the parted waters are not shown and the sea is not colored red. Noah's Ark is shown in Armenia, as a small house on a mountaintop, but the legend notes that these are the remains of the ark, according to "vulgar" opinion among the Armenians.[27] While religious sites continued to be important, they were firmly located in the past rather than the eternal present. A famous event happened here, but it was a long time ago. (In contrast, the mappaemundi gave a dizzying sense of all events occurring simultaneously: the waters of the Red Sea parting, Noah and his family peering out of the windows of the ark, the Tower of Babel still standing.) One exception is Fra Mauro's depiction of the city of Babylon, shown in all its glory, with a long legend telling of its enormous size and wealth. Citing Orosius, Fra Mauro observes that it is not only amazing that human power could build such a city but also that it could be (and was) destroyed. Inside the circle of the impressive walls we return to modern times with the inscription "Babilonia, or Baghdad." It is interesting that there is no Tower of Babel, a fixture on most medieval mappaemundi.

Secular history received the same treatment. Fifteenth-century Ptolemaic maps were burdened with the names of numerous tribes who long since had ceased to inhabit the designated part of the world—or indeed any part of the world—and medieval maps had frequently shown them, too. Fra Mauro's references are more current. Near the northern part of the border between Europe and Asia is a brief inscription: "From this [area of] Gothia came the Goths to Italy." A longer historical note near Scandinavia traces the course of the migration in greater, if somewhat fantastical, detail.[28] Ancient sources are quoted admiringly, but Fra Mauro has a particular problem with Ptolemy, who was cutting such a swath through the intellectual world of the mid-fifteenth century. The superiority of sea charts in many areas to Ptolemy's maps, the archaism of his nomenclature, and the impossibility of constructing a world map on a latitude/longitude structure without sufficient data were some of the problems Fra Mauro faced. "I do not wish to contradict Ptolemy by not following his cosmography," he writes, "for if I wished to respect as far as possible his meridians or parallels or degrees of the northern part of this sphere, I would have had to omit many provinces which Ptolemy never mentioned at the extremes of south and north latitude, said by him to be 'terra incognita,' because in his time they were not known."[29] In the far north, he says, "In his fourth map of Europe Ptolemy names this area 'Scandinaria' and says that it has 18 hours of daylight, which is amazing to me in that all this part of Norway and Sweden was unknown to him."[30] This comment, along with a dozen other critical remarks about Ptolemy, shows that Fra Mauro certainly had access to a Ptolemaic atlas and no doubt had studied it with some care.

The problem of ancient versus modern place-names was also an issue with more recent mapmakers. The old provinces of the Roman Empire had been retained on European maps for a millennium, long after those names had fallen out of everyday use. Fra Mauro announces, "In this work I have chosen by necessity to use modern and common names because truly if I had done otherwise few would have understood me except for some bookish type [qualche literato] who, even nowadays, cannot make himself use the names which are currently employed."[31] For example, in North Africa he writes that he is not going to use the three classical divisions of Mauritania (Cesarensis, Sitifensis, and Tingitana) because they are no longer in use today.[32]

Voices of Experience

His most respected source was the skillful sailors ("i marinari experti"), those who had seen with their own eyes.[33] Their testimony is repeatedly put up against antique authority, which comes off poorly. We know who some of his sources were, as he made good use of Marco Polo for place-names and legends in Cathay, and Nicolò de' Conti for southeast Asia. Marco Polo's adventures had been mapped in a mural at the doge's palace, which was apparently still there in the 1450s. It was ordered to be repainted in 1459 and was destroyed by fire in 1483. (The map visible today was painted in the eighteenth century.) When Ramusio wrote his collection of "Navigations" in the sixteenth century, he averred that Fra Mauro used a map that Marco himself had made in Cathay and brought back with him, but no trace of such a map has ever been found.[34] The Catalan Atlas of 1375 was the first surviving map to make use of Marco Polo's information from the East, and Fra Mauro does the same. He notes that this area was "terra incognita" to Ptolemy and repeats Marco Polo's fulsome description of the wonders of the great city of Cansay (Quinsay)—that it is built upon a lake like Venice but is much larger. He goes off into a flurry of statistics, citing its twelve thousand bridges, its circumference of one hundred miles, and an estimate of its population (900,000) based on the number of hearths, and he concludes with a tribute to its great magnificence and order.[35] He lists cities and provinces, mentions the great size of the Yangtze (Quian) River, and refers to the trade goods to be found there, such as rhubarb, ginger, silk, and porcelain.

While he may not have had Marco Polo's map, Fra Mauro does seem to have had a Portuguese chart with the most up-to-date information on their voyages. He describes his source as a "new map" with "new names for rivers, gulfs, capes and ports, of which I have had a copy."[36] The newest names appear on the west coast of Africa, where Cavo Rosso and Cavo Verde indicate that the mapmaker had information from the voyage of 1445. Venice itself was full of merchant travelers coming and going from exotic places, which is reflected on the map by many references to trade routes and trade goods—spices, silk, precious stones, gold, and silver. The business of making nautical charts flourished in Venice as well. The most unusual bit of information on Fra Mauro's map probably involves the Catalan ship which, in his own day, went off course in the far north and lost its cargo.[37]

Fra Mauro's connections with the intellectual center of Florence may have provided him with more information. For example, he mentions the conquests of the king of Ethiopia in central Africa in 1430, which may have come to him from those who interviewed the Ethiopian delegation at the Council of Florence.[38] In East Africa is a legend explaining that, in mapping this area "virtually unknown to the ancients," he had consulted with monks native to the region, who "with their own hands had drawn all these provinces and cities and rivers and mountains with their names" (fig. 6.3).[39] He makes one other reference to these informants. "The Abassians [Abyssinians] say that they have more territory above the source of the Nile than below, that is, toward us. And they say that they have rivers greater than the Nile, which among us has such a reputation for its great size." He goes on to say that the Nile is increased by the many rivers that flow into it and the rains that fall at the time of their winter, that is, May and June.[40]

On the complex and vexed question of the configuration of the Indian Ocean, the cartographer marshaled the forces on all sides. "Some authorities," he notes, "write that the Indian Ocean is closed like a lake and that the Ocean does not enter it, but Solinus says that the Indian Ocean is navigable from the southern part to the southwest, and I affirm that some ships have gone out and come back by that route. Pliny also confirms this when he says that in his day two ships went out from the Arabian Sea, according to the account which they left, and, loaded up with spices, finally disembarked in Spain and at Gibraltar. This is also confirmed by Facio [Fazio degli Uberti, author of the geographic poem, *Il Dittamondo*] and by those who have experienced this journey, men of great prudence, in agreement with those authorities."[41] It is not clear exactly who these "men of great prudence" were, but he seems to be referring to contemporaries, specifically "those whom the king of Portugal sent with his caravels to seek out and see with their own eyes. These are said to have gone more than 2000 miles southwest beyond the strait of Gibraltar."[42]

Even more intriguing is his reference to a "ship of India," which he calls a *çoncho*, or junk. This vessel had sailed around Africa from east to west in the year 1420, traveling for forty days "beyond the cape of Soffala and the Verde islands to the southwest and west, and by the judgment of their astrologers [astronomers] which guided them they went about 2000 miles."[43] Returning to the "Cape of Diab," the southern tip of Africa, they went ashore and found the giant egg of an even larger bird, one capable of easily lifting an elephant. Gavin Menzies is convinced that the "ship of India" was one of the great Chinese fleet of 1421 and that the tale of its misadventure was brought back to Venice by Nicolò de' Conti, who had encountered the fleet when it passed through India. Conti, like Marco Polo, does describe the huge junks with five or more masts, solid construction, and great carrying capacity but does not say anything about the fleet.[44] This would be an interesting speculation if he went no farther, but unfortunately Menzies identifies the bird as an ostrich, when it is clearly the Rukh of the ancient sailors' tale. Fra Mauro even refers to it as a "chrocho," and says it was "most rapid in its flight." Menzies has the fleet going on to the Cape Verde Islands ("the green islands"), America, and the Straits of Magellan, and discovering Australia and Antarctica before returning home.[45] It seems possible that one of the ships may have gone astray, as Fra Mauro reports, or that it could have been any of the many traffickers in the Indian Ocean who ran into unpleasant weather. That Fra Mauro shows the tip of Africa, albeit divided from the rest of the continent by a thin band of water, and insists that it is circumnavigible, raises irresistible speculation. How did he know this? Or was it a lucky guess?

Other Maps as Sources

Fra Mauro makes sparing references to other maps, but he must have seen a number of them in Venice, in addition to the maps of his collaborator Andrea Bianco. Fortunately, we have preserved in the Vatican the chart that may have been a copy of the source map Fra Mauro says the Portuguese supplied.[46] He apparently had access to the maps of Ptolemy, as Bianco had when he produced his atlas in 1436. The very form and ambition of the Mauro/Bianco map, however, links it to the monumental maps that had had their heyday several centuries earlier. The impulse to make such maps had not vanished. In the mid-fifteenth century, Giovanni Leardo, working in Venice, made several mappaemundi of the traditional type: east at the top, Jerusalem at the center, the inaccessible frigid and torrid zones plainly marked. Other world maps Fra Mauro might have seen were those of Paolino Minorita, Marino Sanudo, and Pietro Vesconte.

Figure 6.3. Detail, southern Africa. Fra Mauro's depiction of Africa shows it to be circumnavigable. On the west coast is a gulf similar to the later discovered Gulf of Guinea, while the southern tip of the continent is divided from the whole by a narrow river. The Red Sea may be seen at the lower left. (Remember south is at the top.) Courtesy of the Biblioteca Nazionale Marciana.

Sanudo made sure that numerous copies of his maps survived and, according to his will, had put several of them into monasteries in Venice. The world map in this group was one of the very first to incorporate the geographical forms and structure of the sea charts.

Another possible source of inspiration was the Catalan school of cartography. Like Fra Mauro, the Catalans melded the sea chart with a complex depiction of the world as a whole, embellished with extensive details on land and colorful images. One of the most ambitious productions of this school, the Catalan world map of Modena, is dated to about 1450 and was made for the Este family of Ferrara. Fra Mauro might have seen this map or one like it (fig. 7.8). The Modena map, unlike the Catalan Atlas of 1375, was circular like Fra Mauro's and the ocean was painted in a similar rippled pattern of blue and white. While the shape of South Africa on the Modena map is quite different, it is divided by a canal as it is on the Venetian map.

Image of Africa

The configuration of Africa on the Fra Mauro map is not completely out of line with that of other mid-fifteenth century maps, which show some kind of extension to the south and east. Ptolemy, of course, had joined Asia and Africa by means of a long southern shore to the Indian Ocean. Fra Mauro's image startles because it seems so close to the modern image, just curved around to the east to accommodate the circular frame. The indentation on the west coast looks remarkably like the Gulf of Guinea, which is puzzling because the Portuguese had not yet gotten that far. Labeled Sinus Ethiopicus, it is more likely that it is the eastward-bearing gulf described in the *Libro de Conoscimiento* and sought by the Portuguese as the body of water which would bring them within a few days' journey of the fabled kingdom of Prester John.[47] It was also believed to be the entrance to the River of Gold and is so marked on the map.[48] Near this point is a notice: "I have heard many times from many people that here there is a column with a pointing hand and an inscription saying that one should go no further. I wish the Portuguese who sail this sea would say if what I have heard is true, because I do not dare to affirm it."[49] This mysterious column and notice ("ne plus ultra") had once been located at the Straits of Gibraltar but now seemed to be moving rapidly southward down the coast.

The southern portion of the African continent, the Cape of Diab, is shown separated from the rest by a narrow river. It has been speculated that this is the island of Madagascar, knowledge of which may have come from Arab sources. The legend here informs us that the passage separating Diab from Abassia (Abyssinia) is bordered on both sides by very high mountains and huge trees, and so is dark and dangerous for ships.[50] On the east coast of Africa we find several names of Arabic origin: Xengibar (Zanzibar), Soffala, Chelue (Kilwa), and Maabase (Mombasa).[51]

Africa also contains a number of features that date back many centuries. The name Ethiopia appears at least seven times, reminding us that this was almost a synonym for southern Africa (see fig. 6.3). In the far south was the powerful and wealthy kingdom of Benichileb, where the inhabitants had the faces of dogs but had never been conquered, not even by the Romans.[52] Somewhere in central Africa dwelt Prester John, with 120 kingdoms under his dominion, speaking sixty different languages. He is further identified as the "King of Abassia," ruling an almost infinite number of people. He goes into battle with a force of a million men, some of whom fight naked while others wear armor made of crocodile skins.[53] It is interesting that there is no reference to his famous piety and virtue.

Like other mapmakers, Fra Mauro is concerned with the course of the Nile and provides a number of inscriptions speculating particularly on its source and course. From the Ethiopians he may have gotten a description of its originating in a huge mountain, out of which the river emerges to form three lakes and then to flow northward to lower Egypt. This configuration also appears on Arabic maps and in Ptolemy's *Geography*. Still, he cannot let go of the idea that the great river also flows into West Africa because, he says, the same animals are found in both places.[54] After giving various information in different legends, which are not quite consistent, Fra Mauro exclaims, "I believe that many marvel at where I have placed the source of the Nile, but surely if they be moved by reason and desire to understand, they will see how much I have done and with more diligence than I can say here, and that I have been motivated to prove this by the clearest evidence I had."[55]

On the West African coast appears a series of names marking the points attained by the Portuguese voyages, as far as Cape Rosso, just south of the River Gambia at 12° 20″ north, reached by Dinis Dias in 1445. Already the sailors had begun to note the position of the polestar low on the horizon. Fra Mauro puts this information at Cape Verde and remarks that the same is true in the same latitudes of India near the Cape of Chomari (Cape Comorin). Here "one loses the north star or the arctic pole, and this is affirmed by all."[56] The Venetian Alvise Cadamosto, sailing in African waters in a Portuguese ship in the mid-1450s, sighted the southern cross and drew an outline of this constellation in his notebook. Fra Mauro's map is dated too early to include information from Cadamosto, who was traveling in Africa in 1455–56, returning to Venice only in the early 1460s. There is one peculiar reference in Africa, however, which is identical with his report. It is a story of silent barter, of salt for gold in Africa, and includes the strange and unlikely custom of the natives "with the big lips" carrying salt in their lips, which "keeps them from putrefying." While the tale of silent barter is an ancient one, dating back at least to Herodotus, and has been said to occur in many parts of the world, this odd detail is unique to Cadamosto—and Fra Mauro.[57]

The Canary Islands, some of which were settled by Castile beginning in the 1420s, are named on the map with the comment that some have thought them to be the Fortunate Islands. Madeira, "rediscovered" by the Portuguese in 1425 and subsequently settled, although the Azores, discovered in 1427, are not. The discovery dates of these Atlantic islands are unclear, and it is puzzling that they appear on sea charts during the fourteenth century and in the *Libro de Conoscimiento*, before their discovery was otherwise recorded.[58]

Although his collaborator and many of his informants were sailors, Fra Mauro gives us relatively little information about navigational techniques. He is more likely to speak about divergence from common Mediterranean sailing practices than to tell us about something that was perhaps so commonly known that it did not seem worth mentioning. In the Baltic Sea ("an area completely unknown to Ptolemy"), he observes that they navigate "without chart or compass but only with lead and line."[59] In the Indian Ocean, they also navigate without a compass, "but they carry an astrologer [astronomer] on board who perches up above with the astrolabe in his hand gives orders how to sail."[60] One deduces that the audience for the map would already be familiar with compass-and-chart sailing in mid-fifteenth-century Europe.

Indian Ocean

The Indian Ocean on the Mauro/Bianco map not only is open to navigation east and west but also is crowded with an array of islands more clearly identified than those in the Catalan Atlas of 1375. Taprobana was the one island that had dominated older maps, and conventional geographical accounts, as well as Ptolemy's atlas, had made it too large. Marco Polo speculated that it had shrunk in recent times due to the rising of the sea.[61] But which of these depicted islands is the real Taprobana? Fra Mauro writes, "Note that Ptolemy, meaning to describe Taprobana, only described Saylam [Ceylon]."[62] Instead, Fra Mauro identifies the large island of Sumatra, farther east, as the true Taprobana. This confusion was to persist for several hundred years, understandable in view of the multitude of islands in this sea. The map takes Marco Polo's figure of 12,700, and adds that "in this sea are many islands of which one cannot make special note because there is no space, but all are inhabited and very fertile in diverse and precious spices and other novelties and are most rich in gold and silver."[63] Japan (Ciampangu) appears for the first time on any European map, even though Marco Polo had alerted the west to its existence 150 years before. We also find two Javas, major (possibly Borneo) and minor (Java); the island of Colombo (apparently an error for Quilon on the southwestern coast of India); Bandan, which was the source of cloves; and Sondai, where nutmeg grows. The polyglot nature of the Indian Ocean meant that most places had multiple names in various languages with transliterations and translations. Today it is still impossible to identify absolutely some of the islands and cities shown. Fra Mauro's sources on this area were probably Arabic, though perhaps filtered through Venetian merchants.[64]

One of these merchants was Nicolò de' Conti, who had spent decades rambling about southeast Asia and among the islands. His account of his travels was incorporated into Poggio Bracciolini's book on the vagaries of fortune. It is difficult to be specific about Fra Mauro's use of Conti, as many of Conti's observations were similar to those made by others, and inconsistencies of nomenclature and spelling make identification chancy. Conti's travels in the Indian Ocean follow Marco Polo's, and many of Conti's remarks follow those of his great predecessor. He did, however, travel further inland, up the Ganges River (he says for fifteen days) and to the great kingdom of Vijayanagar in central India ("300 miles from the sea"). Both he and Fra Mauro called this kingdom Bisenegal, and Conti comments on its great wealth and size and goes into detail about its elaborate religious ceremonies. Here one finds one of several of his descriptions of the custom of sati, the burning of a living wife on the funeral pyre of her husband.[65] Details that Fra Mauro may have taken from him include the description of the city walls built into the surrounding mountains and the estimate of the king's fighting force at 900,000 men.[66] Conti also gives specific information about the spices to be found on the islands of Sondai (Sunda) and Batavia (Banda), nutmeg and cloves, respectively, not to mention beautifully colored parrots who can talk exactly like human beings. Fra Mauro repeats the information about the spices and the parrots on the map.[67] We may assume that Fra Mauro thought Conti a reliable source because he soberly reports the presence in India of "a seven-headed serpent seven feet long," which Conti had described.[68] Conti's glowing account of the abundant resources and numerous people of India finds an echo on the map, where the mapmaker waxes eloquent about the "cities, castles, innumerable variety of people, conditions, customs, very powerful kings, great number of elephants" as well as "precious fruits, woods, grasses and roots," precious stones, and "so many other things that I cannot possibly list them all."[69]

The Indian Ocean was a good place for marvels—not only rare spices, precious stones, aromatic woods, exotic fruits, and other things one could not find at home but also people with peculiar customs. The islands of men and women, long a part of Arabic folklore, had been mentioned by Marco Polo. These sexually segregated islands were located conveniently near to one another so that the men could pay periodic visits in order to replenish the race. The Catalan Atlas, seeing the obvious connections with the old stories of the Amazons, put the queen of the Amazons on one of these islands. Fra Mauro calls the islands Nebila and Mangla and puts them in the far southeast. Of course, most of these people were "idolaters," and some were cannibals and giants. If one went too far, one might sail into the "dense sea" near the "dark islands," from which no sailor had ever returned. Ibn Baṭṭūṭah, the Moroccan traveler, had told of being becalmed in that frightening sea for thirty-seven days before being miraculously saved.[70]

The Portuguese recipients of the original map must have studied this area carefully and, undaunted by the hazards, determined to go after the pearls, spices, scented aloe, brightly colored parrots, and gold reputed to be found there.

The Far North

At the other end of the world, in the far north (also "unknown to Ptolemy") Fra Mauro drew a rather large Scandinavian peninsula and mapped out the ports along the Baltic Sea and the Gulf of Finland.[71] The Baltic is here called the Sinus Germanicus. It lacks the northward extension of the Gulf of Bothnia, making the Scandinavian peninsula much thicker than it is in reality. This depiction probably reflects the greater familiarity sailors would have had with the southern shore and its many rich trading cities. Fra Mauro says the Scandinavians used to be the terror of Europe, but now they are weak and do not have the reputation that they once had.[72] The Laplanders, called Permiani, live farthest north. They live by trapping and selling furs, and in the winter they retreat into Russia and live underground.[73] Greenland is not shown; it is interesting that Fra Mauro does not seem to know the Clavus map of the north. Variants of Iceland appear three times, once on the Danish peninsula (Islandia, shown as an island), once at the far west of the Scandinavian peninsula (Islant), and once (Ixilandia) on an island in the northwest Atlantic. The Hyperborean mountains and the Riphaean mountains, both of fabled antiquity, are shown as well. In this area the mapmaker has tried to combine the traditional picture of the far north with rather spotty contemporary information. A good part of the northern section of the map is taken up with more general texts on the making of the map and the general configuration of the world.

On the Mauro/Bianco map the conflict between visual representation and textual description is persistent, reflecting the unsettled state of world geography in the fifteenth century. New discoveries of distant lands as well as ancient manuscripts and maps had thrown the eternal verities into confusion. Sometimes the text on the map is thoroughly conventional, as in the case of paradise, while the picture is not. Elsewhere the placement is orthodox and the text more radical. One curious example is the connection between Scotland and England on the island of Great Britain. Sea charts had shown the Firths of Forth and Clyde as a channel separating the two nations. Fra Mauro's Britain is presented as a continuous landmass, but in the nearby text he says that Scotland is in fact "separated by mountains and water" from the southern part of the island.[74] This example seems to indicate that the legends were composed after the map was painted, a circumstance that would explain why many of the legends contradict not only the images but sometimes each other. Some of the legends are pasted on bits of paper, inspiring us to imagine that, after the map was supposed to be finished, the mapmaker kept feverishly composing new texts and sticking them onto the surface, as new information came to Venice or he had second thoughts. Although Cattaneo thinks this map was mostly made in the early 1450s, it was still at the monastery, and the date of 1460 on the back suggests that it was declared finished only after the death of the "incomparable cartographer."

The long inscriptions on the map allowed the mapmaker to struggle with some of the cosmological issues paramount in his day. Was the earth habitable throughout? Was there more land on the other side of the globe, or was there only a vast expanse of sea? And what was the proportion of land versus water on the earth? Recent discoveries, bolstered by the miscalculation of the width of the three known continents, reinforced the idea that land might be dominant. However, the Bible said that the land covered six-sevenths of the globe: "On the third day you [God] commanded the waters to be gathered together in a seventh part of the earth; six parts you dried up and kept so that some of them might be planted and cultivated and be of service before you" (Esdras 6:42). Was there an encircling band of ocean at the equator, which was uncrossable due to the great heat? Increasing exploration to the south seemed to demonstrate that there was no such band, nor was the heat intolerable. Again and again, the mapmakers turned to the oral testimony of eyewitnesses to confirm their information.

Ancient authorities, however, were not cast aside lightly, and in fact many conclusions were held in abeyance with the idea that they might still be upheld. A small example is Fra Mauro's reference to the gold-digging ants, denizens of many geographies and world maps since Herodotus's day. Maybe there was some larger animal who looked like an ant, he wrote.[75] The river Don did not flow from the Riphaean mountains, but these legendary mountains, which marked the farthest north, were still shown. The Permiani were savage and their climate harsh, unlike that of the storied Hyperboreans, but perhaps someday the Hyperboreans would be found near the mountains which bore their name.

The Mauro/Bianco map strains at the seams. The growing expanse of the Atlantic Ocean and the newly discovered islands would not fit onto a circular world map without greatly shrinking other features. Africa here curves around to the east, but further explorations would reveal its southward thrust, and on the 1490 map of Henricus Martellus Germanus, the Cape of Good Hope broke through the map's frame. Throughout their map, Mauro and Bianco express confidence that new discoveries will be made which will clarify some remaining uncertainties. The world was not a static image, established some time in the classical past, but a dynamic one, changing before one's very eyes.

In the general caption of his great map, Fra Mauro writes:

> This work, made for the contemplation of this most illustrious signoria [of Venice], has not achieved all that it should, for truly it is not possible for the human intellect without divine assistance to verify everything on this cosmographia or mappamundi, the information on which is more like a taste than the complete satisfaction of one's desire. Thus some will complain because I have not followed Claudius Ptolemy, neither in his form nor in his measures by longitude and by latitude. I have not wished to go to extremes to justify what he cannot justify himself, for in Book II, chapter one, he says that he can speak correctly about those parts of the world which are continually frequented, but of those places which are not so often visited, he does not think it is possible to speak correctly. Understanding that it was not possible for him to verify everything in his cosmography, it being a long and difficult task and life being brief and experience often faulty, so he concedes that with time such a work could be better produced or that one could have more definite information than he has here. Furthermore I say that in my time I have tried to validate written sources with experience, researching for many years and profiting from the experience of trustworthy persons who have seen with their own eyes all I have faithfully put forth here.[76]

CHAPTER SEVEN

The Persistence of Tradition in Fifteenth-Century World Maps

Most fifteenth-century world maps were touched by contemporary upheavals in geographical thought and mapmaking technique. The traditional model of the mappamundi was, however, compelling and continued to be a force in shaping the vision of the mapmaker and, we assume, the map consumer. Conventional images still hold power over us. To take one example, showing students a copy of the Australian-produced world map with south on top produces disorientation, even outrage, and cries of, "Turn it right side up!" Unusual map conformations such as those of Buckminster Fuller or the Peters projection, whatever their merits as maps, meet initial resistance because they do not look like the image of the world we expect to see.

The salient characteristics of the mappamundi are summed up in table 7.1. Obviously, not all maps are identical, but some features, such as the circular shape and the presence of paradise, are widely shared.

The maps with the most compelling reasons for clinging to traditional forms were those that accompanied historical works, such as the maps illustrating Ranulf Higden's history, *The Polychronicon*. Higden served as a monk in the Benedictine Abbey of Saint Werburgh in Chester in northern England from 1299 until his death in 1363/64. He completed the first version of his work in 1327; it seems to have experienced only a local fame. In 1340 he put together a new edition, much rewritten, which found a wider audience. Indeed, in 1352 he was called to London by the royal council—who asked that he bring his chronicle with him—to serve as a historical consultant about "various matters which will be explained to you." Unfortunately, they have not been explained to us, but we can surmise that the chronicler was being called upon to reinforce England's claims to land in France, these years marking the beginning of the

TABLE 7.1.

Fifteenth-century mappaemundi

Traditional features	
East orientation	Bi, H, L, R
Paradise	B, Bi, C, H, L, M, N, O, R, W
Four rivers	Bi, C, H, M, N, O, W
Circular shape	B, Be, Bi, C, L, M, N, O, R, W
12 winds	Be, H, O, W
Jerusalem in center	C, H, L, N, O, R, W, Z
Red Sea shown red, with division	Be, Bi, G, H, L, M, O, W
Monstrous races/animals	All
Pictures	B, Be, Bi, G, H, L, M, R, W, Z
Biblical events and ancient history	B, Be, Bi, G, H, L, N. O, R, W
Gog and Magog enclosed	Bi, G, H, L, M, N, W, Z
Enlarged area for holy land	H, O, R
Prester John	B, Bi, G, L, M, R, W, Z
Abstract geographical forms	B, H, O, R
Traditional, not modern, names	C, H, R
Zones marked uninhabitable	B, Be, L, N, O, W
Narrow rim of ocean	B, C, H, L, O, R (Rüst)
New features	
References to recent events	B
Sea chart forms, esp. in Mediterranean, Atlantic	Bi, C, G, L, M
8 winds and multiples, rhumb lines	Bi, G, L, M
New discoveries, such as Atlantic islands	Bi, G, M
Scale bar	G, M, W

Note: B = Borgia (1430). Be = Bell (1440–50; fragmentary, similar to W, Z). Bi = Bianco (1436). C = Copper map (c. 1485). G = Genoese (1457). H = Higden. L = Leardo (1442–52). M = Catalan/Estense (1450), Modena. N = Nova Cosmographia (1440). O = Olmütz (1450). R = Rudimentum Novitiorum (1475), includes Rüst/Sporer broadsides. W = Walsperger (1448). Z = Zeitz (1470).

Hundred Years' War.[1] Higden continued to add to his chronicle until shortly before his death.

Higden opened his work with an extensive geographical chapter before launching into the history of the world from the beginning of the first age at the "plasmation" of Adam. By the time of the second edition, he decided to include a world map and referred the reader to it.[2] His work must have been immensely popular, as 185 manuscripts survive, all made in England for the collections of cathedral, monastery, and university libraries, as well as for noblemen and wealthy London merchants. In the sixteenth century, Henry VIII kept two copies in his personal library. This wide distribution was doubtless due to the book's emphasis on Britain, which not only occupies a substantial section of the geography but also serves as the major focus for the history. Higden's "six ages of the world" are not divided in the usual way; instead he covers the period from Adam to Christ in a mere two ages. The remaining eras are divided according to events in British history: the invasions of the Saxons, the Danes, and the Normans. *The Polychronicon* was translated into English by John Trevisa in 1387, and in 1480 and 1482 editions were published by the pioneer printer William Caxton. There were three more printed editions by 1527. Readers doubtless liked its structure (six ages, eight aspects of history) as well as its moral themes, for Higden saw history as a handy storehouse of moral exempla.

The map he included—and we are fortunate to have what appears to be the original map in his autograph manuscript of the second edition—was resolutely traditional (fig. 7.1).[3] He adopted an oval form, perhaps based on Hugh of Saint Victor's idea that the world was shaped like Noah's Ark.[4] East is at the top with a vignette of Adam and Eve, standing on either side of the snake-entwined tree in the Garden of Eden. The geographical forms are abstract, showing no influence of the sea chart. The Red Sea is red, as is the Persian Gulf, and the passage of the Hebrews is clearly marked. The islands of the outer ocean are rectangles bearing their names and arranged in an orderly fashion around the edge. The place-names would have been familiar to Isidore, including the Roman provinces from the heyday of the empire. The Caspian Sea is a gulf of the northern ocean, the Nile originates in west Africa and meanders all over the continent before spilling into the Red Sea, and the pillars of Hercules stand guard at the straits of Gibraltar.

There are twenty surviving maps in nineteen Higden manuscripts, as well as several strays that may have been inspired by his history.[5] There are also several manuscripts with blank pages left for a map that was intended but never drawn. Whether Higden had a map designed especially for his book or appropriated a preexisting model, the map makes a good illustration for his text. Nearly all the names on the map appear in the text, most in the geographical chapters of Book I. Their very archaism seems eminently suitable for a work of history. The various copies of the Higden maps are more or less elaborate but have a great consistency of place-names, which show that they were probably copied from one another, even though the artist felt a certain degree of freedom to tinker with the form. The simplest are just an array of place-names scattered in roughly geographical order over a blank oval interior. The most elaborate, and the one that most accurately reflects the text, is found in the British Library. Vividly colored in red and green, it covers two pages and bears a whopping two hundred names.[6] It is the only one of the Higden maps to include legends other than place-names and the requisite "transitus Ebreorum" next to the Red Sea. Most of these are taken directly from the text and include brief descriptions of twenty different members of the monstrous races. This map was made at Ramsey Abbey and appears to have no direct connection to Higden himself, having been made instead

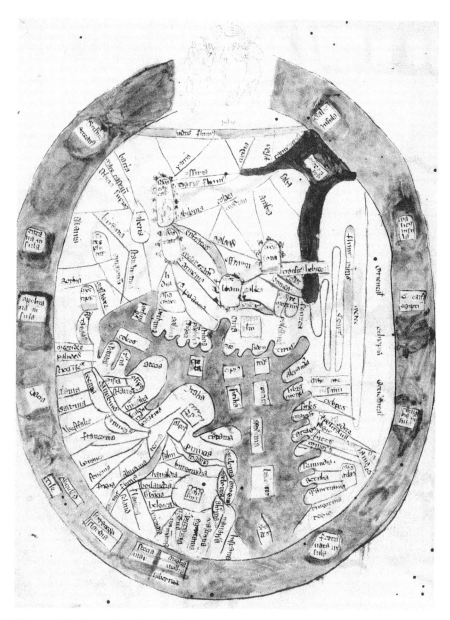

Figure 7.1. World map from Ranulf Higden, *Polychronicon*, late fourteenth century (London, British Library, MS Roy. 14.C.IX, fol. 2v. 46.5 × 34.2 cm / 18.6″ × 13.7″. The world map of Ranulf Higden is oriented to the east. Here Adam and Eve are sketched in, being chased out of paradise by an angel. On the right is the Red Sea, with a wide strip across it showing the passage of the Israelites. The Nile pursues a meandering course in Africa (right) before discharging into the Red Sea. Geographical features are rather abstract, with islands shown as simple rectangles bearing their names. Courtesy of the British Library.

by a careful reader. The Ramsey manuscript also includes a simpler version, more like the original map, now at the Huntington Library in California.[7]

Higden's work continued to be copied in the fifteenth century, from which six manuscripts with maps survive.[8] The later maps resolutely follow the original model with no additions or modifications for changing times. There is also an interesting freestanding map from the late fourteenth century which follows the Higden format. Not specifically made as a book illustration, it was probably posted in the library of Evesham cathedral as an all-purpose readers' aid.[9] By 1450, however, it must have been considered outdated, for the parchment was turned over and an illustrated genealogy drawn on the back.

The popularity and survival of the Higden map into the fifteenth century demonstrates that maps did not have to be up-to-date in order to be valuable to the map-using public. The mappamundi form had stood the test of time and was not about to be quickly discarded. Writing of the location of paradise, Higden commented that the "fame of Paradise had endured for six thousand years and more, from the beginning of time up to our own day. The reputation of a false thing usually falls into oblivion or meets with a contradictory opinion." But paradise has endured. So, too, has the mappamundi.[10]

Rudimentum Novitiorum

The conservatism of the Higden maps can be explained by their fourteenth-century model, but in the late fifteenth century a world map appeared in print which was extremely peculiar and whose antecedents we do not know. Printed in Lübeck in 1475, the map was an illustration for a universal history in Latin entitled *Rudimentum Novitiorum* (a basic book for novices). According to its publisher, Lucas Brandis, it was designed so that "poor men unable to afford a library can have a brief manual always on hand in place of many books." The *pauperes* of whom Brandis spoke may have meant poor preachers or friars minor but also may have referred to a growing number of book owners among the merchant class in a thriving commercial city such as Lübeck.[11] Like Higden, the anonymous author of *Rudimentum* included several lengthy geographical digressions, one on the world and one on the holy land.[12]

The world map is oriented to the east with Asia on top and the other two continents below, T-O style, though the three water divisions (Don, Nile, Mediterranean) are not shown (fig. 7.2). The province of Palestine is more or less in the center, where Jerusalem, unnamed, is represented by a formidable castle. In the west, three columns stand for the Pillars of Hercules at the entrance of the Mediterranean. The map's most striking feature is its conformation of the land

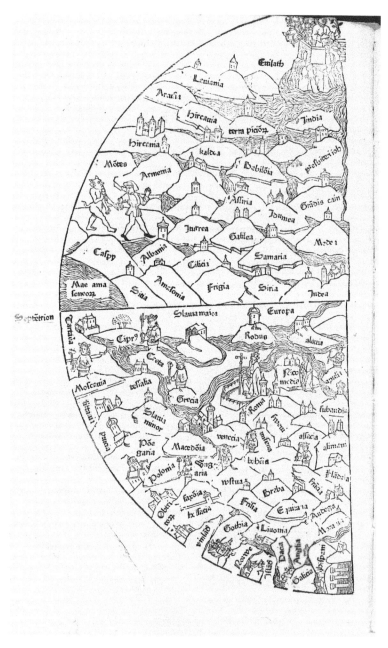

Figure 7.2. Rudimentum Novitiorum (Lübeck), 1475 (Washington, D.C., Library of Congress, Rosenwald Collection, Incun. 1472.R8. 37 cm / 15″ diam.). Like the *Polychronicon*, the *Rudimentum Novitiorum* was a comprehensive handbook for readers. The map appears more earthlike with its hillocks and castles but bears little relation to the physical structure of the world. East is at the top, with the two mysterious men in the walled garden, and the Mediterranean is not shown at all. Courtesy of the Rare Books Division of the Library of Congress.

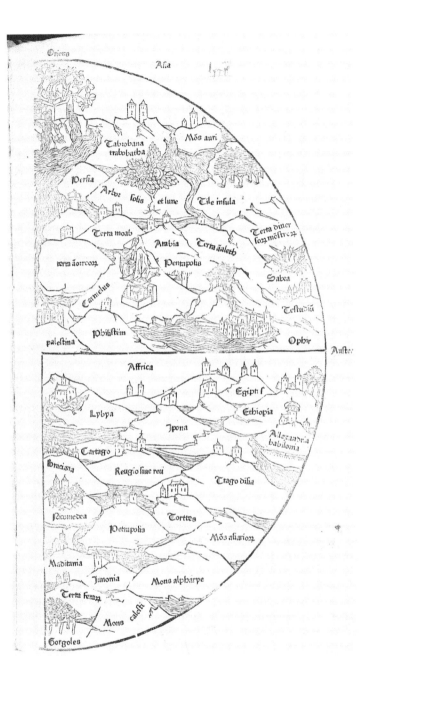

as a succession of hillocks, each surmounted by a castle or tower and bearing the name of a province or territory. It is really a list-map with a decorative base. Rivers flow plentifully among these hillocks but are unlabelled and unidentifiable. There is no encircling ocean, and the only labeled sea is "Mare Amasoneorum" in the north, possibly intended for the Caspian Sea. Geographical locations are somewhat confused. Of many examples, Nicomedia (from Asia Minor) is shown next to the Pope near Rome and again in Africa, perhaps an error for Numidia. India is north of Persia, which is next to Taprobana. There are a number of misspellings, leaving some inscriptions incomprehensible.[13] Printing was, of course, in its infancy and woodblock printing was still rather crude. The poor typesetter, working backwards, probably had some problems with the difficult geographical names.[14]

At the top of the map are two figures in an enclosed garden. They are not, however Adam and Eve, but two men, each holding a branch in his hand, apparently having a conversation. Various identifications have been suggested. Anna-Dorothee von den Brincken thinks they might be a Jew and a Christian having a harmonious discussion, thus symbolizing the unity of the Old and New Testaments. Another idea is that they represent Enoch and Elijah, Old Testament characters who went directly to heaven without suffering death.[15] Paradise is not labeled, but near one of the streams is the word *Evilath* (Havilah), one of the regions watered by the four rivers mentioned in Genesis. Other pictures on the map include two dragons in Libya, crowned kings and a queen in various kingdoms, the Pope in Rome, a figure in the holy land which is perhaps Saint Jerome (a major source for the book), and a phoenix in flames in Africa. A man-eating devil in northern Asia is already in the process of devouring a victim's severed arm.

The *Rudimentum* had only one printing in Germany, and by 1500 its printer was broke. It must have sold fairly well, however, as numerous copies survive today in libraries.[16] We do not know if there was a manuscript version that formed its base, as none has ever been found, but there are a great number of universal histories of this type, which were very popular in the late Middle Ages. Von den Brincken notes that the twenty years before the book's publication are barely covered in the history and opines that the printer updated an older work.[17] The book was, however, translated into French, maps and all, under the title *Mer des Hystoires* in 1488. This work had greater success, as it came out in three more editions: Lyons, 1491; Paris, 1500; and Paris, 1536. The French editor added material on French history and some new illustrations, including a woodcut of the baptism of Clovis. On the recut map a few new images appear—ships, birds, and trees—but the basic format was unchanged.[18]

The Rüst/Sporer Maps

A similar map was circulating in broadsheet in Germany in the late fifteenth century (fig. 7.3). Only a few copies survive, those which happened to get pasted into books, and it is difficult to date them exactly. One bears the name of Hanns Rüst, a printer of playing cards in Augsburg, who was dead by 1485. The block was sold before 1500 to Hans Sporer in Erfurt, who recut it to make some additions, including his own signature symbol, a pair of spurs. The text is in colloquial German. A ribbon floats above the map with the words, "This is the mappamundi of all the lands and kingdoms which there are in the whole world." The maps are round, with east on top, showing Adam and Eve in a walled enclosure. Four rivers flow out of this paradise, and Jerusalem is in the center. The three continents bear the names of the sons of Noah, and the twelve classical winds surround the map. In the encircling ocean a sequence of islands appear, some of which are populated by assorted monsters. The rest of the map is arranged very much like the one in the *Rudimentum*, with castles and buildings shown in circular vignettes above which banners float inscribed with their names. Even some of the details are the same: the devil in northern Europe and three (instead of two) Pillars of Hercules at the western end of the Mediterranean.

Below the Rüst/Sporer map are two circular diagrams, one showing the four elements (water, earth, fire, and air) and the other showing a T-O division, with an urban and a rural landscape above and the sea below. Hugo Hassinger suggests that this vignette is modeled on a north-oriented zone map, with the uninhabitable sea to the south, a European town on the northwest, and the habitable but relatively empty landscape to the east.[19]

Like the *Rudimentum*, the broadsheet map contains mostly information that had been in circulation since the days of Isidore—the Medes and the Persians, the apple smellers, the tree of the sun and moon, the pygmies fighting cranes. An exception is a scattering of German town names on the Sporer version. The geographic organization is somewhat chaotic: Flanders is in the Mediterranean, the Ganges river is between the Tigris and Euphrates, and the city of Augsburg is next door to Jerusalem. The map's legends are somewhat mangled, perhaps due to the woodcut artist or the difficulties in translating from a Latin model map. An interesting source of at least a few of the map's oddities may be German folklore, such as the adventure tales of "Herzog Ernst."[20]

The appeal of these old-fashioned maps is interesting to us. As illustrations for histories, which covered the world from the Creation to the present, or as histories in themselves, they served an obvious function. Their errors probably indicate a relatively uncritical audience, as well as a not very learned author.

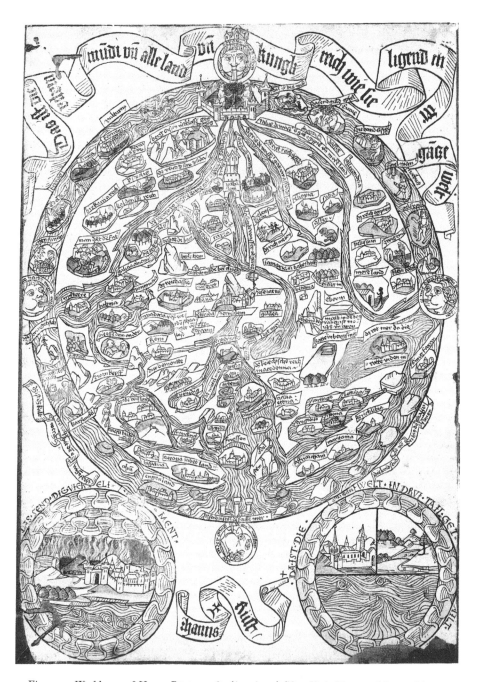

Figure 7.3. World map of Hanns Rüst, c. 1480 (Augsburg) (New York, Pierpont Morgan Library, 19921. 28 cm / 11″ diam.). Made for popular consumption, this map was sold as a single sheet. It is only slightly more realistic than the Rudimentum. The four rivers are shown streaming from paradise at the top, while boats and islands alternate in the ocean rim. The small circles below contain more abstract visions of the world, the four elements to the left and the division between country, town, and the sea to the right. Courtesy of the Pierpont Morgan Library.

Printing had just begun to make books available to a wider, less erudite public. As von den Brincken points out, there are only 1,100 maps in Destombes's catalogue of one thousand years of medieval world maps, while between 1472 (the date of the first printed world map) and 1500 ten thousand world maps were printed in Germany alone.[21] Doubtless many of the new map customers had not had an opportunity to look at a map before, much less to possess one.

Borgia XVI Map

Germany is also thought to be the origin of a large, freestanding map of the world engraved and enameled on copper (fig. 7.4). A circular map with south on the top, it is a striking work of art populated by lively cartoon figures—an elephant with a trunk like a trumpet, dog-heads worshiping their king, a Norwegian mounted on an elk, and a Tartar camp with wagons and tents drawn into a circle. In the northeast, Amazons are bending their bows, while in eastern Europe a battle between the Teutonic knights and the pagan Lithuanians is in full swing. Small ships sail upon the seas, which are ornamented with a decorative pattern of ripples. Some thirty-eight nail holes placed over the surface may have been for the attachment of symbols of some type, as they seem far too many for mere support.[22] Around the rim are twenty-four small diamonds spaced equally and numbered from i to xii, twice. These are unlikely to represent winds, as twelve is usually the maximum for the classical wind structure. It is possible that the mapmaker was copying from a model that framed the world with a calendar and that these numbers represent the hours of the day.

The geographical forms are abstract and show no influence of the sea charts. Italy is a large rectangular block, and a solid Greece shows no evidence of the Gulf of Corinth. The Caspian is a gulf of the ocean, which surrounds the entire land in a narrow band. The Indian Ocean is noted to have seven thousand islands but there is room only for "Travobana" and seven other tiny unnamed islands. The Red Sea cannot be red, as there is no color on the map, and it is not labeled, but it does show the passage of the Israelites—"Transeunt filiorum Israel." Other standard features include paradise, slightly to the south of east by the mouth of the Ganges. Adam and Eve are placed on either side of an angel in a walled-off enclosure labeled "locus deliciarum." The far north and far south are marked as uninhabitable zones. There is considerable biblical and historical content, from Noah's Ark, the Tower of Babel, and the Trojan War, to the conquests of Alexander and Hannibal.

In addition to the southern orientation, there are several unusual features. The traditional four rivers of paradise are not shown. Jerusalem is not on the

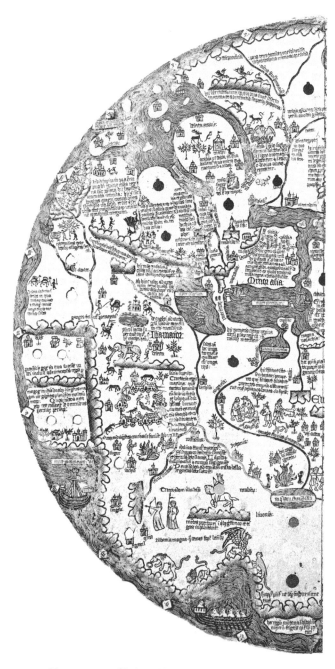

Figure 7.4. Fifteenth century world map, c. 1430 (Vatican City, Biblioteca Apostolica Vaticana, MS Borgia XVI. 63 cm × 25″ diam.). South is at the top of this map, and Europe takes up a great deal of space at the lower right. Geographical forms in the Mediterranean are abstractly drawn. The map is enlivened with cartoon drawings of humans and animals, from a Tartar

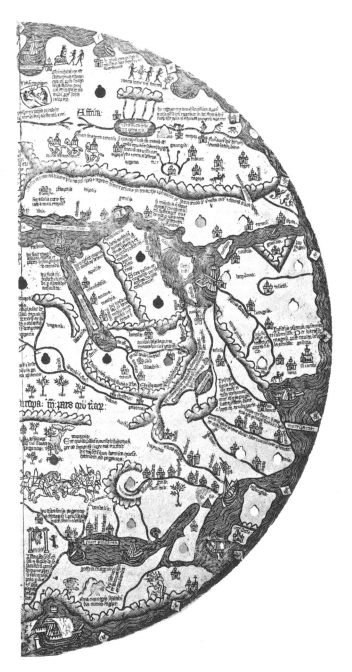

camp at the left to a pair of dog-heads before their king in the upper right. In the lower center, the pagan Lithuanians battle the Teutonic knights. Courtesy of the Biblioteca Apostolica Vaticana.

map at all, and the holy land ("Siria, terra sancta") is not in the center. Europe is greatly enlarged, taking up much more space than the usual one-quarter allotted on the T-O maps. The name of Rome does not appear, but the city is represented by an inscription reading "sedes apostolica et imperialis per vii-c annos." Scholars have tried to date the map by this reference to seven hundred years. In 751, Pepin, the founder of the Carolingian dynasty, received the blessing of the church in return for his acknowledgement of the Pope as his overlord. If the inscription refers to this event, which established the temporal power of the papal states, the map would date to around 1450. Other interpreters have read the inscription as "seven years," demoting the *c* to a mere cedilla, which could imply a reference to the ending of the Great Schism in 1417, thus dating the map in the 1420s.[23] The last historical event recorded on the map occurred in 1402.

The map is rich in classical history as well as events from the reign of Charlemagne. We see his campaign against the "infidels" in Spain and the place where Roland and the twelve peers were ambushed. The map also shows more current happenings. Among these are the fighting between the Teutonic knights and the Lithuanians which ended in the Battle of Tannenburg in 1410 (not shown), the capture of John II the French king at the Battle of Poitiers (1356), and the battles of Nicopolis (1396) and "Savastra" (1402). In the first of these, an army of crusaders were defeated by the Turkish sultan Bayezid, and, as the map notes, many were killed. The second inscription could refer to the city of Sivas, known since antiquity as Sebastea, which was sacked by Tamerlane on his way to Ankara, where Bayezid was defeated and taken captive. This event probably extended the life of the tottering Byzantine Empire for another half-century.

Material drawn from Marco Polo dominates China, which is presented, frozen in time, as it was at the end of the thirteenth century. The names of cities, the desert of Lop, the fact that it took four months to travel by caravan from Urgenj to Cathay, and the number of islands in the Indian Ocean all come from Polo's text. The map's connections with the Catalan school are most apparent in Africa, with its characteristic form for the Atlas Mountains, including the pass through which the caravans went "to the land of the blacks." A sequence of names in the Sahara, its description as a "sea of sand," the presence of Prester John and his Christian kingdom, and the wide "River of Gold" can all be found on Catalan predecessors. "Musamelli," the fourteenth-century monarch of Ghana, is shown still reigning in central Africa. There has been some speculation that the Borgia map was influenced by an Arab model because of its southern orientation, its reference to climates, and its names in Africa, but the climates were common geographical coin by the fifteenth century, and most of the names could also have been derived from Catalan sources.

Several exotic animals (elephants, dragons, lions, the phoenix) appear on the map, but its maker is clearly more interested in ethnography. The energetic little stick figures cavorting over the map represent the dog-heads, Amazons, cannibals, and idol-worshippers. The lazy king, a descendant of the "long-nailed mandarin," is being waited upon by obsequious slaves, here transposed to Africa from China.[24] In a particularly poignant scene, we see the Tartars, who are said to be so impoverished that they sell their own children into slavery, "just as Christians do their cattle."

We know nothing about the provenance or authorship of the Borgia map, which turned up in an antique shop in the late eighteenth century and then entered the collection of Cardinal Borgia. Its script identifies it as south German, but it is impossible to say much about its author. The noticeable emphasis on history and the traditional nomenclature suggests that it was designed as a historical map, perhaps for use in a library or a school. The mapmaker emphasizes the orderly sequence of the four empires (Babylon, Alexander, Carthage, Rome), a contrast to the turbulent state of Italy in the fifteenth century, and he comments that "Italy, beautiful, fertile, strong and proud, from lack of a single lord, has no justice."[25] Another prominent theme is the struggle between Christians and pagans. Although, oddly enough, the Crusades of the eleventh, twelfth and thirteenth centuries are not mentioned, there are special references to Charlemagne's campaign in Spain, to contemporary crusading (in northeastern Europe, in Africa, at Nicopolis), and to the future threat of Gog and Magog, here specifically identified as Jews.[26] The map is clearly a carefully made and thoughtful production, crafted in a medium designed to endure. While the creator's choices are not always the ones we should have made, they are not the result of sloppiness or error.

Olmütz Map

The Olmütz map is another rather conservative world map made in southern Germany around 1450.[27] It was bound in a volume of religious writings but does not appear connected to any of these, so we really do not know the purpose for which it was made. This small map (150 millimeters or six inches in diameter) is oriented to the west, but other than that, its features are the traditional ones. Paradise is in the east with the four rivers flowing from it. The inscription notes that it is guarded by venomous serpents and an angel with a flaming sword. The Red Sea is colored red and shown divided by the passage of the fleeing Hebrews: "The Red Sea which the children of Israel crossed with dry feet."[28] Jerusalem is at the center, and the area of the holy land is much enlarged, including a very large Sea of Galilee and Dead Sea with the five drowned cities. The geographical forms are abstract, showing no influence from the sea charts. For example, the Mediterranean shore of southern Europe is a gently undulating line with no Italian or Greek peninsulas. A narrow band of ocean surrounds the round earth, with a small patch added in the northwest to accommodate the island of Hibernia. Outside are the names and descriptions of the classical twelve winds. There are no pictures on the map, but several of the monstrous races are named in inscriptions. Gog and Magog are shown twice, once in north Asia and once in India. Scott Westrem speculates that the mapmaker was copying from an east- or north-oriented map and got the directions mixed up.[29] Place-names are largely traditional, but two islands in the Indian Ocean, "Insula piperum" and "Tarmelim," recall the Sanudo/Vesconte map. Notations on the side and faint arcs drawn over the map indicate nine different climates, from that of Meroe in the south to a ninth climate located north of Denmark and labeled "uninhabitable because of the cold." The map is on paper with a watermark that dates it to the mid-fifteenth century. Otherwise, it could have been designed at almost any time in the later Middle Ages.

The Maps of the Nova Cosmographia

In the early fifteenth century, there was a scientific school in Austria centered around the Augustinian monastery of Klosterneuberg. Its denizens were especially interested in astronomy and geography and assembled a large collection of figures for latitude and longitude. They also discussed various ways of constructing maps and globes to scale. The surviving world maps most closely related to this school are that of Andreas Walsperger, and the Zeitz and Bell maps, made between 1440 and 1470.[30] The first two survive more or less intact, while the Bell map is a fragment, about 40 percent of the whole, showing Africa, the Mediterranean, southern Europe, and Asia Minor. In addition to these three maps, the scholar Dana Durand has reconstructed a map he calls the Nova Cosmographia from the extensive notes and tables in one of the monastery's manuscripts. The information was compiled by Brother Fredericus between 1447 and 1455; Durand believes that it was taken from a preexisting map.[31] What is most interesting about this project is the conservatism of the world maps produced. It seems that the so-called Vienna-Klosterneuberg school, while theoretically interested in accurate measurement and placement, was not ready to abandon the more important features of the traditional medieval world map (fig. 7.5).[32]

World maps from this school are circular and cover the eastern hemisphere from pole to pole rather than cutting off somewhere above the equator as most earlier mappaemundi do. They are oriented to the south, for reasons not made clear, yet paradise with its four rivers is still in the East. Jerusalem is in the center of the land mass, creating problems, as Fra Mauro noted, in not leaving enough space in Asia for the discoveries of the past two centuries. Africa curves around to the east, to form the southern shore of the Indian Ocean, in the style of Arabic maps and of many European maps since Sanudo. Monsters of the usual types abound, and these German mapmakers even added one of their own native species, trolls, found in the far north and prone to mischievous habits. The picturesque monsters account for most of the illustrations on the maps, although paradise on the Walsperger map is indicated by an elaborately turreted castle. The Zeitz map has no pictures but carefully reproduces the descriptive text for about twenty monstrosities. The Bell map is the most heavily illustrated; in the surviving fragment, one sees numerous castles, a wind head wearing a stylish fifteenth-century cap, as well as a goat-headed man, a lion-headed one, and a figure with no head at all.

Ptolemy was an important name to the German school, but it seems as though his work was not very well understood. An inscription on all the maps reads, "Ptolemy placed the inhabited world as 180° from north to south." Of course, Ptolemy did no such thing. His maps showed the world only as far as 16° south of the equator, and he opined that the rest was probably uninhabited or uninhabitable. The 180° on his maps was the measure of longitude from the Fortunate Islands to the east coast of China. The Walsperger map comes with a statement of purpose, announcing that it has been "made geometrically from the cosmography of Ptolemy proportionally according to longitude and latitude and the divisions of the climates." Yet latitude and longitude do not appear on the map. Konrad Kretschmer, who discovered the Walsperger map back in the 1890s, said with disgust, "Not a single line of the map is from Ptolemy!"[33] This is not quite fair, as random features from Ptolemy appear, such as the "Aurea Kersonesus" (Golden Chersonese) in the Far East, an enclosed Caspian Sea—though a Caspian gulf is also retained—and the stone tower that marks the border with India. On the other hand, Heinrich Winter blames the pernicious

Figure 7.5. (*overleaf*) Nova Cosmographia, c. 1440 (reconstruction by Dana B. Durand from table of locations in Munich, Bayerische Staatsbibliothek, CLM 14583, fols. 236r–277v. *The Vienna-Klosterneuberg Map Corpus* [Leiden: E. J. Brill, 1952]). Durand has put north at the top of this map, although the other maps of this school are south-oriented. Paradise is at the east, and the Indian Ocean is open to the east, with Africa forming its southern coast. An oddity is that the British Isles are confused with the Fortunate Islands (usually the Canaries) in the upper left. The radiating lines on the map were used to locate places according to a novel system of coordinates.

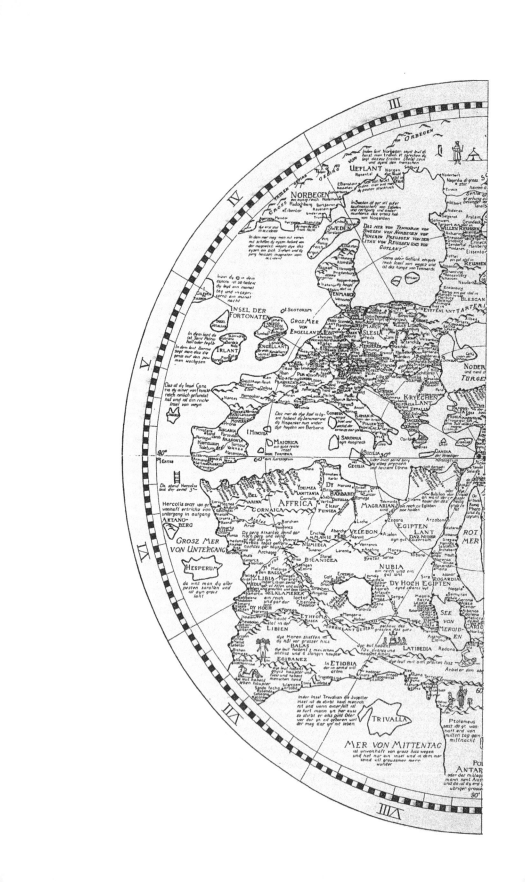

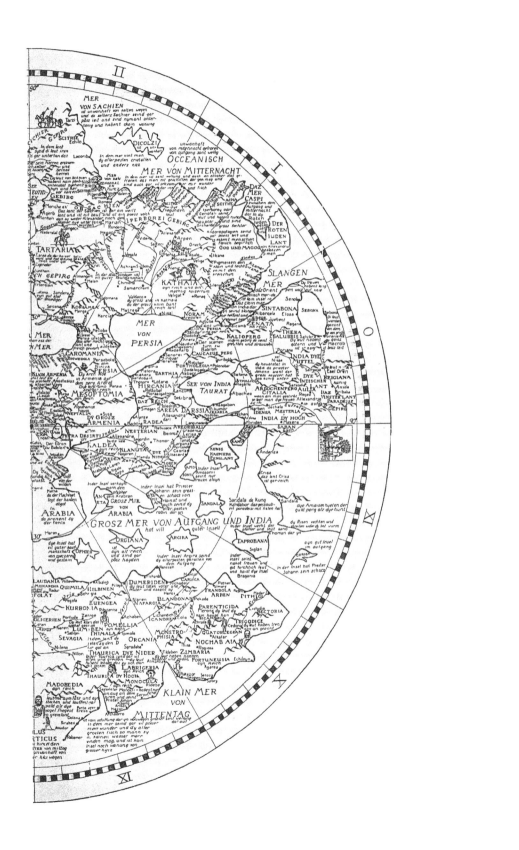

influence of Ptolemy for the maps' abandonment of the accurate geographic forms of the contemporary sea charts.[34]

The Vienna-Klosterneuberg geographers developed a novel method of map construction based on the circle. Placing the foot of the compass on the center point of their map (Jerusalem for a world map), they drew a circle which they divided into 360 degrees. Then they drew six diameters, also calibrated, dividing the circle into twelve sections. Rotating a calibrated strip of paper over their map, they tried to locate various points with precision, either by copying an accurate map or by using known distance measurements. Their coordinates for each place included a normal latitude figure, but the longitude was based on the rotating calibrated diameter. Each place receives a *signa* number, based on the pie section in which it is found, a *gradus* and *minuta* indicating its position within the section. The idea of organizing a map in this fashion had been the basis of Burchard of Mount Sion's description of the holy land in the late thirteenth century, although it is not clear that he ever constructed such a map. Apparently, it was also suggested by the German mathematical genius Regiomontanus in a letter to his patron Cardinal Bessarion. Vestiges of the diameters can be seen on the Walsperger map, and Durand draws them on his reconstruction, based on the figures in Brother Fredericus's chart.[35]

The idea of scale was important to the mapmakers, even if they were not quite able to carry it off. In the Munich manuscript, a short treatise entitled "Instrumentum de distantiis," asserts that the new map should enable its readers to take the most direct route from one region or city to another, and that he should be able to see the correct distances on the map "without any mental strain."[36] Walsperger echoes this intention, saying that if one wants to know how far it is from one place to another, one should place the foot of the compass at one point and extend the other foot to the desired goal. Then moving the open compass to the scale on the map, one should be able to tell how many German miles lie between them. Unfortunately, this hopeful text is a sham, and the map does not appear to be drawn to any scale at all. There are also numerous errors of location—the positions of cities are reversed and so on—that one hopes would-be travelers did not rely entirely on the map.[37]

The Walsperger and Bell maps are placed within a cosmographical diagram showing the elements surrounding the earth (air and fire), the spheres of the planets, stars, angels, and primum mobile. Special emphasis is given to the signs of the zodiac and the twelve winds and their characteristics. Walsperger also included at the bottom of the map in the far north the lowest circle of all, hell, in the "belly of the earth." The schema of the earth is also structured in reference to the seven climates, from Meroë south of the equator to England in the north.

Of course, this arrangement does not include most of the southern hemisphere, much less the south pole, and a mysterious inscription in the text adapted by Durand indicates that this area is uninhabitable, due to the extreme heat. Walsperger agrees that it is uninhabitable but mentions instead the numerous monsters, animal and human, that live in the region.

The maps are in Latin, except for Durand's reconstruction, which he perversely presents in Bavarian dialect and orients to the north, probably for our convenience as all the maps in this group are south oriented. Many classical names are retained, such as the archaic islands of Crise, Argire, and Ophir in the Indian Ocean. In northern Europe, particularly in Germany, the mapmakers are more up-to-date, showing that they were willing, if not eager, to include modern data when they had it on hand. This area of the map is much richer in nomenclature than any other fifteenth-century world map. One does not have to go very far afield, though, before the names become problematic. On the Baltic coast, northeast of Riga, appears a series of names that sound Germanic enough (Globurg, Kneussen, Balenor) but that Durand believes to be sheer invention.[38]

These maps share some unique features, showing their intimate connection. One of these is the island of Dicolzi in the far northeast. All we hear about this place is that it is *unwanhaft* (uninhabitable). Another is the Island of Jupiter located in the Atlantic near southern Africa. The maps' inscriptions vary somewhat, but the idea is that no one ever dies on this island: "in qua nullus hominis moritur," says Walsperger. The island of the immortals had a previous existence among the imagined islands of the Atlantic, but it was usually located in the north and sometimes was attached to Ireland. It appears on the Dulcert chart of 1339 and on the Catalan Atlas, with the added detail that no woman can give birth on this island. Here the phenomenon is rationalized; when people get old or sick, they are quietly taken off the island to die elsewhere, and women in labor also are removed to maintain the island's reputation. As for Jupiter, Kretschmer opines that it might be an error for Junonia, the island of Juno, which, according to Pliny, was one of the Fortunate Islands.[39] The many ideal features of the Fortunate Islands, which Isidore says were believed to be the site of an earthly paradise, make it not unlikely that immortality could be added to their attractions. The Fortunate Islands themselves are confused with the British Isles in the Nova Cosmographia, while the island of Jupiter is placed off the West African coast.

A curious inscription is found on Zandala or Sandala, an island in the Indian Ocean. This island, we learn, was once under the rule of King Kandabor and was believed to be paradise. The source for this bit of information is John Mandeville, who retold the story of the Old Man of the Mountain, who created a paradise for his assassins-in-training, complete with compliant virgins. The stronghold of the Assassins, a sect of the Shiite Muslims, was established around 1090 in Iran, south of the Caspian Sea, and was destroyed by the Mongol army in the mid-thirteenth century. Mandeville, following Odoric, transplanted this kingdom from the Middle East to a vague location in India near the kingdom of Prester John, also on the map. This "false paradise" must have had a compelling urge to move closer to the "real" paradise.[40]

The configuration of Asia on the maps of the Nova Cosmographia is unusual: East Asia is almost completely cut off from the rest of the continent by a deeply indented gulf on the north, the Sea of Serpents, and another gulf, Ptolemy's Sinus Magnus, reaching north from the Indian Ocean. This continental fragment is not only the location of the genuine earthly paradise but also accommodates a host of other fantastic peoples and places, such as dragon-fighting giants, gold-digging ants, bearded women, the trees of the sun and the moon, and the kingdom of Prester John. Walsperger adds Mount Vaus, home of an astronomical observatory where the three kings first sighted the Star of Bethlehem before setting off on their journey.[41] Nearby in the Indian Ocean is the island where one of the kings, Caspar, lived. Brother Fredericus puts instructions in his chart to aid the mapmaker, including descriptions of how the monsters should be drawn and such directions as "turn here" or "sharp corner." For the island of Taprobana, he writes, "Make it as broad as you wish with the paint, yet not completely covering the sea."[42]

Both Asia and Africa show a number of place-names that appear to be imaginary, along with others that are well attested. In Africa, for example, the names along the coast would be familiar from any sea chart, but around the lake in the center of the continent, which is the source of the Nile, the only recognizable names are Algol, Altair, and Vega, which seem to have wandered in from a star chart. Several river names are taken from Greek gods, such as Iris and Eolus. Durand suggests that the cosmographer "was afraid of blank spaces on the map" and filled it with made-up names.[43]

We do not know the context in which these maps were produced, with one exception. The Zeitz map was made for a Ptolemaic atlas, as one of the "tabulae modernae." The atlas includes the usual run of regional maps, plus a more straightforward Ptolemaic world map. The Walsperger map was for a long time bound with Pietro Vesconte's atlas in the Vatican Library, but this seems to have been merely a convenient place to store it. The Bell map is the largest and most elaborately decorated of the three, twenty inches in diameter, but it survives only in part, and we have no idea of its context. Durand's reconstruction of the ur-map follows the text in a geographical manuscript, but it is not clear that the map itself was ever made in the fifteenth century, although Durand believes that it was. If it did exist, it may have been a separate item, thus vulnerable to the disappearance and destruction to which single-sheet maps are prone.

Compared to Catalan and Italian world maps, the geographic productions of the Vienna-Klosterneuberg school show little influence from the sea charts, as far as geographical forms are concerned, although Walsperger claims to have used them. In addition to his claims to have drawn on the work of Ptolemy, he boasts in his caption that the map is based on "a true and complete chart of the navigation of the seas." Place-names along the Mediterranean and Black Sea coasts are similar to those on sea charts, but the shapes of these bodies of water are only roughly recognizable. The Sea of Azov appears twice, once adjacent to the Black Sea, or "Mare La Taniorum" (Sea of Tana), and once far to the north, connected by the river Don and labeled Palus Meotidis. Durand suggests that the mapmaker has conflated Meroe in Egypt, which he has transformed from an island into a lake, and the Meotide swamps of the Sea of Azov. These two lakes make a satisfyingly symmetrical pattern on the map.[44] There is also a table of coordinates in the manuscript for a *Schyfkarte*, or sea chart, which Durand obligingly reconstructs.[45] The radiating lines and the use of coordinates based on them are somewhat reminiscent of the marine charts, but it is von den Brincken's opinion that sea charts were not known in Germany at this time.[46] Certainly, the cartographers at Klosterneuberg had at least a textual portolan, if not a chart, but the world map makes only limited use of it. Outside the limits of the chart they were forced to improvise: for example, no Italian sea chart from this period shows the coast of the Baltic in any detail.

A refinement of Walsperger's map is the red dot placed by each city controlled by Christians; pagan-ruled sites get a black dot. Eastern Asia is, interestingly enough, all red dots except for Zandala, the fake paradise. Red dots predominate in Europe up to the river Don but occasionally appear elsewhere. There is one lone red dot on the West African coast, but there is no place-name given—could this be a relic of the Portuguese voyages? Another red dot is found on Taprobana and the island where Saint Thomas is thought to be buried, and another marks the stone tower on the narrow land bridge to east Asia. Reality is recognized in the holy land, where all dots are black, and Russian orthodoxy is apparently considered beyond the pale, as all places in Russia are also marked with black. This feature is reminiscent of the sea charts, which found it useful to put religiously keyed symbols or flags on the various ports. It was better for sailors not to meet with any surprises.

Durand attributes the peculiarities of the maps of this school to "the characteristic mental processes of late medieval scholarship: the mixture of criticism and credulity, tradition and innovation, scientific intention and wanton fantasy."[47] Such a judgment is too facile, for there are many more accurate and thoughtful late medieval maps than the Nova Cosmographia, and one would hesitate to attribute any apparently serious cartographic enterprise to "wanton fantasy." It was, in fact, very difficult to reconcile the various strands of geographical information, as we have seen in Fra Mauro's sophisticated debate on his world map. The contrast between his skeptical treatment of monsters, however, and the Vienna-Klosterneuberg school's wholehearted acceptance of dozens of these is striking. We cannot escape the perception that we are looking at a backwater school of geography, the same people who would have provided an appreciative audience for the travels of Johannes Witte von Hesse, who sailed "around the world" in the late-fourteenth century, visiting paradise, purgatory, the kingdom of Prester John, and the grave of the miracle-working Saint Thomas.[48] It is particularly curious to think of Walsperger working in Constance, which had been a hotbed of the most up-to-date geographical thinking in 1414–18, the time of the council. Now, thirty years later, he had no better map to copy.

The cartographers of the Vienna-Klosterneuberg school are interesting because of their brave attempts to establish mapping on a more scientific basis, and they were more successful in constructing a regional map of Germany. These landlocked Germans were plainly intending to do for land travel what the chart makers had done for travel by sea. When it came to the world map, however, not only was their data faulty but their need to present the world in its usual format—with its many traditional features, from paradise to cannibals—won out over potential reform. The Nova Cosmographia is thus the old cosmography, despite its manifestos and good intentions.

The World of Giovanni Leardo

Germany may have been somewhat removed from the sources of up-to-date cartographic knowledge, but a world map every bit as traditional as those of the Nova Cosmographia was produced in midcentury Venice, in the very shadow of Fra Mauro. The cartographer was Giovanni Leardo, and in the mid-fifteenth century he made at least four maps, three of which survive.[49] "Mappa Mundi—Figura Mundi" is the title of the map, though the rest of the text is in Venetian dialect (fig. 7.6). The maps are each on a single sheet and placed in a cosmological frame. Below is a lengthy text, beginning with God's creation of both "created and uncreated things" [*sic*], and going on to give the diameters of the earth, the other three elements (water, air, fire), and eventually that of the orbits of the

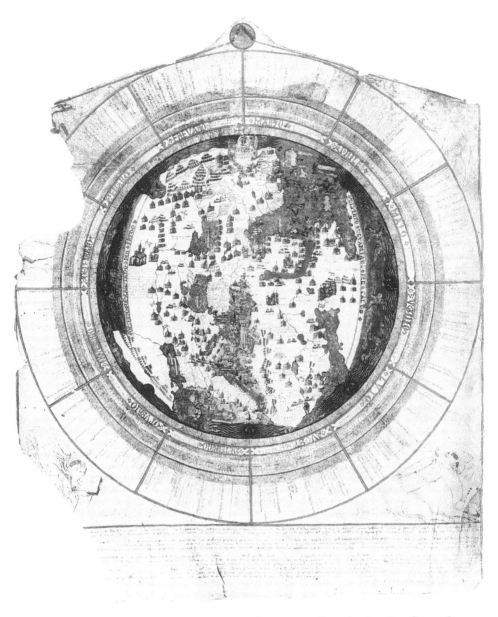

Figure 7.6. World map of Giovanni Leardo, 1452 (Milwaukee, Wisconsin, American Geographical Society. Page is 73 × 60 cm / 29″ × 23″. Map is 12.5″ diam.). One of several maps made
by Giovanni Leardo, this is oriented to the east with the terrestrial paradise shown as a forti-
fied city with the four rivers flowing from beneath its walls. The two polar zones to the left
and right are described as uninhabitable. Around the map are the names of the months and a
calendar of saints' days. Symbols of the four evangelists appear in the corners. Courtesy of the
American Geographical Society Library, University of Wisconsin–Milwaukee Libraries.

planets, stars, and the outermost crystalline sphere. Instead of these spheres surrounding the world map, the mapmaker provides an array of circular calendars that would aid one in calculating the date of Easter. In the corners are portraits of the four evangelists.

The maps vary among themselves, but all are oriented with east and paradise at the top and Jerusalem in the center. The Red Sea is red, Gog and Magog are confined in the northeast, along with griffins and cannibals, and Noah's Ark and Mt. Sinai are marked. Within this conventional structure, however, Leardo has found it possible to incorporate some modern material. The Mediterranean and Black Seas are drawn according to the model of the sea chart, though the map's small size does not allow for many place-names, and the winds are eight, in seaman fashion, rather than the classical twelve. Ptolemaic names appear, especially in Africa, where the Nile flows from the Mountains of the Moon, located in West Africa. The earth is shown from the pole to the equator, with the north polar region marked red on the Verona and Vicenza maps, green on the American Geological Society (AGS) map, and noted as uninhabitable due to the cold. The southern region is described as uninhabitable due to the heat and is colored red on all three maps. Africa is indented by a deep gulf on either side, while the southern part of the continent is fanned out as it is on the Catalan Modena map (see below), producing an Indian Ocean with an opening to the east. Prester John inhabits a large castle on this southeastern extension of Africa. The Persian Gulf runs from east to west, cutting off the Arabian peninsula on the north, while the Red Sea has a right-angle bend in it, oriented north-south in Egypt and turning to run into the Indian Ocean from the west. A large Scandinavian peninsula lies along the northwestern section of the map, taking up so much room that the British Isles are pushed down to a location off the western coast of France. There are several references to the spice trade in the Indian Ocean, and in the Sahara are some of the towns that began appearing on Catalan maps in the previous century. The AGS map, which was studied intensively by Wright, is badly worn and some of the inscriptions are now illegible.[50]

The *computus* and theological framework of the Leardo maps show their purpose: to illustrate the totality of God's creation in both space and time. Modern features are included where that can be done without violating the integrity of the model, but the basic format is the traditional one. We do not know the purpose for which the maps were made. One was apparently owned by the Trevisano family in Venice, and in the eighteenth century Giovanni degli Agostini remarked, "Giovanni Leardo: This man lived shortly before the middle of the fifteenth century, and he delighted in geography and spheres. In the Trevisan Library was preserved a planisphere by him on parchment on which could be seen delineated the whole terraqueous globe with all the signs and celestial constellations, beneath which, according to his assertion, every part is placed . . . It is curious to see how in his time, when not many discoveries had been made and navigation was so little advanced, the positions of the provinces and of the seas were conceived."[51] The Trevisano family, deeply involved in the Indian trade in the next century, must have soon found this map to be an antiquarian curiosity.

The Copper Map

Two examples survive of a map printed from a copper plate, the so-called Copper Map, made in the late fifteenth century. The small (17.5 centimeters in diameter), circular map shows the three continents of the inhabited world surrounded by a narrow band of ocean. The map is unsigned, though most researchers have guessed that its origin is Venice, since that was the only place where copper-plate printing was being done at the time. It is also untitled and removed from its context, so we do not know what its purpose was. The Mediterranean, Black Sea, and Atlantic coasts show the influence of the sea charts, as does the northern orientation of the map. An interesting feature is that the Caspian is shown for the first time in its correct north-south orientation. The other characteristics of the map are resolutely traditional. From paradise in the Far East stream the four rivers, the Nile following a southern course through an eastward extension of the African continent, to flow through its delta into the Mediterranean. The river is joined by tributaries, which flow, Ptolemy-style, from mountains in central Africa. The names are all in Latin and most are of the Roman era, such as the two Mauritanias in Africa and the land of the Colchi east of the Black Sea (Pontus Euxinus). In the northwest, however, the mapmaker has added a Scandinavian peninsula labeled "Norbegia, Suetia quae est Gottia, Helandia." Further north, another peninsula is named Engrovelant, an error for "Engronelant," or probably Greenland. Thus, the map is an amalgam of Ptolemy, the sea chart, and a traditional mappamundi, reflecting, as Tony Campbell says, "the cartographic bewilderment of the late fifteenth century."[52]

The Genoese World Map of 1457

Another world map that displays the tension between ancient, medieval, and modern knowledge, but with a more critical sense, is the so-called Genoese map, now in Florence (fig. 7.7).[53] A Genoese flag in the upper northwest corner of the map has established its origin, along with the coat of arms of the Spinolas, a prominent Genoese mercantile family. Oddly, there is no city view of Genoa ("Janua"), while

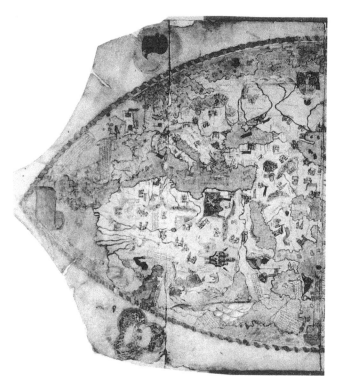

Figure 7.7. World map, 1457 (Florence, Biblioteca Nazionale Centrale, Port. 1. 82 × 41 cm / 32″ × 16″. North is at the top of this interesting map. The northern shore of the Indian Ocean shows the influence of Ptolemy as does the large island of Taprobana, but the ocean is open to the south and east, and Africa is circumnavigable. Decorated with kings, fantastic creatures, and animals, the map nevertheless represents the most up-to-date thinking on world geography. By concession of the Ministero per i Beni e le Attività Culturali della Repubblica Italiana.

Venice is represented by an impressive set of buildings. An unusual feature of this map is its oval shape. It shows the islands of the Atlantic (Azores, Canaries, Madeira), the south coast of Africa, the east coast of China, and the northern sea above Asia and Europe. The map is richly illustrated and was made not for practical use but for display, probably in the library of the Spinola family.

A text on the map states its purpose: "Hec est vera cosmographorum cum marino accordata de(scri)cio quorundam frivolis narracionibus reiectis 1457" (This is a true description in agreement with Marinos, having rejected the frivolous tales of certain cosmographers: 1457). Much scholarly ink has been spilled over the meaning of this caption, particularly "cum marino." It has been read to refer to Marino Sanudo's map, to the knowledge of sailors (though ungram-

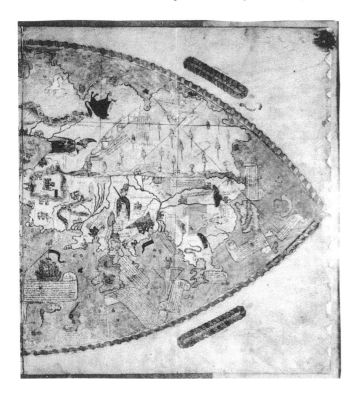

matically), and to Marinos of Tyre. In support of the last idea, it is noteworthy that the map appears to extend 225° by 87°, the extent of the habitable world, according to him. What we now know about Marinos is mostly from Ptolemy's criticism of him, but the Arab geographer al-Ma'sūdī reported in the tenth century that he had seen Marinos's geographical treatise.[54] Gaetano Ferro suggests that the map copyist left out the word *charta* before *marino* (still ungrammatical), which would make the inscription read: "This is the true description according to the marine chart, having rejected the frivolous tales of certain cosmographers."[55] Certainly, the map tries to pull together not only the medieval tradition (however frivolous) but also Ptolemaic geography and the world of the nautical charts.

The abandonment of the circular form enabled the cartographer to show features that were increasingly squeezed on fifteenth-century mappaemundi. Jerusalem does not appear in the center of the map but is considerably to the west of a center located south of the Caspian Sea. The highly pictorial character of this well-preserved map leads us to think of it as more traditional than it actually is. Oversized crowned or turbaned kings, monstrous and simply exotic animals, an elephant bearing an elaborate howdah, and scary sea monsters associate with more scientific signs, such as flags and city symbols. Inscriptions on the map, as well as recently discovered geographical features, however, proclaim the map to be a document of cartographic thinking similar, if not as large and ambitious, to that of Fra Mauro.

The oval form is not unknown among medieval maps. Hugh of Saint Victor had described the world as being the shape of Noah's Ark, and Ranulf Higden's world maps were oval. A standard way of describing the earth, from Bede to the Catalan Atlas, was to compare it to an egg. The main purpose of the analogy seems to have been to describe the various spheres surrounding the earth (egg white, shell), but the idea of an egg shape could have been derived from these works.[56] Another possibility is that the oval form represents the *mandorla,* or nimbus, which surrounded Christ in many medieval works of art.[57] In the fourteenth-century didactic poem, "Il Dittamondo," Fazio degli Uberti, described the inhabited world as long and narrow ("lungo e stretto") like an almond ("mandorla"), with no apparent religious significance.[58] Ptolemy's maps, while not exactly oval, were wider from east to west than they were high (north to south). For a fifteenth-century mapmaker, this form made convenient room for discoveries in the Atlantic and in Asia. By the end of the century, the circular form was becoming impractical, and once the Americas were added to world maps, it was gone almost completely.

Ptolemy is cited by name in several inscriptions, and there is evidence of his influence in the representation of Africa (Ethiopia, the source of the Nile), an enclosed Caspian Sea (Mar de Sara), the southern coast of Asia, and the Golden Chersonese, not named but identified by a legend noting that it is particularly rich in gold and precious stones. The Indian Ocean, however, is open to the east and south, the mapmaker suggesting that Africa is circumnavigible and that the torrid zone is passable. An inscription reads: "Beyond the equator Ptolemy reports an unknown land, but Pomponius and many others, doubting this, [say] that the passage of sailors from this place to the Indies might be possible, and tell of many who have gone through these parts from India to Spain, especially Pomponius in his last chapter."[59] The map also relies on a modern source, the travel report of Nicolò de' Conti, who returned to Italy from Asia in 1439. The map shows the city of Bizungalia (Vijayanagar) in India, the land of Macina (Siam), Xilana (Ceylon), Cimiteria (Sumatra), the Javas major and minor (here Java and Borneo), and the islands of Sanday and Bandam (apparently the Moluccas). Longer legends take material from Conti on the funeral practice of wife burning ("if they refuse out of fear, they are forced to do it"), the cultivation of pepper, the collection of human heads in Sumatra, the sea-tight compartments of Chinese junks, the practice of tattooing, and the availability of spices and multicolored parrots.

Paradise does not appear on the map, and an inscription in southeastern Africa tells us why: "In this region some depict the earthly paradise. Others say it is beyond the Indias to the east. But since this is a description of the cosmographers, who make no mention of it, it is omitted from this narration." Who are the cosmographers, who also appeared in the caption cited above but in a more negative context? The reference seems to be to the geographers of the classical world. The title of Ptolemy's work, as it circulated through Italy in the fifteenth century, was usually given as *Cosmographia* rather than *Geography*, a less familiar Greek term. Certainly, neither Ptolemy, Pomponius Mela, nor Pliny had given a location for paradise, save for fantasies about the Fortunate Islands, and the Genoese mapmaker appears to associate classical with scientific geography.

Of course, the cosmographers had no objection to monsters—even Ptolemy mentions a few, although he did not put them on his maps. It is interesting that the maker of the Genoese map mentions peculiar customs (cannibalism, people who have no names) but no "monstrous races," that is, people with aberrant physical characteristics, other than the pygmies. On the other hand, he is happy to include pictures of bizarre animals. In the Indian Ocean are shown a mermaid and a fish with a devil's head, while on land nearby is a snake with a human head. In northern Asia is a very large griffin, while a couple of dragons appear in Ethiopia. These fantastic creatures join other wonderful but real animals, such as a giraffe, a leopard, a crocodile, two monkeys, and a swordfish.

Some standard features of medieval mappaemundi are also present. Gog and Magog are enclosed in northeast Asia, Noah's Ark rests on a mountain range in Armenia, and the Red Sea is red, though there is no text about the passage of the Israelites through its waters. We can also find Mount Sinai, and the tomb of Saint Thomas in India. Prester John is represented behind a wall, protecting him from the future rampages of Gog and Magog. As we might expect from a Genoese map, good use has been made of the nautical chart as well. The Mediterranean, Black, and Atlantic coasts reflect the forms. A partially finished network of rhumb lines appears on the map, and on the right are two scale bars, though they are more a sign of intention than reality.

The Genoese map became the center of controversy in the 1940s when the Italian scholar Sebastiano Crinò, suggested that this was a copy of the map that Paolo Toscanelli sent to the Portuguese court in 1474, touting the possibility of a sea route to India via the Atlantic.[60] The evidence is purely circumstantial, though the map would have been more encouraging to anyone hoping to cir-

Figure 7.8. Catalan world map, 1450 (Biblioteca Estense / Modena, MS C.G.A. 1. 113 cm / 45″ diam. This map is drawn so that one can walk around it, reading the inscriptions from each direction. It was probably meant to be mounted, as it is today, on a flat table. In the Catalan tradition, it includes colorful images. Its most unusual feature is the large southern addition to the continent of Africa. Sailors might take courage, however, from the canal that is shown linking the Gulf of Guinea to the Indian Ocean. Courtesy of the Ministero per i Beni e le Attività Culturali.

cumnavigate Africa, and Crinò's thesis has had no other advocates. (Toscanelli's letter and map are discussed in chapter 8.)

The Catalan Map of Modena

In the Biblioteca Estense in Modena is a large, circular world map (fig. 7.8). It was first recorded in a 1488 library inventory in the collection of the Este family and after various adventures was restored to Modena in the late nineteenth century. It is now mounted under glass on a specially designed table. It is written almost entirely in the Catalan language and displays a number of features of that school of mapmakers.[61]

The map has proved difficult to date. The presence of the place-names Cape Verde and Cape Rosso on the West African coast indicate that it incorporates the discoveries of the Portuguese through 1446, but almost everything else about the map could have been assembled much earlier. Indeed it is very close to the Catalan Atlas of 1375, some of the legends being nearly the same, word for word. The usual date now assigned is between 1450 and 1460.

The map shows the legacy of the marine charts in the coastal outlines of the west, and it sports an incomplete set of rhumb lines with sixteen centers around the rim for the sixteen chief nautical winds. There are also two scale bars at the edge. The orientation is north-south, in the manner of marine charts, so that one can read the place-names and legends while moving around the circular map. Its large size and elaborate decoration show that it was intended for display, not for sea-faring. The Catalan cartographer is anonymous, although Marcel Destombes has suggested it may have been Petrus Roselli, a chart maker known to have been active in Majorca in the 1460s.[62] If so, this would have been the only world map he made, and no one else has followed up on this idea with any enthusiasm.

Many elements of the traditional mappamundi persist here, including the Red Sea, colored red and marked with the crossing of the Hebrews, and Noah's Ark in Armenia. Paradise is present, with Adam and Eve kneeling on either side of the tree, but has been moved to East Africa on the equator. A single river flows from it into a lake, from which the four rivers emerge, one going on to form the Nile. The inscription lists the traditional names of the rivers. The garden is surrounded by a ring of fire, and to the north of paradise are six mountains, the "mountains of diamond," which guard the sacred enclosure. In northeast Asia we find reference to Gog and Magog, described as cannibals who will join the Antichrist. A lone trumpeter, stationed there by Alexander the Great, will warn the world of their advent. The large illustrations that appeared on the Catalan

Atlas are not here—indeed much of Asia is left almost blank. The elaborate caravan of the Catalan Atlas is reduced to two men on horseback.

The Modena map also shows some significant deviation from the mappa-mundi. Jerusalem is not in the center and has no city vignette; it is simply marked "San Sepulcra" and located on the River Jordan. Other than the coastal cities, only the Dead Sea (Mar Gomora), Iudea, and the Jordan are mentioned. The shape of Africa on this map is unique, and it is much enlarged in relation to Europe and Asia. Below the Gulf of Guinea, which nearly cuts the continent in two, is a large crescent-shaped appendage extending to the east and forming a southern shore for the Indian Ocean. A thin canal across its narrow waist implies a passage between the Atlantic and Indian Oceans. The southern landmass, which may be intended for a separate continent, has no place-names or pictures, demonstrating remarkable restraint on the part of the artist. An inscription is its only decoration: "Africa begins at the river Nile in Egypt and ends at Gutzola in the west, including all Barbaria and divides it from the south." Gutzola is shown on the Moroccan coast just south of Safi.[63] Near Cape Verde we are told, "At this cape is the end of the land of the west part of Africa. This line is at the equator on which the sun stays continually, making twelve hours of night and twelve of day."[64] Nearby is an island labeled "illa de cades: Here Hercules placed his two columns."[65] So the pillars of Hercules have slipped down the coast and will eventually disappear completely. Ptolemy's Mountains of the Moon are placed on the north shore of the Gulf of Guinea. Five rivers flow from their base into a large lake, from which one river runs west to the coast and is labeled as the river of Gold, meant for the Niger.

Africa contains half a dozen reigning monarchs, from Musamelli to Prester John, sitting in splendor in their royal tents. The mapmaker omits the usual array of monsters in Africa, and the only animal depicted is a camel with a rider, sedately proceeding along the caravan route to the sea. The Saharan cities that appeared on the Catalan Atlas also appear here; among them are Siguilmese, Tenduch, Tagort, Buda, and Melli. The far north in Europe and Asia is more frightening than Africa, showing a naked giant pursuing a fox, a nine-headed idol being adored by two worshippers, and a strange hanging head, which appears on several other fifteenth-century world maps. The Caspian Sea is enclosed, but there are two unlabeled gulfs in the northern ocean. China is pretty much the world of Marco Polo; the Great Khan is still ruling there. To the south the Indian Ocean is greatly enlarged and full of brightly colored islands, but only three are named: Silan, Trapobana, and Java. A Chinese junk, identified in a legend, sails through the water, menaced by three half-human figures: one part fish, one part bird, and one part horse. South Asia lacks a definite Indian peninsula and shows no trace of the Golden Chersonese.

The entire map has been shifted to the east in its circular frame, thus making more room in the Atlantic for its islands. The Azores, Canaries, and Madeiras are shown. Next to the Canaries, a long Latin text, drawn from Isidore and the voyage of Saint Brendan, describes the Fortunate Islands of antique fame. Plato's tale of Atlantis is recalled near an island labeled "illa de gentils"; it was once as large as all Africa but now, by the will of God, is covered with water.[66] In the north is a group of colorful islands marked, "These islands are called 'islandes,'" which may be a reference to Iceland.[67] West of Ireland can be found the islands of Mam and Brezill.

The combination of archaism and modernism is an outstanding characteristic of this map, and it is interesting to note that the cultured and humanistic Duke of Ferrara, Ercole d'Este, the owner of this map, also had in his library a copy of Ptolemy's *Geography*, edited by Nicholas Germanus. No evidence of Ptolemy's influence on this map can be discovered. The duke owned a copy of Mandeville's *Travels* as well, which he must have treasured, as there survives a letter he wrote demanding its return from a borrower.[68]

The Mappamundi of Andrea Bianco

Andrea Bianco's mappamundi has already been discussed in the introduction, but it is worth returning to him briefly here to put his work in the context of the other fifteenth-century world maps (see fig. I.3). Made in Venice in 1436, the world map forms part of an atlas, which also includes nautical instructions, a series of sea charts, and a Ptolemaic world map.[69] The circular map is oriented to the east, surrounded by a blue rim representing not the ocean (which is green) but the heavens, as we can see from the stars painted on it. The landmass of the earth is shrunk considerably within its frame in order to increase the size of the ocean and to include the polar regions. Even so, a bit of land in East Asia protrudes into the frame. To the south of this promontory lies a long gulf, and on the peninsula nearby is the Garden of Eden, with Adam and Eve standing on either side of the tree, and the four rivers flowing to the west. Below is another garden, where God is instructing Adam, alas, to no avail. The Indian Ocean is open to the east and crowded with islands, while Africa extends far to the east, bounding the ocean on the south. The two polar regions are not inviting. In the north, marked off by a half-circle, we read that it is terribly cold and that everyone born in that region is a savage. To the east are the enclosed peoples of Gog and Magog. The south pole is described as "nidus alli malion" (nest of all evil), and there is a man hanging from a gallows, as well as several sea monsters.[70]

Not surprisingly, the map borrows the coastal forms from the sea charts, and the vernacular winds, whose lines divide it into eight pie-shaped sections. Despite dire warnings, a small ship drawn with loving detail is advancing bravely around the southern coast of Africa. Because of this map's early date, there is no record of the Portuguese voyages, but in Bianco's 1448 chart, their progress in west Africa is duly noted. An unusual feature of this map is the large number of human figures on it. In southern Africa there is a procession of dog-heads bearing a banner. Asia is almost entirely taken up by an array of enthroned kings, flanked by what appear to be their entourages. The holy land is illustrated by Mary holding the infant Jesus on her lap with the three wise men in attendance, and Jesus is also shown being baptized by John in the river Jordan. It is almost as though all of Bianco's pent-up creativity is unleashed after drawing the more restrained sea charts.

The circular medieval mappaemundi, oriented to the east, had placed the world in its universal context, frequently including the structure of the four elements, the nine spheres, and the ordering of time by the motions of the moon and the sun. The three continents of the then-known world were associated with the biblical distribution of the lands to the sons of Noah. The classical heritage was marked by the surrounding twelve winds, as well as the dominance of geographical names harking back to ancient times. While current events and newly founded cities were not excluded from the maps, historical sites were equally important: the cities of Troy and Carthage, the sites of ancient battles, the exploits of Alexander, the progress of the empires from east to west. Of special significance was biblical history, and the most complete mappaemundi covered it all, from the Creation through the Incarnation to the Last Judgment. Important sites such as the barns of Joseph, the sites connected with the life of Christ, the missions of the apostles, and the bishopric of Saint Augustine, set forth the sacred story in spatial terms. The Red Sea, bearing a text on the passages of the Israelites and/or the drowning of Pharoah's army, was shown still divided, in order to indicate the eternal present of these spiritually significant events.

The later medieval mappaemundi had a strong sense of center and edges. Around 1200 Jerusalem began to be put in the center of world maps, and the significance of this position was well appreciated. Around the rim were found monstrous races and exotic animals, part of the wonder of creation not often seen at home. Many of these creatures were inherited from classical writers. Both the form and the content of the maps appeared to discourage exploration, by separating "uninhabitable" zones and filling them with terrifying monsters. Mapmakers pushed and pulled at the medieval world map form, trying to accommodate events and places they saw as significant. Travels to Asia in the thirteenth century, which culminated in the work of Marco Polo, filled the depictions of that continent with new names and a new awareness of its geographical features and distances. Polo also contributed important information about the Indian Ocean with its islands, spices, and vibrant, lucrative trade. This information remained fairly static over the next two centuries until Europeans once again took up travel and trade in the Indian Ocean and the Far East. In the fourteenth century the rediscovery of the Canary Islands, followed by Madeira and the Azores, began to stretch cartographic space to the west.

Then in the mid-fifteenth century the Portuguese voyages down the African coast expanded the size of the known world to the south and scuttled the idea of an impassable equatorial zone. By the late fifteenth century, the old circular map form with its narrow rim of ocean did not have enough space to fit all this in, and the implication that there were parts of the world where humans could not travel, either because they were too hot or cold for human habitation or were forbidden by God, was less convincing. Enterprising travelers to the north and south reported busy human activity in the so-called uninhabitable zones. Although some charlatans claimed to have seen paradise from a distance or to have heard the roar of the four rivers departing, others told of searching in vain and finding more mundane places on the alleged spot.

The fifteenth-century mapmakers discussed in this chapter did what they could to retain the beloved and familiar features of their world, while making their maps as up-to-date as possible. Peter Barber calls these "hybrid" maps and points out that discriminating sixteenth-century collectors, such as Henry VIII and Ferdinand Columbus, had examples in their libraries.[71] Many makers of maps (Olmütz, Borgia, Leardo, Catalan Modena, Rudimentum Novitiorum and maps of the Nova Cosmographia, Fra Mauro) kept the traditional circular form of the world, while a few others, such as the Genoese map, dispensed with it. Jerusalem continued to be the center for some maps, but others (Fra Mauro, Catalan Modena, and Genoese) expanded to the east and north. Paradise continued to appear on most maps in the east with its four rivers. The original explanation, other than biblical authority, was that the sources of these great rivers were unknown, but in the fifteenth century we find Florentine humanists questioning Ethiopian monks about the real source of the Nile. The influx of new classical works (such as those by Ptolemy, Pomponius Mela, and Strabo) and the critical reading of these in connection with the preexisting library staples led to questions about the reality of this divine hydrology. Only the Genoese mapmaker denied paradise any place on the map, while Fra Mauro hedged on the question, and the Catalan mapmaker relocated it in equatorial Africa.

The influence of the marine charts had already in the fourteenth century led to modification of the abstract geographical shapes of the medieval world maps. The charts also brought with them a sense of direction and the idea of scaled distance, though it was not yet possible to represent this on a global scale. The twelve winds of classical antiquity were jettisoned in favor of the sea-going winds, in multiples of eight and adapted to the use of the compass. Inevitably, the evidence of sailors about coasts, distances, and geographical features began to have an impact on the content of maps, even though some of these came along with tall tales of sea monsters, sluggish seas, and magical islands.

The eastern orientation was oddly enough the first traditional feature to go; possibly Arabic influence caused more maps to be oriented to the south. Only the Leardo maps, the Rudimentum Novitiorum, the German broadsheets, and the Bianco map kept their eastern bias. The sea charts tended to use a north orientation, based on the pointing of the compass, but the larger charts often had no clear orientation so that the user could rotate them and read off the coastal names from various angles.

The holy land had been greatly enlarged on the thirteenth-century mappaemundi, reflecting its importance and the number of places the mapmaker wanted to include. Beginning with Vesconte, the holy land shrank drastically to put it in proportion with the rest of the world; after all, it was not a very big country, physically speaking. Some cartographers (Vesconte and Sanudo, the Rudimentum Novitiorum maker) made up for this insult by drawing a separate map of the holy land, which could include all its interesting features.

The monstrous races and animals were to have a long history on maps. "Here be dragons" is a phrase which has actually been found on only one map, the Lenox Globe of 1503–7, where "Hc sunt dracones" appears on the southeastern coast of Asia.[72] Even if the phrase itself were absent, maps continued to show monstrous animals, particularly at sea, into the seventeenth century. As world maps developed to include the great expanses of the ocean, this empty space cried out for embellishment. Genuine monsters such as whales or giant squids were joined by enormous sea serpents and the god Neptune, armed with a trident, rising from the waters.[73] The monstrous human races moved from the edges to the borders of maps, where they survived into the nineteenth century, sometimes transformed into individuals of various races in colorful "native dress."[74]

One effect of the Ptolemaic atlases was the influence of their sober character. Most editions produced maps that showed simply geographic forms and place-names with no pictorial embellishment. The vivid effect of the densely illustrated medieval map was replaced by a more scientific-looking production, but (not to worry) the need for fantasy was supplied by elaborate cartouches and borders, packed with mythological figures, personified continents, ferocious animals, colorful natives, and monsters of various sorts, not to mention flowery dedications and descriptions. An early example: the world map of the *Nuremberg Chronicle* of 1493 was a straightforward Ptolemy-style map, but around its edges the sons of Noah embraced their respective continents, while down the sides marched a selection from the monstrous races.[75]

World maps are never of much practical use, as the scale is too small. The pleasure and reward one gets from regarding them is more philosophical or theological. Consider the many posters made of the twentieth-century photograph of the earth seen from space. One version is captioned "Love your mother," emphasizing the beauty and fragility of our small blue planet. Circled with strands of clouds, the world in this photo has no political boundaries, a salient feature of most maps. To the Middle Ages, the world map was part of a larger cosmos, the center of a gigantic nest of transparent spheres, bearing planets, stars, and angels, and eventually the ultimate sphere of God himself. The structure of the earth and the arrangement of places on its surface were not merely physical questions. The earth, as God's creation, bore important messages for the human race. Names, geographical shapes, and the history of each place were imbued with many layers of meaning. It was the task of the medieval mapmaker to present this whole as clearly and beautifully as possible. By the last half of the fifteenth century, the influence of the marine charts and the *Geography* of Ptolemy began to transform the way space was mapped, but it was not so easy to dismantle the received wisdom of the medieval period. Although mapmakers could incorporate the coastlines from marine charts, they did not have the capacity to measure the entire world. Astronomically determined coordinates were not precise enough for distances on the ground, and the variable length of a degree of longitude posed an, as yet, insuperable obstacle. Even determining latitude was no simple matter. In addition, mapmakers were reluctant to abandon the rich historical/theological understanding that had shaped their perception of the world for so long.

The medieval mappamundi was a powerful statement of medieval culture and beliefs, but the question remains, how had the image and idea been spread throughout Europe? We do not know how many large, public mappaemundi were made—there were certainly a great many more than survive today—but still they could not be found in every city or monastery. Even supposing someone traveled to Hereford Cathedral and looked at the map there, how long would he study it, and how much would he take away in his mind? Maps in books would be seen only by the readers, which were never very numerous. For example, we are not sure that any mapmaker was influenced by the fascinating maps of Matthew Paris, for the simple reason that they were locked away in the monastery at Saint Alban's, and few people saw them. What we can gather from the material included in maps is that many of the commonplaces on them, such as their structure, came from popular texts. A small library including Solinus, Isidore, Orosius, and possibly one of the later scholastic writers, such as Honorius Augustodunensis would have supplied most of the information that appeared on medieval maps. If these books included a map as well, even a simple framework, as a number of Isidore's works did, all to the good. Some of these books were used in schools, with the teacher reading aloud to the students, and thus the information was passed along. When Chaucer says that Rosamunde was round as a "mappemonde," or when we learn that the Spanish composer Juan Cornago based his mass on a popular song, entitled "I'ho visto il mappamondo," we assume that their hearers had some idea of what they were talking about.[76]

In the mid-fifteenth century, the mappamundi was still holding its own, but in the last twenty years of the century it began to give way. Long before the Pinta, the Niña, and the Santa Maria sailed out of the Palos harbor, the ancient form was burst apart, and space had already been created on the map for new discoveries at all points of the compass.

CHAPTER EIGHT

The Transformation of the World Map

In the 1480s the Columbus brothers, Bartholomew and Christopher, were making the rounds of the courts of Europe, looking for a royal sponsor for their expedition to the Orient. They proposed a route heading west across the Atlantic Ocean, setting out from the Canary Islands for Japan. Based on his calculation of the length of a degree of longitude, Christopher Columbus estimated the distance from the Canaries to Japan at 2,400 nautical miles. The actual distance is closer to 10,600 miles, and, of course, there is a large obstacle in the way—the American continents. This optimistic outlook for the transatlantic journey was based on Columbus's own calculations of the length of a degree of longitude at the equator (56 2/3 Italian miles), which produced a terrestrial circumference of 20,400 miles or less than 19,000 statute miles, instead of the modern figure of 24,900.[1] In addition, an overestimation of the length of the inhabited world (Spain to China) brought Asia still closer, as did the placing of Japan 1,500 miles from the coast, based on Marco Polo's account. These last two estimates were not unique to Columbus, but still the ad hoc committee of scholars convened by the King of Portugal in 1483 turned down his proposal. We no longer have their reasons, but one scholar surmises that Columbus was asking for too much money.[2] Of course, the Portuguese were already heavily invested in the project of reaching India by going around Africa, but King João II sent out at least one surreptitious expedition to the west.

Royal patronage was important to Columbus not only for the money but also because he hoped for the political privileges and social promotion that only a monarch could give him. His ambition drove him from Portugal to Spain, back to Portugal, where he says he witnessed the return of Bartholomeo Dias in December 1488 from his voyage round the southern tip of Africa and back to Spain again. Meeting with a rebuff from another learned council, Columbus had started off to try his luck in France when King Ferdinand of Aragon called him back. The final expulsion of the Moors from Granada in 1492 had freed up potential funds, and in addition the Portuguese monopoly of the African routes made an alternative route appealing to Spain. One million maravedis were granted for the voyage, and an additional 140,000 maravedis for Columbus's personal salary.[3] His idea of a practical western route to Asia was about to be tested.

Geographical Theory

In his popular nineteenth-century biography of Columbus, Washington Irving attempted to recreate the scene at the University of Salamanca in which the brave Columbus, "a simple mariner," faced a group of Roman Catholic know-nothings, some of whom believed the earth to be flat.[4] It makes for a dramatic encounter, and subsequent, less scrupulous authors, enchanted by the heroism of Columbus, have passed the story on in an embellished form. The concept of the spherical earth, however, was a commonplace in Europe since at least the third century BCE and had been transmitted to the Middle Ages through various authors such as Macrobius and Martianus Capella. Early medieval writers simply recorded the idea as authoritative, but in the twelfth century, after the recovery of Aristotle's works, writers began to expand on the evidence for the fact. Aristotle had marshaled a number of arguments to prove the earth was round, such as the curved shadow on the moon during a lunar eclipse and the fact that the polestar is seen lower in the sky as one travels south.[5] Later medieval writers mulled over these ideas and added to them, testing the hypothesis by their own observations.[6]

That being said, a number of issues remained. One was the relative proportion of land and water on the earth's surface. Given that knowledge was limited to the northern section of the eastern hemisphere, discussions of what occupied the rest of the earth were necessarily theoretical. There were several possibilities. One was that the *ecumene* and its three continents were the entire land surface and that the rest of the earth was water. Another theory, found principally in Macrobius, was that there was a similarly sized landmass in each quarter of the globe, but they were not accessible to Europeans due to the impassability of the equatorial zone and the hostile environment of the ocean to the west and east. All these ideas were brought up at Salamanca and probably had been discussed at Lisbon as well.

There were other theories, however, partly fueled by the imaginative literature of antiquity. The most famous of these was Plato's Atlantis. If, in fact, it had now sunk beneath the sea, the story suggested that there might be other lands as well, or perhaps it had not sunk deep enough and was an enormous sandbar, preventing further travel to the west. The existence of other lands posed theological problems. If inaccessible lands existed, how could they be populated by the descendants of Adam? And, if Christ came to save all mankind, how could he have left out the entire population of the other continents? This question was to come back with a vengeance at the end of the fifteenth century.

The Middle Ages had supplied its share of fictional islands, such as those encountered by Saint Brendan and his Irish monks in their travels in the Atlantic. Saint Brendan's islands, in various forms, appeared on many medieval maps, sometimes identified with the Fortunate Islands.[7] The island of the Seven Bishops or the Seven Cities, also called Antilia, was supposed to be the refuge for Portuguese clerics fleeing the Muslim invasion in the eighth century. The Antilles appear on sea charts in the fourteenth and fifteenth centuries, and in the late fifteenth century the Portuguese king dispatched sailors to look for this fabled land. Another imagined island was that of Brazil, whose name is perhaps derived from Irish legend. It appears on maps west of Ireland, often as a perfect circle divided in half by what appears to be a canal. We hear of this island being sought diligently by the sailors of Bristol in the late fifteenth century. Both the Antilles and Brazil eventually achieved their own reality in the naming of places in the New World.[8]

As for the relative proportion of land and water, there was the evidence of 2 Esdras 6:42, in which the prophet Ezra said that on the third day God "commanded the waters to be gathered together in a seventh part of the earth; six parts you dried up and kept so that some of them might be planted and cultivated and be of service before you." Esdras was not a canonical book, but the credentials of the prophet Ezra were unimpeachable. There were also the classical authorities. Aristotle had suggested that there was not much space between the Pillars of Hercules and India, thus dramatically reducing the size of the ocean.[9] Pliny had also described the great extent of India to the south and east.[10] Very weighty as well was Saint Jerome's testimony about the length of the voyage of Solomon's ships, which took three years to go and return from the fabled city of Ophir.[11] Even today no one is quite sure where Ophir was, but, judging by the riches the fleet brought back, it was likely in India. The length of the voyage suggested an extensive Asian continent. In the *Geography*, Ptolemy had stated that the width of the *ecumene* was one-half the circumference of the earth or 180°.[12] In his popular work *Ymago Mundi* (1410), Pierre d'Ailly examined all these theories and concluded that the land area of the globe was greater than that of the sea, an opinion enthusiastically approved by Columbus, who annotated his copy with extensive notes.[13]

While there was, as yet, no way to know what was on the other side of the globe, there was the question of the size of the known world, land and water together. The trouble was that it was not all that known. Considering that the worthy Ptolemy overestimated the size of his local sea, the Mediterranean, by 20°, it is not surprising that Europeans at the end of the Middle Ages had only the vaguest notion of the extent of land to the East. Ptolemy's 180° estimate of the longitude of the *ecumene* (the Fortunate Islands to the Cattigara peninsula) had been preceded by Marinos's guess of 225°. Marinos, whose work was preserved only by Ptolemy's criticisms of it, met with newfound popularity in the later Middle Ages by would-be explorers such as Columbus, as Marinos's estimate dramatically shortened the western route to Asia. Another reason for the expansion of the picture of the *ecumene* was that mapmakers had added Marco Polo's Asia to Ptolemy's, including the offshore island of Japan.

As for the size of the earth as a whole, Eratosthenes had worked out a method for determining the length of a degree and, thus, the circumference. His estimate was 700 stades to a degree and 252,000 stades for the whole. If a stade is one-eighth of a mile (and this is problematic), this would result in a degree of eighty-seven and one-half miles and a total circumference of 31,500. The other possibility handed down from antiquity was a smaller earth, with fifty-six and two-thirds miles to a degree and a total circumference of 20,400 miles. Columbus reported that he tested this calculation himself, measuring a degree of longitude at the equator by taking observations at the fort of Sao Jorge de Mina in Africa. The problem was that Elmina is actually not at the equator at all, but about 5° north.[14] Columbus was using Italian miles, which were 4,810 or 4,860 feet in length, rather than the 5,280-foot long statute mile, and the resulting circumference of 20,400 Italian miles was less than 19,000 statute miles, which is 32 percent smaller than the actual size.[15]

Columbus's attempt to measure a degree, however faulty, is an indication of a new approach to geography in the fifteenth century, one we have already seen in practice in the work of mapmakers such as Fra Mauro. Eyewitness accounts or personal experience challenged ancient assumptions. In particular, voyages to the newly discovered Atlantic islands indicated that the ocean was not uninhabitable after all, while the Portuguese expeditions along the West African coast revealed that the so-called torrid zone was not boiling hot and could be safely navigated. In his copy of d'Ailly, Columbus remarks on the return of Bartolomeo Dias to Lisbon from his epic journey around the Cape of Good Hope in 1488. Dias brought the chart he had made and showed it to the king, estimating that he had reached a point 45° south of the equator. "In quibus omnibus, interfui," says Columbus (At all these events, I was present).[16] This shows, Columbus goes on to say, that the sea is navigable in all parts and that navigation is not impeded by the heat of the torrid zone. Dias's ten-degree error in calculating the latitude of the Cape of Good Hope is evidence of the imprecise nature of astronomical observation at this time.[17] Even the equator was incorrectly located. Ptolemy had located it 10° south of the Canary Islands, or about 15° too far north, while the Catalan world map of 1450 puts it off Cape Verde, or 10° north of its true location.[18]

Toscanelli's Letter

While Columbus was certainly an original thinker, he did not arrive at his conclusions alone. We know from the list of the books in his library that he found encouragement for his project in works as varied as those of Pierre d'Ailly, Marco Polo, Aeneas Sylvius (Pope Pius II), and Pliny the Elder. He also owned a copy of the 1478 Rome edition of Ptolemy with its maps. Among the documents that inspired him was a letter from the Florentine savant Paolo dal Pozzo Toscanelli (1397–1482) in response to a request from Alfonso V, king of Portugal (1438–81). Toscanelli was one of those versatile intellectuals of the Renaissance, a medical doctor, an astronomer, a geographer, and a translator from the Greek, whose day job was working for the Medici bank. In the 1420s and 1430s, he had been a regular participant in the seminars held in Florence at the Camaldolese Convent of Santa Maria degli Angeli. The Portuguese court seems to have consulted him as early as 1459, when he borrowed a map from a friend to show their ambassadors.[19] In 1474 he wrote the letter responding to the king's request and directed to his counselor Ferdinand Martins, a canon of Lisbon cathedral whom Toscanelli had met in Italy. Some years later, between 1479 and 1482, Columbus said that he also wrote to Toscanelli and received a copy of the same letter. The letter survives in three versions, one from the biography of Columbus by his son Ferdinand, one in the Las Casas transcription of Columbus's papers, and one written on the flyleaf of Columbus's copy of Aeneas Sylvius's universal history. There are variations among these versions, written in Italian, Spanish, and Latin, respectively, but the substance is the same. Toscanelli had been requested to give his opinion about the best route to the land of spices. Along with his letter, he enclosed a chart, marked with lines of longitude and latitude, showing the distance from Lisbon to Quinsay in China as "26 spaces" or 6,500 miles (the actual distance about 12,000 miles), while from "Antilia" to Cipangu (Japan) it was a mere "10 spaces" or 2,500 miles. "The distance across unknown seas, therefore, is not great," Toscanelli concluded. The chart unfortunately does not survive, although modern historians have attempted to reconstruct it, and his description accords well enough with several surviving cartographical monuments, such as the Behaim globe and some of the world maps of Henricus Martellus Germanus.[20]

Toscanelli's information about China and Japan was derived from Marco Polo. Henry Vignaud scoffs at his dated information about the East, but there had been little European travel in Asia since the early fourteenth century, and the two-hundred-year-old information of Marco Polo remained the most authoritative source. Even Nicolò de' Conti's information about Cathay is nothing more than a recounting of Polo.[21] Columbus himself left home bearing letters to the "Grand Khan" from the Catholic kings of Spain, intending to deliver them in Quinsay. Toscanelli stressed the riches of the East—pepper, precious stones, aromatic spices, gold and silver—but he pointed out that the East was rich in knowledge of the sciences and the arts as well. In a later letter to Columbus, he wrote that there were Christians in the East who were eager to make contact with their coreligionists. Toscanelli died in 1482, ten years before the voyage was made. The authenticity of the letter has been challenged by Vignaud. He argues that it is not a very impressive document from someone as learned as Toscanelli, which does ring true, but the triple evidence of the letter's existence seems too powerful to discard.[22]

The Changing World Map

The world as pictured on the map had changed significantly in the preceding century, making it a world open to exploration. First of all, there was the shape. The round world of the mappamundi, surrounded by a narrow ocean rim, privileged the land over the sea. There was no sailing room in the Atlantic, no place to put the newly discovered and important islands, not to mention those hypothetical islands as yet undiscovered. The maps made for Ptolemy's *Geography* had taken an oblong shape, stretching the length of the known world from east to west. This form could be indefinitely extended as new lands were found in either direction. After the discoveries in the Atlantic in the fourteenth and fifteenth centuries, mapmakers needed either to shrink the *ecumene* within the circular map to make room for the Azores, Madeiras, and Canaries, or to extend the map to the west. The use of maps had also changed. No longer an object of pure scholarly or philosophical contemplation, the map was a tool to be used, and, as such, it was important to put the latest information upon it.[23] In his presentations to various courts and royal commissions, Columbus always produced a map as evidence of the practicality of his scheme.

It would be nice to start this discussion with the Vinland Map as the earliest map to show a new conception of the world. This interesting document was recently discovered in a fifteenth-century manuscript, which also contains the "Tartar Relation," an account of the Pian di Carpini mission to Asia in the thirteenth century, and an excerpt from the encyclopedic work of Vincent of Beauvais. The map, on a rectangular double folio, is oriented to the north. It shows an Indian Ocean open to the east and several islands east of Asia, which are taken to indicate Japan.[24] More important, to the northwest are shown the islands of Iceland ("Isolanda Ibernica"), Greenland, and Vinland ("Vinlanda insula"), with a note on the discovery of the last "by Bjarni and Leif." When the map first came to light in the 1960s, it produced great excitement as the first evidence that the Norsemen had mapped their voyages and discoveries in the North Atlantic. About the same time the map surfaced, the Norwegian archaeologists, Helge and Anne Stine Ingstad, discovered what appeared to the remains of a genuine Viking settlement at L'Anse aux Meadows in Newfoundland. The authenticity of the site is now generally accepted, but the map has been less fortunate, and many now believe it to be a clever modern forgery. This opinion is based partly on technical grounds (the composition of the ink) and partly on geographical grounds (how could anyone in the fifteenth century have known that icebound Greenland was actually an island?). Greenland, when it appears on other fifteenth-century maps, was usually drawn as a peninsula attached to northern Europe.[25] Vinland can be found in a few other medieval geographical texts and maps, as an island somewhere to the north of Europe. Whether this is an echo of the Norse discovery is not entirely clear.[26]

The "Columbus Map"

Three extremely interesting late fifteenth-century maps give us a picture of the world as it was seen on the eve of Columbus's voyage. These are the so-called Columbus map in the Bibliothèque Nationale in Paris, the world maps of Henricus Martellus Germanus, and the globe of Martin Behaim. The first of these gained notoriety when it was proposed by Charles de la Roncière that it had been made by Columbus and actually taken along on his voyage (fig. 8.1). Columbus's career as a mapmaker rests on flimsy evidence, as does that of his brother Bartholomew.[27] Christopher Columbus probably did not learn this skill in Genoa, where the once-thriving map trade had fallen on hard times, for in 1472 he was describing himself as a "lanerius" or wool-worker.[28] In Lisbon he may have sold maps as a dealer, and, as a sailor, he naturally was familiar with the use of maps. He may have commissioned maps to be made and added information from his own experience. In his notes to d'Ailly's *Imago Mundi,* he makes several references to "our maps," but these could just as well be maps in

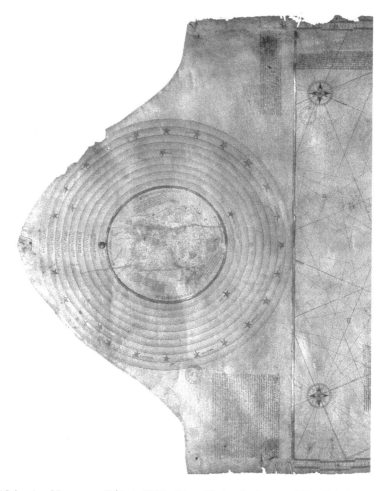

Figure 8.1. "Columbus Map, c. 1492" (Paris, Bibliothèque Nationale, Cartes et Plans, Rés Ge. AA 562. 82 × 41 cm / 28″ × 44″. This map, which has been the subject of much controversy, cannot be definitely linked to Columbus. It pairs a small circular world map (left) with a sea chart, which gives an accurate outline of the known world. Texts above and below the world map are taken from Pierre d'Ailly and discuss the structure of the universe. Courtesy of the Bibliothèque Nationale de France.

his possession as maps he had made himself.[29] As Gaetano Ferro has pointed out, the only surviving map that can be securely assigned to Columbus is his sketch map of Haiti.

The Paris map, unsigned, was acquired by the library in 1848 from an unknown source.[30] The main part of the large sheet of parchment is taken up with a sea chart of the known world, from the Congo River in Africa in the southwest to the Tanais River north of the Black Sea to the northeast. It also shows

Iceland and the Red Sea. In the upper left-hand corner is a circular world map, surrounded by the celestial spheres, with two texts taken from Pierre d'Ailly describing the composition of the universe. The small (twenty centimeters in diameter, or eight inches) world map displays some 250 place-names, and unlike the sea chart, shows all of Africa, including the results of Dias's voyage—the discovery of the Cape of Good Hope and the premise that one could sail around Africa into the Indian Ocean. These details indicate that the map was drawn after 1488 and presumably before 1493. The mappamundi also shows paradise as an island east of the Asian continent. The place-names and inscriptions are in Latin.

De la Roncière presented this map with great éclat at a meeting of the International Congress of Geographers in Cairo in April 1925.[31] In support of his thesis, he cited the prominent vignette for the city of Genoa and a reference to the Genoese discoverer of the Cape Verde Islands. Marcel Destombes agrees that the script on the map is Genoese.[32] De la Roncière says that the map also reveals the preoccupations of a commercial traveler, such as Columbus was, showing where wheat could be grown and where it needed to be imported, and citing the availability of other trade goods, such as ostrich plumes, cotton, sugar, pepper, civet, and parrots. The maker of the map also displays much interest in places where gold might be found. The mappamundi makes it clear that the voyage to the Indies going around Africa, though feasible, was very long, a view reinforced by an inscription describing the three-year voyage of Solomon's fleet, going and returning from Ophir. This subject had greatly interested Columbus and was mentioned in a note he wrote in his copy of d'Ailly's book. De la Roncière also points to the Island of the Seven Cities, the Antilles, accompanied on the sea chart by a description of an accidental encounter there; the cabin boys of a Spanish ship, scrubbing their kitchen utensils on the beach, found silver in the sand.[33] De la Roncière thinks that this story inspired Columbus, and there is evidence from his journal that he had on board a map that showed the Antilles, one of the goals of his expedition.[34] Also important to a would-be explorer, the map makes it clear that the earth is habitable and the sea navigable in the equatorial and polar zones. Finally, de la Roncière refers to Columbus's notes in d'Ailly's book pointing specifically to "our maps on paper where there is a sphere."[35] Although sea atlases sometimes contained world maps, the configuration of the "Columbus map" with its round world map on the same sheet, is unusual. Summing up, de la Roncière says, "A more or less felicitous mishmash of archaic terms and modern nomenclature, of theory and reality, of established facts and discoveries assumed before they happened, this strange model is that of the maps of Christopher Columbus."[36]

More recently, "all passion spent," Monique Pelletier reviewed the evidence for the "Columbus map," confirming that it is in fact a showcase for many of Columbus's ideas. She describes it as a fine presentation copy, perhaps designed for unveiling before the Spanish monarchs at the final, and victorious, hearing at Santa Fe. It is difficult, however, to explain away the absence of Cipangu (Japan) on the world map, since that island was key to Columbus's route. Instead, several islands identified with St. Brendan float vaguely in the ocean northeast of Asia. All we can conclude is that St. Brendan's adventures, usually located in the North Atlantic, have drifted to the North Pacific, perhaps another piece of evidence of the "small sea between Spain and India."[37] There is no hard evidence to link it with Columbus, either as author or owner.

Henricus Martellus Germanus

Very little is known about Henricus Martellus Germanus, a mapmaker working in Florence from 1480 to 1496. He spent at least part of that time in the workshop of Francesco Rosselli (c. 1445–before 1527). Although he used the sobriquet Germanus, Martellus is not a German name. The Martelli were a prominent merchant family in Florence, and Enrico Martello sounds Italian enough. The one piece of information he gives us about himself, that he has traveled extensively, suggests a career in business.[38] His surviving work includes two editions of Ptolemy's *Geography,* a large wall map now at Yale University, and five editions of the *Insularium Illustratum* of Cristoforo Buondelmonti. Roberto Almagià has pronounced that Henricus Martellus was an excellent draftsman, who drew upon the latest information and improved the maps he adapted for his collections.[39]

His earliest known work was a *Geography* of Ptolemy, which is dated just after 1480, based on a reference to the Turkish conquest of Otranto in that year.[40] This includes the standard set of twenty-seven maps, including a world map, all in the Ptolemaic model, with the regional maps using Nicholas Germanus's trapezoidal projection. His other Ptolemy is later, sometime before 1496, when Camillo Maria Vitelli, the patron for whom it was made, died. This is the grandiose MS Magliabechiano XIII.16 in the National Library of Florence. The world map in this codex is almost identical with the Vatican Ptolemy map, but the regional maps have been made on the cylindrical projection, and there are many more of them. In addition to the first tabulae modernae, which appeared in the work of Nicholas Germanus and Piero del Massaio, Henricus Martellus made modern maps of Mediterranean islands, Asia Minor, northern Europe, and the British Isles. There is also a nautical map of the north African coast. In a prefatory note, the mapmaker announces that his "most splendid" *Cosmographia* contains regional maps of the present day and shows all the ports and coasts newly discovered by the Portuguese.[41]

In between these two works, Henricus Martellus produced at least five editions of Cristoforo Buondelmonti's *Insularium Illustratum.* Buondelmonti, a canon of the Florence cathedral, set out on an Aegean tour around 1414, visiting every island. His book, completed in 1420, featured a set of maps along with notes on the most interesting mythological, historical, and geographical features of each island. It was dedicated to Cardinal Giordano Orsini and must have been popular, as it was recopied a number of times.[42] Henricus Martellus takes the basic format but adds new text and other maps, including a world map, sea

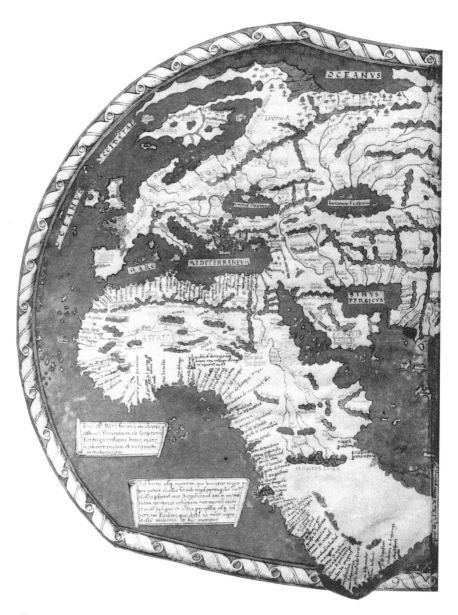

Figure 8.2. World map of Henricus Martellus Germanus, c. 1490 (London, British Library, Add. MS 15,760, fols. 68v–69r, 47 × 30 cm / 18.8″ × 12″. Henricus Martellus made maps for Ptolemaic atlases, as well as for various editions of Christoforo Buondelmonti's *Insularium*, or "Book of Islands," from which this map is taken. Some Ptolemaic features are retained, such as the absence of the Indian peninsula, but the Indian Ocean has been opened up. He retains the southern extension of the eastern coast, a feature repeated on many subsequent world maps and called the Dragon's Tail. The names on the African coast reflect the progress of the Portuguese voyages up to 1488. Courtesy of the British Library.

charts, and maps of islands beyond the Aegean, such as Sicily, Sardinia, England, Ireland, and, in one edition, Taprobana and Cipangu.[43]

His world maps in the *Insularium* build on the Ptolemaic model but are modernized (fig. 8.2). For example, the Mediterranean on the world maps in the *Geography* is 62° long, while on the *Insularium* maps it follows the dimensions on the nautical charts and is 50°, closer to the modern figure of 42°.[44] On the world map in his *Insularium* at the British Library, he shows the results of Bartholomew Dias's rounding of the Cape of Good Hope in 1488, listing the names of the various ports and landmarks along the way.[45] "Here is the true modern form of Africa between the Mediterranean Sea and the southern Ocean, according to the Portuguese description," says an inscription. Modern Africa is so long, however, that it breaks through the frame at the lower edge of the map.

The Indian Ocean is open to the south, but Henry Martellus retains its eastern coast, which Ptolemy had called Sinae, home of the fish-eating Ethiopians. This was the edge of Ptolemy's map—he did not show the full extent of Asia, noting that "unknown lands" lay beyond. On the Martellus map there is a long peninsula to the east of the Golden Chersonese (Indochina), featuring the mysterious port of Cattigara. Unlike Ptolemy's Asia, Martellus's version has an eastern coast with the island-bedecked ocean beyond and additional names taken from Marco Polo. The peninsula of Cattigara was copied on most of the world maps of the first half of the sixteenth century, as the shape and extent of Asia was being sorted out. Called the Tigerleg or the Dragon's Tail, there has been an unconvincing attempt to identify it as an early map of South America, turning Ptolemy's "Sinus Magnus" into the Pacific Ocean.[46]

Martellus retains other Ptolemaic features, such as the appearance of the northern coast of the Indian Ocean with a flattened India and a huge Taprobana. In the north, Greenland is a long, skinny peninsula attached to Europe and north of Scandinavia, a concept derived from the Claudius Clavus map of 1427. The Mediterranean and Black Seas and the Atlantic coasts are taken from sea charts, while the east coast of Africa, as yet unexplored, follows Ptolemy's design. From this we can see that in the *Insularia* Martellus used modern information when he had it, incorporating it into a classical format. If the two Ptolemy atlases can be dated as the first and last of his works, it is interesting to see that he reverts to the pure Ptolemaic form for the world in the later edition, with the lengthened Mediterranean and the closed Indian Ocean. Clearly, Ptolemy maps were considered as illustrations for a revered and ancient text and did not have to include the latest news.

Although Martellus's world maps in the Ptolemy manuscripts show an extension of the inhabited world as 180°, those in the *Insularium* make it 220° or more, an attractive feature to those who dreamed of reaching Asia by sailing west. One of the *Insularium* manuscripts, that at the Laurentian Library in Florence seems to be a working copy, as it has many cross-outs and corrections on the maps and in the texts.[47] Here the habitable world is extended to 265°, with northern Asia coming right up against the right-hand border.

The largest of the Martellus world maps is at Yale University (43 × 76 inches). Long in the possession of a family from Lucca, it went to Austria in the nineteenth century and was bought for Yale in 1961. It shows a great deal more ocean to the east of Asia with several large islands in it, including Japan in the far northeast. It also has the islands of Madagascar and Zanzibar off the coast of Africa. The map is surrounded by a dozen puffing classical wind heads, and for the first time in a non-Ptolemaic map, there is a latitude and longitude scale on the side. Of course, this is mostly symbolic, as neither Henricus Martellus nor anyone else had a complete set of accurate coordinates.

The work of Henricus Martellus Germanus epitomizes the best of European cartography at the end of the fifteenth century. His intelligent effort to reconcile modern discoveries with traditional knowledge, especially Ptolemy, provides a workable model for the world map in a time of rapid change. The role of experience was clearly important to Martellus, for in his copy of the Laurentian *Insularium* he includes a verse about his own travels, noting that he has been traveling around for many years, and adding that it is worthwhile but difficult to "set a white sail upon the stormy sea." He suggests that the reader might prefer to stay safely at home, and learn about the world through his book.[48]

Rosselli's Printed World Map

An unsigned, undated, printed world map attributed to Francesco Rosselli, demonstrates many characteristics of the maps of Henricus Martellus.[49] Rosselli, trained as a painter, had a shop in Florence in the later 1480s and 1490s which specialized in copper engraving, and there remains an inventory of the contents of the shop from 1527, when his heir Alessandro died.[50] The shape of Africa, the "Dragon's Tail" peninsula, the northern peninsula of Greenland, and the combination of nautical chart and Ptolemaic features resemble the work of Martellus. The islands in the Indian and Pacific Oceans, including Madagascar and Zanzibar, are similar to those on the Yale map, though the Rosselli map does not show such a large extension of the ocean to the east, and Japan is not depicted. On the west coast of Africa, the toponymy is not exactly the same. Henricus Martellus gives more names, mostly taken from the Portuguese. Many of Rosselli's names are similar, but they are in a slightly different order, and each map has names that the other lacks, showing that one is not an exact copy of the other.

The confusion over dating the map began with Rosselli's inscription near a drawing of the columns erected by Bartholomew Dias on the Great Fish River, marking the progress of his voyage. Rosselli writes, "Huc usque ad ultimam columpnam pervenerunt nautes lusitani 1498" (Here as far as the last column

Portuguese sailors came in 1498). Tony Campbell opines that this is a simple error and that the date was meant to read 1488, which seems to be the case. The map bears a caption: "Placed here is the entire form of the whole (world) which is surrounded by the ocean sea, with the part of upper India, discovered after the time of Ptolemy, and with the part of Africa which Portuguese sailors traversed completely in our times."[51] The comment about India probably refers to information from Marco Polo, though the overall configuration of India on the map is resolutely Ptolemaic. Campbell suggests 1492 as the best guess for the date of this map and notes that, with the Contarini map of 1506, Rosselli was responsible for the last printed map before Columbus's discoveries and the first published afterward.[52]

In his world maps, Rosselli used three different projections in an attempt to portray accurately the changing world in which he worked. The first, Ptolemy's homeotheric projection, made room for the extension of Africa to the south. The second, a conical projection that he used for the 1506 world map, showed Columbus's discoveries attached to East Asia. Rosselli's last world map was drawn as an oval, a form that continued to be used by mapmakers for some years, as it was considered most able to show the entire surface of the globe.[53]

The Globe of Martin Behaim (1492)

In his letter to the king of Portugal, Toscanelli had commented that it was much easier to show the advantages of the proposed western route to Asia on a globe, which suggests that he actually possessed or had easy access to one. Globes were made in the ancient world, though none survive, and there is written evidence of several constructed earlier in the fifteenth century. Among these are a terrestrial globe thought to have been made by the astronomer and instrument maker Jean Fusoris about 1432 to accompany his treatise on the sphere.[54] Somewhat later, between 1440 and 1444, a globe was made for Philip the Good, Duke of Burgundy. Originally attributed to the Flemish painter Jan van Eyck, it has recently been argued that it was the work of Guillaume Hobit.[55] By the end of the fifteenth century, the German humanist Conrad Celtis was apparently using globes as visual aids for his lectures on geography at the University of Vienna.[56]

The oldest surviving terrestrial globe is that of Martin Behaim, made in the annus mirabilis of 1492 (fig. 8.3). This date appears on the globe itself; however, it shows no discoveries after 1485, the voyage of Dias not being represented. Behaim, born in Nuremberg in 1459, had an adventurous and somewhat checkered career.[57] He went into business, being apprenticed to a cloth merchant in Mech-

Figure 8.3. Globe of Martin Behaim, c. 1492 (Nuremberg, Germanisches Nationalmuseum. 50.7 cm / 20″ diam.). The oldest surviving terrestrial globe, Behaim's work appeared on the very eve of Columbus's voyage across the Atlantic and is a graphic representation of the relative ease of such a journey. The globe is brightly colored, ornamented with pictures, and burdened with lengthy inscriptions on trade, history, and the life of its maker. Courtesy of the Germanisches Nationalmuseum.

lin in 1476. Business took him to the Frankfurt fairs, to Antwerp, and eventually in 1484 to Lisbon. He claimed to have studied with his great compatriot Regiomontanus, and apparently this claim earned him a position as a participant in a mathematical conference at court in 1484–85. The conference was dedicated to determining latitude by the position of the sun, as Portuguese sailors were now traveling so far south that it was no longer possible to see the polestar. Behaim was also knighted by the king of Portugal in 1484. Behaim tells us that he made several voyages along the African coast. He went as far as the Azores, where he married the daughter of an official, and possibly got as far as the Guinea coast. In 1490 he returned to Nuremberg to settle his parents' estate. The city council, impressed by his claims of extensive travel and mathematical skill, commissioned him to design a globe, which was completed in 1492. The actual drawing was done by Georg Glockendon. After this Behaim returned to the Azores and eventually died in poverty in Lisbon in 1507.

Behaim seems to have been something of a con artist. The globe, however, is an undoubted triumph, twenty inches in diameter (507 millimeters), painted with dark blue seas, tan-colored lands, green forests, gold and silver mountains, along with red and black inscriptions and numerous pictures. It is a wonderful mix of traditional geographical lore and more recent discoveries. He seems to think Ptolemy, Marco Polo, John Mandeville, and the Portuguese navigators are all on a par as sources for his work. The globe is marked by a meridian and the equator, graduated in degrees, and by the tropic and polar circles. The ecliptic, shown on the globe, bears the signs of the zodiac.[58] Like the maps of Henricus Martellus, Behaim's globe shows a circumnavigable Africa, though with a noticeable eastward extension at the south, an extended Eurasia (236°), and a Dragon's Tail peninsula in East Asia. Most striking, and as could only be made clear on a globe as Toscanelli had said, is the relatively short distance between Europe and Asia, particularly Japan, which is portrayed as a very large island off the eastern coast on the Tropic of Cancer, considerably south of its true location, with an inscription lauding its pepper forests and gold supplies. A handy stepping-stone is Antilia, shown here as two islands and located halfway across the ocean. Behaim himself was so inspired by this vision that he applied to Emperor Maximilian to sponsor him on a westward voyage in 1493, not realizing that Columbus was already on his way home.

The globe is highly illustrated and packed with paragraphs of text, all in German. On the map one can find forty-eight flags, fifteen coats of arms, forty-eight kings in tents or on thrones, four saints (Peter, Paul, Matthew, and James), eleven ships, as well as fish, seals, sea serpents, mermen. and a mermaid. On land are elephants, a leopard, a bear, a camel, an ostrich, a parrot, and two Sciapods in central Africa. Inscriptions describe sirens, satyrs, unicorns, dog-heads, cannibals, and idol worshippers.[59]

The message of the globe is unabashedly commercial, as would appeal to the city fathers of Nuremberg. Gold, precious stones, spices, and furs await the adventurous merchant. The longest inscription, which occupies the prominent midoceanic position between Europe and Asia, recounts the course of the spice trade from Java Major in the Far East until the cargo finally reaches western Europe. The route can be followed on his globe going from Java Major in the Far East, to Seilan (probably Sumatra) to the Golden Chersonese (Indochina) to Taprobana, than on to Aden, Cairo, Venice, Germany, England, and France. Behaim makes the point that customs duties are paid twelve times in the course of this voyage, as well as profits being taken, resulting in very high prices. Obviously, eliminating these middlemen would be a very profitable course of action. Behaim attributes his information to Bartholomew of Florence, a person not otherwise known, and his account is made even more dubious by having Bartholomew report to Pope Eugenius IV in 1424, as Eugenius did not assume office until 1431. Bartholomew may be an error for Nicolò de' Conti, who did travel many years in the East, returning to Italy in 1439, when Eugenius IV was Pope. There is, however, no trace of Conti's wanderings in Southeast Asia shown on the globe.

Like other mapmakers of the late fifteenth century, Behaim draws on a variety of traditions for the features of his globe. Ptolemy provides information for Asia (land of the Seres, the Sacae, Scythia, Sinus Magnus, though not Cattigara) and Africa (the mountains of the moon, Agisimba, the Garamantes). Alexander the Great is named, with the trees of the sun and the moon, his victory over Darius, and the altar he built. The map has little biblical content, but we can find the Red Sea painted red, Noah's Ark, Havilah, and the city of Ophir. The holy land is not enlarged and thus has only a handful of names. Jerusalem is marked with a picture of a kneeling saint. From Marco Polo come many legends and place-names in Asia, including Cathay, the desert of Lop (which Behaim describes as a "forest"), Karakorum, Kambalu, Quinsai, and Tartary. There are few more modern references, but there is one on the destruction of Tamerlane, which happened around 1400. Prester John is in Asia, and several paragraphs are devoted to his power. "Og and Magog" are on the map, but there is no reference to their enclosure or their future depredations. Missing from the globe is the earthly paradise. There is also no southern continent, though Behaim does place the Antipodes on the island of "Candyn," located in the ocean southeast of China. He says that these inhabitants are placed "foot against foot with respect to our land," that they do not see the polestar and that they have night when we are in daylight. "All this is because God has created the world together with the water of a round shape, as described by John Mandeville, in the third part of his voyage."[60]

Behaim said that he had traveled to Africa, crossing the equator and rounding the southern cape. A lengthy inscription recounts a voyage made in 1484 to "King Furfur's land," 1,200 miles from Portugal, and another 1,100 miles farther on where cinnamon was found. His map is a doubtful witness. Monte Negro, which Henricus Martellus shows at 13° S is here placed below the Tropic of Capricorn, where the coast turns abruptly to the east. The names on the coast are scrambled and, in many cases, unidentifiable, including a river that Behaim names for himself, the privilege of a mapmaker. There is no Cape of Good Hope, though Behaim's Cabo Negro is more or less in the right position. "Here," he says, "were set up the columns of the King of Portugal on January 18 of the year of the Lord 1485." There is no reference to the voyage of Dias, which seems to be unknown to Behaim. His latest references are to the expedition of Diogo Cão (1484–85).

Behaim notes that his father-in-law, Sir Jobst von Hürter, a Fleming, is the governor of the island of Fayal in the Azores. In his brief account on the map, he makes a number of surprising mistakes, saying the islands were first settled in 1466 (settlement began in 1439) and that sugar is grown there (it is not, though it was a main product of Madeira). Behaim places two flags on Fayal, one with the arms of Nuremberg and one with his own.

Despite Behaim's numerous errors, self-advertisement, and anachronisms, his globe made a point that could not be made nearly so well on a flat map; it showed the way open from western Europe to the riches of Asia. "Let none doubt," writes our mapmaker, "the simple arrangement of the world, and that every part may be reached in ships, as is here to be seen."[61]

The maps discussed in this chapter, made at the close of the fifteenth century, demonstrate the state of cartography at a crucial moment in history. Certain elements of the traditional world map have been abandoned or modified, such as the circular form, retained by only one of our three maps. The mapping of the ocean space has been greatly expanded. No longer is the ocean a simple rim around the edge of the known world, but an open, explorable highway, sprinkled with friendly islands and others yet to be discovered. No place is marked uninhabitable, including the formerly forbidding polar and torrid zones; instead, the whole world is open to human adventuring and settlement. The supernatural element is greatly reduced. The small sphere on the Columbus map is the most traditional, retaining a space for paradise, but paradise was now to become rare on world maps. Even Gog and Magog, who dominated the northeast corner of maps with their apocalyptic threat, disappear. Behaim still has them on his globe, as "Og and Magog," but now they are just a group under the khan's control rather than the servants of the Antichrist. Maps are now more commonly oriented to the north, based on the use of the compass and the common arrangement in many sea charts. Martellus's Yale map and Behaim's globe include latitude and longitude scale, although they do not mean very much as yet. The sorts of observations, however faulty, that the Portuguese made in their southward journeys were attempts to fill in the world picture with precise coordinates. It would be several centuries before this was even remotely possible, but the idea was firmly implanted.

There is a clear agreement among our maps about certain important features, such as the circumnavigability of Africa, the extension of Eurasia (although the estimates vary), and the corresponding shortening of the distance from Spain to Asia, and the open Indian Ocean. The question of which map was made or seen or taken on the voyage by Columbus, which has exercised so many scholarly minds, may be put aside, as the maps' agreement represents a kind of consensus about the shape of the world which the great explorer may have encountered in a number of places.[62] Another interesting characteristic of these maps is their internationalism. We have a German working in Florence, a Genoese in Portugal and Spain, another German from the Azores, and a Florentine corresponding with the Portuguese court. Francesco Rosselli even spent a significant term in Hungary, working for the enlightened King Matthias Corvinus. The supposed policy of secrecy on the part of the Portuguese government is exploded by the rapid appearance of their discoveries on maps throughout Europe.

The game of trying to find "Christopher Columbus's map" will probably be played indefinitely, but the search is complicated by the time that has passed, the poor survival rate of cartographical documents, and the intense publicity that has surrounded Columbus ever since 1493.[63] As has been shown in this chapter, there were several maps that illustrated his ideas on the eve of his first voyage, at least insofar as we can know what these were. His "errors" about the size and shape of the world were not his alone but were widely shared.

By the end of the century, the voyages of Columbus and Vasco da Gama would enlarge European knowledge of world geography to an unprecedented degree. What would mapmakers do to keep up with the dizzying rate of change? The story of the gradual emergence of the Americas, the mapping of the Indian Ocean, the circumnavigation of the world by Magellan, and the eventual discovery of Australia and New Zealand has been told many times, and it is an enthralling tale. But it is also important to remember that these developments came at a point when mapmakers, those who pictured the earth's surface, were ready and willing to adapt to changes. There was no ferocious clinging to traditional shapes and distances but a quick response to the reports that poured in from all sides. If the American discoveries adhered to the coast of Asia for a while, that was a reasonable assumption, though soon to be overturned by further exploration. Even the ancient wisdom of Ptolemy was thrown overboard with surprising swiftness. By 1500 world maps in Ptolemy atlases were showing an open Indian Ocean.

The kind of self-confidence that propelled Columbus and his captains and crew onto an unknown ocean had infected a whole class of Europeans. Fra Mauro was already giving greater credence to his navigators than to musty old authorities. The mapmakers, drawn from a class of artisans, businessmen, and navigators were part of this cultural frame of mind. While some sailors may have gone to the Americas before, either by accident or design, and certainly the Vikings did go, it was only in the late fifteenth century that Europeans were able to map where they had gone and so were able to come back and to go again. Theories of geography, eagerly discussed in humanist circles in the fifteenth century, were now to confront hard-core reality—and yield.

The World Map Transformed

Before America was discovered, there was a place to put it on the map. In the two centuries before 1492, from the era of the great mappaemundi of the thirteenth century to the world maps at the end of the fifteenth century, mapmaking was transformed. The closed circular shape of the medieval mappamundi, where the three known continents filled most of the space, leaving only a narrow rim for the ocean, gave way to other forms, like the oval and the rectangle, which allowed for indefinite expansion. New information appeared on maps. Modern names supplanted those handed down from the past, such as the terminology on the Hereford mappamundi, which is mostly from the era of the Roman Empire. Geographical spaces far from Europe took on new definition and content, incorporating the reports of travelers. At first this was a slow process—it took almost a century (1295–1375) for Marco Polo's Cathay to appear on a world map, but the African voyage of Bartholomew Dias in 1488 was recorded after just two years. Finally, discoveries of Columbus and his successors introduced previously unknown regions to the European mind. The first map to show Columbus's voyage appeared in 1500, a mere seven years after the fleet returned home from its first venture. While there may have been other Europeans who had landed in the New World in the past, this was the first journey to be mapped, however incorrectly at first, and integrated into European consciousness as new space.[1]

The transformation of the world map brought losses as well as gains. The very ambition of the mappamundi, which provided a veritable encyclopedia of human knowledge and belief about the world, yielded to a plain representation of physical space. The transformed map answered only the question, "where?" while the old map had tried to answer the questions, "what?" "when?" and even "why?" Conservative mapmakers and map users were loath to give up the old form and content, and continued to make and purchase quite traditional mappaemundi. The atlas of Andrea Bianco demonstrates that several different styles of world map could be presented simultaneously, while the broadsheet maps made in Germany toward the end of the century peddled an old-fashioned image of the world for ordinary folk to buy.

Content

The mappamundi imbued space with meaning, reminding the viewer that the earth was created by God and that sacred history was playing itself out on this great stage. The medieval period was torn between two views of the world: as the scene of sin and temptation but also as God's creation, a good and beautiful thing. The Hereford map (see fig. 1.3) illustrates this well, with its grim evocation of death (MORS) encircling a document bursting with life: energetic monsters, towered cities, beasts, and humans all engaged in their myriad activities. Some mappaemundi stressed the larger framework in which the world was situated by encircling the map with explanatory texts or embedding it in a cosmic diagram. On the Ebstorf map, for example, a number of long inscriptions are arranged around the world, expounding on the form of the heavens, the shape of the earth, the division of the continents, and the members of the animal kingdom. Other mappaemundi appeared surrounded by the heavenly spheres or by elaborate calendars, showing the calculation of time by the motions of the sun and moon. Recent work by Brigitte Englisch suggests that early medieval world maps were constructed with an underlying geometrical framework, witness of the mathematical perfection of the divine creation.

Medieval mappaemundi contained places that could never be entered, explored, or mapped precisely. While the mappamundi was certainly intended to portray terrestrial space, it invariably included certain "unearthly" details, such as the terrestrial paradise, believed to exist in real space but not space that could be entered by any living human being since the Fall. Similarly, the monstrous races, never actually seen by anyone, were arrayed in conveniently inconvenient locations, and zones, marked uninhabitable due to heat or cold, found their places on edges of the map. When a territory is labeled *Terra Incognita*, we may ask if the unknown quality is a permanent one or if we are merely waiting, like Joseph Conrad's narrator in the *Heart of Darkness,* to explore those intriguing "blank spaces on the map." One of the important changes in mapping in the fourteenth and fifteenth centuries was the elimination of such inaccessible and impossible places in favor of the world as a continuum of homogeneous space, open in all directions to human travel. Christian belief about paradise made it difficult to do away with it altogether, although the Genoese map of 1457 suggested this as a reasonable course of action, and Columbus and his successors continued to search for paradise, hoping to put it on the map, as had been done for the "uninhabitable" zones of the equator and the poles. But by its very nature, paradise proved to be unmappable, and, as further exploration revealed more about the disparate sources of its four rivers, it was increasingly relegated to a marginal illustration outside the map.

Another characteristic of the mappamundi was the charting of time, the unfolding of human history in space. The geographical names that appeared tended to be ancient and traditional ones. The Hereford map labels the regions of the world with the names of the late Roman Empire, such as the three Mauritanias in Africa. Tribes, such as the Massagetae, the Goths, and the Colchi, long since moved on or become extinct, are shown in their original homelands. By the fifteenth century, mapmakers were endeavoring to illustrate the state of the world in their own day. Most striking on the map of Henry Martellus Germanus are the series of names on the west coast of Africa, marking the progress of the Portuguese voyages. This region was unknown to the maker of the Hereford map and had been left correspondingly vague. "This is the true form of modern Africa," Martellus writes on his map.

Important places that had disappeared, such as the sunken cities of the Dead Sea or the destroyed Tower of Babel, appeared on the mappamundi. Noah's Ark sits on a mountaintop in Armenia, not as a relic, but with Noah and his family still inside, waiting to disembark. Hippo is not just the former bishopric of St. Augustine, but we can see Augustine standing inside his church, while Abraham looks out from his house in Ur of Chaldea. The faithless Israelites are kneeling to worship the golden calf after losing patience with Moses in the Arabian desert. It may be this feature that the canon of Pisa is missing when he complains about the marine chart: it is too spare, too unadorned, and too many important places are missing, just because they no longer exist.[2]

The meaning of space was also an element in the construction of the mappamundi. The holy land was not just any old scrap of land but radiated spiritual energy from the many important events of religious history that had transpired there. Jerusalem, Bethlehem, and the River Jordan were all real, physical places but something more as well. No wonder the mapmakers enlarged this space in tribute to its great significance. The symbolism of every physical object was a fundamental concept in the Middle Ages, as one can see in the bestiaries that appear on the larger mappaemundi, reminding the viewer of the moral lessons that Nature can teach. By the sixteenth century much of this material had been moved off the map, and a separate map was provided for the holy land.

Such works as the Nuremberg Chronicle provided plentiful textual and pictorial supplements to expound upon such topics, but more and more the world maps themselves were unadorned.

The cultural agreement of the High Middle Ages, as expressed by the mappaemundi and the great encyclopedic compilations of the late thirteenth century, had begun to disintegrate in the fifteenth century. In geography, the dominance of authority over experience was threatened, as was the absolute primacy of spiritual goals. Roger Bacon's approach to geography always came back to religious purposes, often by way of analogy. Actually going there was less important than the "spiritual roads" one might travel in contemplating God's world. Two hundred years later, geography had become a different subject. Although no one questions the religious passion of Columbus, for example, he sought out physical "roads," looking for power, profit, and fame, instead of staying in the monastery of La Rábida, contemplating the beauties of a mappamundi.[3] The recovery of classical books by writers such as Ptolemy and Strabo brought concealed baggage. These works did not merely add to the mounting pile of authorities, but the methods by which they were composed had an impact on their readers.[4] Inspired by classical models, humanists began to make observations of their own. Augustine's distrust of the classical heritage had stimulated him to urge his followers to take what was consistent with Christian belief and to leave the rest. The fifteenth-century classical revival produced uncensored texts that raised disturbing questions. If ancient works contradicted one another, how was one to decide the truth? Perhaps one should consult one of Fra Mauro's "trustworthy men, those who have seen with their own eyes."

The Form of the Map

Toward the end of the fifteenth century, mapmakers began to experiment with new shapes. The circular map of the three continents had always been something of an abstraction, but now it raised definite problems. How could one include the thousands of islands in the Indian and Chinese Oceans, not to mention the fewer but better known islands of the Atlantic? An oval or rectangular form allowed for expansion in all directions, making room for the ever-lengthening coastline of Africa, the northward extension of the Scandinavian peninsula, and the coast of China, no longer quite as unknown as it had been in Ptolemy's day. Fifteenth-century mapmakers found themselves breaking through the frame, trying to depict the geography they now knew. The earliest maps to show the American lands, the map of Juan de la Cosa of 1500 and the Cantino map of 1502, used the rectangular form of the sea chart, stretching it to the west to encompass the new territories. Ptolemy's chapters on projection gave several alternative forms that a world map might take. Abandoning the circle, however, was also to abandon a form of deep symbolic significance and perfection.

Mappaemundi had nearly all been oriented to the east, but after 1500 most mapmakers chose north for the top of their world maps. They were undoubtedly influenced by the sea charts, which were often oriented to the north, reflecting the use of the north-pointing compass in their construction. Abandoning the eastern orientation was an important change, for east was the direction of Creation and the Garden of Eden, and the course of human history shown on the mappamundi moved down it from east to west. Changing the map's orientation was another step in the direction of secularization.

In the thirteenth century, Jerusalem had moved to the center of the mappaemundi, but it lost that position on later world maps. The greater extent of Asia pushed the holy city from a physically central location, and, as maps shied away from depicting spiritual meaning, Jerusalem was displaced. Later maps have no clearly defined central point, physical or spiritual, and it is interesting to note that on Henry Martellus Germanus's map of 1490 (see fig. 8.2), Jerusalem is not even there.

Not only the outer form but the shapes within the map were altered. The sea chart and the mappamundi experienced a parallel existence for several centuries, perhaps longer than we now know, but eventually they began to influence one another. It is noteworthy that the first known map to combine the two forms was commissioned by Marino Sanudo (fig. 3.1), a Venetian businessman who had traveled widely and was familiar with the excellence of the charts used by seamen. There was really nothing incompatible with drawing an accurate outline of Italy within the confines of the mappamundi, as we can see in several fourteenth-century mappaemundi that willingly gave the Mediterranean a more recognizable shape. The Catalan Atlas is one production that pulled together the two traditions without doing violence to either one. It was only when the same mapping principles were applied to the whole world that it became problematic, as the smooth outline of the old *ecumene* was disrupted.

Although the sea charts tended to be sober, allowing no pictorial fantasy to mar the line of the coasts or the order of the ports arrayed along them, it did not take long for colorful illustrations to take over the interior spaces. The hometown of the mapmaker—Genoa, Venice, Majorca—got an elaborate symbol, while the Sahara desert was soon populated with camels, elephants, and kings enthroned in their royal tents. Eventually, as physical details began to accumulate about inland trade routes and geographical features, these pictures needed to be moved out of the way. Possibly it was also a matter of taste, as some sixteenth-century maps, particularly those made for wealthy patrons, continued to be beautifully illustrated, with tropical fauna and flora in Brazil, for instance. An example is the fabulous Atlas Miller of Paris (1519), which is something of a throwback with its gilding, many colors, and vivid images.[5]

The Use of Maps

Map usage is one of the most difficult topics to pin down, as documentary evidence is elusive. Local maps were probably always practical tools—a survey of one's estates, a diagram of the town's water supply or fortifications. Measuring might be as simple as pacing off a boundary or the distance between milestones, but it was an integral part of this form of mapping. On the other hand, world maps were designed for contemplation or study, rather than for practical use. In the early Middle Ages, we know from the books in which world maps appear that their appeal was to scholars and teachers. Almost always written in Latin, the maps were surrounded by literary, ancient historical, and biblical references or by scientific treatises on the structure of the universe. Marine charts, a type of local map, had a clear practical purpose and were based on concrete facts, such as distance and direction. The charts were wisely limited to known coasts and seas and did not attempt to map the entire world. We are uncertain about the extent to which they were used aboard. Sailors probably spent more time looking out at the sea, searching for coastal landmarks and watching out for hazards, than they did poring over a map, which few could read anyway. The charts' main venue may have been the boardroom, helping merchants plan their trading voyages and estimate when the profits on them would come in, like Antonio in *The Merchant of Venice*, "piring in maps for ports, and piers, and roads."[6] The demands of commerce made physical accuracy imperative. A map could be a beautiful work of art, but nothing should stand in the way of the true shape and proportion of land and sea.

Merchants in late medieval society were not exactly riffraff. While Thomas Aquinas may have doubted the morality of their profession, the Venetians were quite certain about its value and prestige. Someone like Marino Sanudo moved in ecclesiastical as well as mercantile circles and had received a classical education of which he was rather proud. He was ideally situated to be a go-between for the clerical and mercantile classes, bringing the two traditions together.

Another use of maps, which is obvious to us, is for military strategy. In his book Sanudo stresses the value of accurate geographical information recorded on maps to the success of his proposed crusade. The fifteenth-century voyages of exploration were, in a sense, military campaigns. The Portuguese did not go abroad to contemplate the exotic flora and fauna of the African coast or the wonders of India but to seize opportunities for trade and for power. The columns erected to mark their progress were soon followed by forts intended to impress their control over the hinterlands. Vasco da Gama's voyage to India was quickly followed by gunboats, which blasted away at resisting coastal cities, upsetting the balance of forces (Arab, Chinese, Malaysian, Indian, Javanese, etc.) that had traded there for centuries. In the Atlantic the brutal extermination of the Canarian people was a rehearsal for the scene to be enacted in the larger theater of the Americas. Although, of course, it is perfectly possible to conquer and destroy without a map, the connection between mapping and possessing, whether a country estate or a continent, is an intimate one.[7]

The Combined Form

The last great attempt to make a mappamundi was that of Fra Mauro in 1450 (see fig. 6.1). He held on to the circular form, but he incorporated into it as accurate a physical picture of the world as he possibly could. It is significant that his collaborator was an experienced seaman and chart maker, and this partnership of monk and sailor produced an eloquent document of mid-fifteenth-century cartographic consciousness. The texts on the map bear witness to the struggles between antique authority and contemporary experience which took place in the monastery of San Michele da Murano in the Venetian lagoon. The mapmakers' resistance to Ptolemy was not only due to lack of sufficient data to make a Ptolemy-style world map; such a map lacked important qualities of the medieval mappamundi, such as the deeper meaning of terrestrial space. Could the virtues of the narrative mappamundi be retained in the modern world? Monsters, alas, had to go, and Jerusalem moved from the center of the world but only after a long, torturous explanation. The great size, expense, and decoration of this work and its public placement make it a part of the mappamundi tradition, while its content, especially its argumentative texts, and its Venetian dialect show it to be moving in a different direction. We imagine that the Venetian Senate contemplated the world on the map not in a mood of scholarly detachment but in eagerness to see the scene of trade routes, markets, and opportunities for making money.

Fra Mauro's map teemed with unresolved questions. The "hybrid" maps of the fifteenth century also strove to unite the two forms. The "Columbus" chart of 1490 (see fig. 8.1) combined a sea chart with a small circular mappamundi on the same sheet. Behaim's famous globe (see fig. 8.4.) was basically a mappa-mundi, but its very shape pointed the way to exciting new developments, such as the possible circumnavigation of the world.

Why Did Maps Change?

In the late Middle Ages several factors affected the perceived shape of the world. The revival of the classics had far-reaching implications for European culture. For mapmaking, better information on world geography was found in the works of Strabo, Pomponius Mela, Pliny, and Ptolemy, but the method that these ancient geographical writers had used was also important. Ptolemy conceived of the world as a single and continuous entity that could be properly depicted only by the use of mathematics. He emphasized that knowledge of the world must be assembled from actual experience, from surveying and astronomical observations, whenever possible.[8] Although he might be made into an "authority" by his readers, he set forth in his own text the means for challenging his work, and we know that the ink was hardly dry on the first Latin translation before aspects of his world picture were being questioned. The essay of Cardinal Fillastre illustrates the confusion of this time, as he struggles to unify conflicting classical sources, as well as bring them into line with Christian doctrine. Almost immediately, atlases of maps made according to Ptolemy's instructions were supplemented with new local and world maps. These *tabulae modernae* accepted, rejected, and revised the original Ptolemaic maps but invariably followed his model of unornamented, measured space.

It is important to remember that the "geometric rationalization of space" that began in the fifteenth century had little in the way of new technology to support the process.[9] Astronomical navigation, other than the rough, centuries-old observation of the polestar and its "guards," was in its infancy. Errors of ten degrees of latitude and more of longitude, as well as miscalculations of hundreds of miles were common, for the various sighting devices (astrolabe, quadrant, cross-staff) were inaccurate and the practitioners unskilled. The revolutionary maps of the late fifteenth century were made, as were Columbus's voyages, by the dead reckoning of old salts, equipped with no more than the compass, log, and sounding line. Once the concept of measured mapping was understood and its desirability established, there was a strong incentive to develop better tools to create accurate maps. Still, it was not until the eighteenth century that adequate equipment was available to the mariner to determine latitude and longitude at sea.

In the fifteenth century Europeans began once more to lift their sights—and their sails—to the greater world around them. Ambitious merchants and evangelizing missionaries were lured into distant places, while curious geographers at home questioned travelers eagerly about hitherto little-known lands. We see this interest in the ongoing seminar on geography held in Florence in the middle of the century, where Ethiopian monks, a "North Indian" ambassador, and the vagabond merchant Nicolò de' Conti became the center of attention. Eventually, experience would overwhelm the strictures of antique tradition, although these died very hard. A burst of exploration into the Atlantic and the settlement of the nearest island groups pushed Europeans into unfamiliar territory. The great series of Portuguese voyages was driven by a desire to find the gold mines of central Africa—and financed by trade in slaves—but one result was an ambitious mapping project that would change the form of the world. Not only did Africa extend much farther to the south than had been previously assumed, but it was inhabited at the equator and beyond, while the Atlantic and Indian Oceans flowed together at the Cape of Good Hope. These discoveries were recorded on world maps and immediately raised questions. What else was out there to discover? With some of their most cherished ideas about the shape of the world scuttled, mapmakers were less surprised to find two continents previously unknown to them located on the other side of the world. The maps of first half of the sixteenth century show the gradual detachment of the newfound lands from Asia and the realization of their independent existence. Building on the transformation of the world map before 1492, post-Columbian mapmakers were now poised to consider contemporary discoveries, to entertain new ideas about the configuration of the world, and to try and put these on the map.

Introduction • Andrea Bianco's Three Maps

1. A beautiful facsimile edition was made by Piero Falchetta, *Andreas Biancho de Veneciis me fecit M.CCCC.XXX.VJ* (Venice: Arsenal Editrice, 1993). Made for the customers of the Banco San Marco, it is now out of print.

2. For a full discussion of this issue, see Tony Campbell, "Portolan Charts from the Late Thirteenth Century to 1500," in *Cartography in Prehistoric, Ancient, and Medieval Europe and the Mediterranean*, ed. David Woodward and J. B. Harley, vol. 1 of *History of Cartography*, (Chicago: University of Chicago, 1987), 432–33, 425n.

3. Fra Mauro map, inscription in Russia. See Tullia Gasparrini Leporace, *Il Mappamondo di Fra Mauro* (Rome: Istituto Poligrafico dello Stato, 1956), tav. 40, p. 62.

4. E. G. R. Taylor has a reproduction of Bianco's circular diagram and a brief explanation of the *marteloio*, as these tables are called. See *The Haven-Finding Art* (1956; reprint, New York: American Elsevier, 1971), plate 8, p. 117.

5. See, e.g., the essay by Cardinal Guillaume Fillastre on Ptolemy and Pomponius Mela, Ms. Arch. S. Pietro 31, Biblioteca Apostolica Vaticana, Vatican City. Critical edition by Patrick Gautier Dalché in *Humanisme et Culture Géographique à l'Époque du Concile de Constance*, ed. Didier Marcotte (Turnhout: Brepols, 2002), 293–356.

6. This was the map included in Cardinal Fillastre's copy of Ptolemy, now, MS 441, fols. 184v–185r, Bibliothèque Municipale, Nancy.

7. Isidore, *Etymologiae*, ed. W. M. Lindsay (Oxford: Clarendon, 1911), 13.17, "De sinibus maris." Isidore's information came from the Roman geographers, such as Pliny, *Natural History*, trans. H. Rackham (Cambridge, MA: Harvard University Press, 1942), 6.36, pp. 363–65.

8. Dana B. Durand, *The Vienna-Klosterneuberg Map Corpus* (Leiden: E. J. Brill, 1952), 177–79 and throughout.

One • The World View of the Mappamundi in the Thirteenth Century

1. Richard of Haldingham or Sleaford, described as the one who commissioned the map, died in 1278. It is probable that the map was brought to Hereford from Lincoln by another Richard, perhaps a relative. See Paul D. A. Harvey, *Mappa Mundi* (London: British Library, 1996),

7–11, for a discussion of this knotty question. Malcolm Parkes dates the map on paleographical evidence to c. 1290–1310 in "The Mappa Mundi at Hereford: The Handwriting and Copying of the Text," in *The Hereford World Map: Medieval Maps and Their Context: Proceedings of the Mappa Mundi Conference, 1999*, ed. Paul D. A. Harvey (London: British Library, 2006).

2. Macrobius, *Commentary on the Dream of Scipio*, trans. William H. Stahl (New York: Columbia University Press, 1952).

3. Patrick Gautier Dalché argues that the simplicity of these diagrams may indicate that they are medieval glosses, added to assist the reader, and not original to the work ("The Classical Heritage of Medieval Cartography: The Known and the Unknown," lecture delivered at the University of British Columbia, October 29, 2005).

4. Brigitte Englisch, *Ordo Orbis Terrae: Die Weltsicht in den Mappae Mundi des Frühen und Hohen Mittelalters*, in *Orbis Mediaevalis: Vorstellungswelten des Mittelalters*, Bd. 3 (Berlin: Akademie Verlag, 2002).

5. Cod. 324, Nationalbibliothek, Vienna; Ekkehard Weber, *Tabula Peutingeriana: Codex Vindobonensis 324* (Graz: Akademisch Druck-und Verlagsanst, 1976), includes a facsimile of the map.

6. Patrick Gautier Dalché, "La Trasmissione Medievale e Rinascimentale della Tabula Peutingeriana," in *Tabula Peutingeriana: Le Antiche Vie del Mondo*, ed. Francisco Prontera (Florence: Leo Olschki, 2003), 43–53.

7. Emily Albu, "Imperial Geography and the Medieval Peutinger Map," *Imago Mundi* 57, no. 2 (2005): 136–48.

8. Alessandro Scafi, "Mapping Eden: Cartographies of Earthly Paradise," in *Mappings*, ed. Denis E. Cosgrove (London: Reaktion, 1999), 51–70.

9. For various meanings and uses of the term, see David Woodward, "Medieval Mappaemundi," in *Cartography in Prehistoric, Ancient, and Medieval Europe and the Mediterranean*, ed. David Woodward and J. B. Harley, vol. 1 of *History of Cartography* (Chicago: University of Chicago Press, 1987), 286–88.

10. Add. MS 28681, fols. 9r–9v, British Library, London.

11. A complete list of world maps, including lost ones, may be found in Harley and Woodward, *History of Cartography*, 1:359–68. Add to this list: Duchy of Cornwall, Aslake, Chalivoy-Milon, San Pedro de Rocas, and two maps at Reims described by Patrick Gautier Dalché in "L'oeuvre géographique du Cardinal Fillastre," in *Humanisme et Culture Géographique à l'Époque du Concile de Constance*, ed. Didier Marcotte (Turnhout: Brepols, 2002), 307, 336. For the map at Chalivoy-Milon, see Marcia Kupfer, "The Lost Mappamundi at Chalivoy-Milon," *Speculum* 66, no. 2 (June 1996): 286–310.

12. This note appears on a rough copy he made of the world map at Westminster, MS 26, p. 284, Corpus Christi College, Cambridge.

13. Jeffrey B. Russell, *Inventing the Flat Earth* (New York: Praeger, 1997); Rudolf Simek, *Heaven and Earth in the Middle Ages*, trans. Angela Hall (Woodbridge, Suffolk: Boydell, 1996), chap. 3; Wesley M. Stevens, "Figure of the Earth in Isidore's De Natura Rerum," *Isis* 71 (1980): 268–77.

14. For *continents* as a modern term, see Martin W. Lewis and Kären E. Wigen, *The Myth of Continents: A Critique of Metageography* (Berkeley and Los Angeles: University of California Press, 1997). On the issue of the continent of Africa, see Francesc Relaño, *The Shaping of Africa* (Burlington, VT: Ashgate, 2001), 2, 16.

15. James M. Scott, *Geography in Early Judaism and Christianity: The Book of Jubilees* (Cambridge: Cambridge University Press, 2002). Scott reprints the relevant text in chap. 2, pp. 28–32.

16. John Williams disagrees, asserting that the body of water is the "interior ocean" and that the so-called fourth continent is actually part of Africa ("Isidore, Orosius, and the Beatus Map," *Imago Mundi* 49 (1997): 7–32, esp. 17–23).

17. Simek, *Heaven and Earth*, 125.

18. Presumably these are "mille passuum" or Roman miles. This works out to about 303 by 292 English miles, which is too small. Scott D. Westrem, *The Hereford Map: A Transcription and Translation of the Legends with Commentary* (Turnhout: Brepols, 2001), item no. 552.

19. Alfred W. Crosby, *Measure of Reality: Quantification and Western Society, 1250–1600* (Cambridge: Cambridge University Press, 1997), esp. chap 5.

20. MS 2*, Corpus Christi College, Oxford.

21. David Woodward and Herbert Howe, "Roger Bacon on Geography and Cartography," in *Roger Bacon and the Sciences: Commemorative Essays,* ed. Jeremiah Hackett (Leiden: E. J. Brill, 1997), 208–10.

22. Dante Alighieri, *The Divine Comedy: Inferno*, trans. John D. Sinclair (New York: Oxford University Press, 1939), canto 26.

23. Stephen McKenzie, "Conquest Landmarks and the Medieval World Image" (University of Adelaide, Ph.D. diss., 2000). Examples of such maps include: CLM 14731, fol. 83v (German, 12th cent.), Bayerische Staatsbibliothek, Munich; *Imago Mundi* of Honorius Augustodunensis (13th cent.), MS e Mus. 223, fol. 185, Bodleian Library, Oxford. On the far south and north, see also Danielle LeCoq, "Place et Fonction du Désert dans la Représentation du Monde au Moyen Age," *Révue des Sciences Humaines* 2 (April/June 2000), 15–112.

24. "Ista est Hierusalem. In medio gentium posui eam et in circuitu eius terras." *Biblia Sacra Vulgata* (Stuttgart: Deutsche Bibelgesellschaft, 1983).

25. Isidore, *Etymologiae*, ed. W. M. Lindsay (Oxford: Clarendon, 1911), XIV.iii.21.

26. Ingrid Baumgärtner, "Die Wahrnehmung Jerusalems auf Mittelalterlichen Weltkarten," in *Jerusalem im Hoch-und Spätmittelalter*, ed. Dieter Bauer, Klaus Herbers, and Nikolas Jaspert (Frankfort/New York: Campus, 2001), 271–333, esp. 294, 309–10; also Woodward, "Medieval Mappaemundi," 340–42; Anna-Dorothee von den Brincken, "Jerusalem: A Historical as Well as an Eschatological Place on Medieval Mappae Mundi," in Harvey, *The Hereford World Map.*

27. See chapter 6 on the Fra Mauro map of 1450.

28. A handy transcription of all the marginal texts on the map can be found in Paul D. A. Harvey, *Mappa Mundi: The Hereford World Map* (London: British Library, 1996), p. 54, as well as in Westrem, *The Hereford Map.*

29. Graham Haslam, "The Duchy of Cornwall Map Fragment," in *Géographie du Monde au Moyen Âge et à la Renaissance*, ed. Monique Pelletier (Paris: Éditions du C.T.H.S., 1989), illustration on p. 42.

30. These texts are interpreted by Konrad Miller, *Mappaemundi: Die Ältesten Weltkarten* (Stuttgart: J. Roth, 1896), 5:10. See also Armin Wolf, "News on the Ebstorf World Map," in Pelletier, *Géographie*, 67.

31. Add. MS 28681, fols. 9r–v British Library, London. For reproductions of these maps, see Evelyn Edson, *Mapping Time and Space: How Medieval Mapmakers Viewed Their World* (London: British Library, 1997), plate VI and fig. 7.1.

32. Roger Mason thinks it may be a self-portrait of Richard, who commissioned the map (presentation at the Hereford Mappamundi Conference, Hereford, July 1999). Valerie Flint has proposed that it refers to a hunting-rights dispute involving Thomas of Cantilupe, former bishop of the diocese. She thinks the drawing has a double meaning as part of a campaign by the succeeding bishop to get Thomas canonized. The rider (Thomas) gestures at the world while moving away from it, symbolizing the view that the world is a metaphor for life's pilgrimage. See Valerie Flint, "The Hereford Map: Two Scenes and a Border," *Transactions of the Royal Historical Society,* 6th ser., 8 (1998): 19–44. On these theories, see Westrem, *The Hereford Map,* p. xxv.

33. The map thought to be the closest to the original version is MS 1 in the collection of the Cathedral at Burgo de Osma, Spain. The way the apostles are pictured, however, suggests reliquaries, or their traditional burial sites, rather than preaching. For a reproduction, see John Williams, *The Illustrated Beatus* (London: Harvey Miller, 1998), IV, fig. 5.

34. Westrem, *The Hereford Map,* item nos. 776 and 785.

35. G. R. Crone, "New Light on the Hereford Map," *Geographical Journal* 131, no. 4 (December 1965): 447–62, esp. 451–52.

36. Westrem, *The Hereford Map,* no. 725.

37. David Abulafia and Nora Berends, eds., *Medieval Frontiers: Concepts and Practices* (Aldershot: Ashgate, 2002).

38. Gervase of Tilbury, *Otia Imperialia,* ed. and trans. S. E. Banks and J. W. Binns (Oxford: Clarendon, 2002), 527.

39. Westrem, *The Hereford Map,* no. 141.

40. The standard work on this subject is John B. Friedman, *The Monstrous Races in Medieval Art and Thought* (Cambridge, MA: Harvard University Press, 1981), esp. chap. 3 on world maps.

41. Or Psylli, according to Solinus, *The Excellent and Pleasant Worke: Collectanea Rerum Memorabilium,* trans. Arthur Golding (1587; reprint, Gainesville, Fl: Scholars' Facsimiles and Reprints, 1955), 27.42. Westrem, *The Hereford Map,* no. 969.

42. Westrem, *The Hereford Map,* no. 202; Solinus, *The Excellent and Pleasant Worke,* 25.1–5.

43. Westrem, *The Hereford Map,* no. 95. The text in the Christian *Physiologus* completes the inscription: "A casu testamenti salvatur homo vere Christi" (Man is truly saved by the testament of Christ). Naomi Reed Kline has an excellent chapter on the influence of bestiaries in *Maps of Medieval Thought* (Woodbridge: Boydell, 2001), chap. 4. See also Margriet Hoogvliet, "Animals in Context: Beasts on the Hereford Map and Medieval Natural History," in Harvey, *The Hereford World Map,* 153–65.

44. Friedman has a most interesting account of the role of dog-heads in Christian thought (*The Monstrous Races,* 61–75).

45. The marvels are depicted in MS Cotton Tiberius B.V-1, British Library, London. See also Alexander, *Epistola ad Aristotelem,* ed. W. Walter Boer (Meisenheim: Anton Hain, 1973).

46. For Modena, see Jeanne Fox-Friedman, "Visions of the World: Romanesque Art of Northern Italy and the Hereford Mappamundi," in Harvey, *The Hereford World Map,* 137–51. Kline, *Maps of Medieval Thought,* 133–38.

47. Westrem, *The Hereford Map,* no. 138.

48. Ibid., no. 157.

49. Ibid., no. 142.

50. Ibid., no. 536.

51. See Harvey, *Mappa Mundi*, 41–53; and Westrem, *The Hereford Map*, xxvii–xxxvii, for further discussion of sources.

52. Natalia Losovsky, *The Earth is Our Book: Geographical Knowledge in the Latin West ca. 400–1000* (Ann Arbor: University of Michigan Press, 2000).

53. These are described by Jerome, *Liber de Situ et Nominibus Locorum Hebraicorum*, in *Onomastica Sacra: Studia et Sumptibus Alterum Edita*, ed. Pauli Lagarde (Göttingen: Horstmann, 1887), 150, 155, 156, 186.

54. Scafi, "Mapping Eden," 58–59.

55. Add. MS 10049, fols. 64 r–v British Library, London; Miller, *Mappaemundi*, 3:1–21; Harvey, "The 12th Century Jerome Maps of Asia and Palestine," Seventeenth Annual Conference on the History of Cartography, Lisbon, July 10, 1997. Harvey has made a thorough physical examination of these maps, noting numerous erasures and corrections. He concludes that there is no plausible connection to Jerome.

56. Paulus Orosius, *Seven Books of History Against the Pagans*, ed. and trans. Roy J. Deferrari (Washington, DC: Catholic University Press, 1964).

57. Harvey, *Mappa Mundi*, 45.

58. Losovsky, *The Earth Is Our Book*, 78.

59. Ibid., 54; Edson, *Mapping Time and Space*, 46–50.

60. Patrick Gautier Dalché, "Décrire le monde et situer les lieux au XIIe siècle: L'Expositio mappe mundi et la généalogie de la mappemonde de Hereford," *Mélanges de l'ecole Française de Rome: Moyen Age* 113, no. 1, (2001): 343–409; discussed by Westrem, *The Hereford Map*, xxxiv–xxxvii. The manuscripts discussed are MS latin 3123, fols. 126r–131v, Bibliothèque Nationale Française, Paris, and MS 344, fols. 52va.–56va., Bibliothèque Municipale, Valenciennes. The *Expositio* has now appeared as one of three documents attributed to Roger of Hoveden in a critical edition by Patrick Gautier Dalché: *Du Yorkshire à l'Inde: Une "Géographie" Urbaine et Maritime de la Fin du XIIe Siècle (Roger de Howden?)* (Geneva: Droz, 2005).

61. Gautier Dalché, "Décrire le monde," 380.

62. Roger Bacon, *Opus Majus*, ed. John Henry Bridges (Oxford: Clarendon, 1897; reprint, Frankfurt-am-Main: Minerva, 1964). On Rubruck, see pp. 303, 354–55; on Sallust as a source for African geography, see p. 315; on the making of a grid-based map, see pp. 296–300.

63. See Bacon, *Opus Majus*, 183–87, on the spiritual purposes of studying geography.

64. At the 1999 conference, participating scholars were given the rare privilege of spending twenty minutes with the map, outside the case, and allowed the use of a ladder.

65. Marcia Kupfer, "Medieval World Maps: Embedded Images, Interpretive Frames," *Word and Image* 10, no. 3 (July–Sept. 1994): 262–88.

66. Kline, *Maps of Medieval Thought*, 76–78, proposed this idea. Dan Terkla and Dominic Harbour have since found physical evidence consistent with its original mounting in the north transept, alongside the tomb of the recently sainted bishop. See Dan Terkla, "The Original Placement of the Hereford Mappa Mundi," *Imago Mundi* 56, no. 2 (2004): 131–51.

67. Patrick Gautier Dalché, *La "Descriptio Mappe Mundi" de Hugues de Saint-Victor* (Paris: Etudes Augustiniennes, 1988), 100–107, gives evidence of Hugh of Saint Victor lecturing before a mappamundi.

68. "Que scilicet non parvam prestat legentibus utilitatem, viantibus directionem rerumque viarum gratissime speculationis directionem (or dilectionem)," Miller, *Mappaemundi*, 5:8; dis-

cussed in Catherine Delano-Smith and Roger Kain, *English Maps* (London: British Library, 1999), 31.

69. Gautier Dalché, "Décrire le monde," 348; Gautier Dalché, *Du Yorkshire à l'Inde*, has all three of these texts.

70. C. S. Lewis, *Discarded Image* (Cambridge: Cambridge University Press, 1967), 98–100, 111–12, on the sense of meaning in the medieval universe.

Two • *Marine Charts and Sailing Directions*

1. Edward L. Stevenson, *Portolan Charts: Their Origin and Characteristics* (New York: Hispanic Society of America, 1911), 2.

2. Rés. Ge B 1118, Cartes et Plans, Bibliothèque Nationale, Paris.

3. Inv. 100, Biblioteca dell'Accademia Etrusca, Cortona.

4. But see Tony Campbell, "Portolan Charts from the Late Thirteenth Century to 1500," in *Cartography in Prehistoric, Ancient, and Medieval Europe and the Mediterranean*, ed. J. B. Harley and David Woodward, vol. 1 of *History of Cartography* (Chicago: University of Chicgo Press, 1987), 371–463, esp. 406, where he dates Carignano to 1329–30.

5. Campbell, "Portolan Charts," 390. On marine charts, see especially the excellent chapter in Harley and Woodward, *Cartography*; see also Jonathan T. Lanman, *On the Origin of Portolan Charts* (Chicago: Newberry Library, 1987); George Kish, *La Carte: Image des Civilisations* (Paris: Seuil, 1980); Eva G. R. Taylor, *The Haven-Finding Art: A History of Navigation from Odysseus to Captain Cook*, rev. ed. (New York: Elsevier Publications, 1971); Theobald Fischer, *Sammlung Mittelalterlicher Welt-und Seekarten: Italienischen Ursprungs aus Italienischen Bibliotheken und Archiven* (Venice: F. Ongania, 1886); and Konrad Kretschmer, *Die Italienischen Portolane des Mittelalters: Ein Beitrag zur Geschichte der Kartographie und Nautik* (1909; reprint, Hildesheim: G. Olms, 1962). Good illustrations can be found in Michel Mollat du Jourdin and Monique de la Roncière, with Marie-Madeleine Azard, Isabelle Raynaud-Nguyen, and Marie Antoinette Vannereau, *Sea Charts of the Early Explorers* (London: Thames and Hudson, 1984) mostly from French collections, and Paul D. A. Harvey, *Medieval Maps* (London: British Library, 1991).

6. Campbell, "Portolan Charts," 388–89.

7. Ibid., 390; Piero Falchetta, "Marinai, Mercanti, Cartografi, Pittori: Ricerche sulla Cartografia Nautica a Venezia (sec. XIV–XV)," *Ateneo Veneto: Rivista di Scienze, Lettere ed Arti*, n.s., 183, no. 33 (1995): 74–76.

8. W. G. L. Randles, "De la Carte-Portulan Méditerranéenne à la Carte Marine du Monde des Grandes Découvertes: La Crise de la Cartographie au XVIe Siècle," in *Géographie du Monde au Moyen Age et à la Renaissance*, ed. Monique Pelletier (Paris: International Cartographic Association, 1989), 25–32.

9. For examples of rhumb-line charts from the sixteenth and seventeenth centuries, see Mollat du Jourdin and de la Roncière, *Sea Charts*.

10. Although see Campbell, "Portolan Charts," 382n107, for the earlier existence of Aigues Mortes.

11. Campbell, "Portolan Charts," 398–401.

12. Bacchisio R. Motzo, ed., *Il compasso da navigare*, in Facoltà di Lettere e Filosofia, Università di Cagliari Annali 8 (1947), xxxviii.

13. Mollat du Jourdin and de la Roncière, *Sea Charts*, plate 11 and p. 204.

14. Taylor, *The Haven-Finding Art*, 169.

15. The first reference I have found to the use of the term *portolan* is in a late fifteenth-century pilot book, printed in Venice: "Lo libro chimado portolano" Taylor Collection #30, Beinecke Library, Yale University.

16. MS Hamilton 396, Staatsbibliothek Preussischer Kulturbesitz, Berlin; Motzo, *Il compasso*, xxvi.

17. On Compasso, see Taylor, *The Haven-Finding Art*, 104–9; Campbell, "Portolan Charts," 382–83; and Patrick Gautier Dalché, *Carte Marine et Portulan au XIIe Siècle: Le Liber de Existencia Riveriarum et Forma Maris Nostri Mediterranei (Pise, c. 1200)*, in *Collection de l'École Française de Rome* 203 (Rome: École Française de Rome, 1995), 39–44, and appendices 3–6. Gautier Dalché reads the text as *conpasso*.

18. Motzo, *Il compasso*, 60–61. "La Licia è porto con catena e à entrata da maestro. En bocca de lo porto da tramontana à una torre. La predicta Gloriata è bono ponedore da tramontana. De la dicta Licia a Valenia XV millara per meczo zorno." "Suri è porto, et à pluzori escolli da maestro, li quali scolli devete tucti largare da meczo dì."

19. Motzo, *Il compasso*, li.

20. Ibid., xl, xliv, li.

21. Campbell, "Portolan Charts," 382–83; Simonetta Conti, "Portolano e Carta Nautica: Confronto Toponomastica," in *Imago et Mensura Mundi*, ed. Carla Clivia Marzoli, Atti del IX Congresso Internazionale di Storia della Cartografia (Rome: Enciclopedia Italiana, 1985), 1:60.

22. Cl. XI, 87 (7353), Biblioteca Marciana, Venice; reprinted in Gautier Dalché, *Carte Marine*, app. 1, pp. 181–82.

23. MS 506, Beinecke Library, Yale University. The list is in the Venetian dialect and also includes ports on the Black Sea and the eastern Mediterranean islands.

24. Annalisa Conterio, "L'Arte del Navegar: Cultura, Formazione Professionale ed Esperienza dell'Uomo di Mari Veneziano nel XV Secolo," in *L'Uomo e il Mare nella Civiltà Occidentale: Da Ulisse a Cristoforo Colombo*, Atti della Società Ligure di Storia Patria, n.s., 32 (1992): 187–225.

25. These traverses are also known as *peleio* or *pileggio*. See Gautier Dalché, *Liber*, 79–80, and maps at the back of the book.

26. Gautier Dalché, *Carte Marine*, 44. Numerous references to *pieleghi* (variously spelled) can be found in a pilot book in the Taylor Collection (#30), Beinecke Library, Yale University. This is a text printed in Venice in 1490. There is no title, but the opening line is "Questa e una opera necessaria a tutti li naviganti" (This is an essential book for all sailors).

27. Jonathan Lanman, *On the Origin of Portolan Charts* (Chicago: Newberry Library, 1987).

28. Lionel Casson, *Periplus Maris Erythraei* (Princeton, NJ: Princeton University Press, 1989), 59.

29. Konrad Miller, *Itineraria Romana: Römische Reisewege an der Hand der Tabula Peutingeriana* (1916; reprint, Rome: Bretschneider, 1964), lxvii–lxviii.

30. Francesco Prontera, "Períploi: Sulla Tradizione della Geografica Nautica presso i Greci," in *L'Uomo e il Mare nella Civiltà Occidentale: Da Ulisse a Cristoforo Colombo*, Atti della Società Ligure di Storia Patria, n.s., 32, (1992): fasc. 2, p. 39; excerpts in A. E. Nordenskiold, *Periplus: An Essay on the Early History of Charts and Sailing Directions*, trans. Francis A. Bather (Stockholm: P. O. Norstedt and Söner, 1897), 11–14.

31. Benedetto Cotrugli, *De Navigatione Liber* (1464–65), MS 557, fol. 43v, Taylor Collection, Beinecke Library, Yale University.

32. Gautier Dalché, *Carte Marine*, 43.

33. Prontera, "Períploi," 33, 36.

34. Gautier Dalché, *Carte Marine*, 43.

35. Anna Comnena, *The Alexiad*, trans. E. R. A. Sewter, (Baltimore: Penguin, 1969), 13.7, p. 415.

36. G. L. Huxley, "A Porphyrogenitan Portolan," *Greek, Roman and Byzantine Studies*, 17 (1976): 295–300, describes a brief tenth-century document on sea mileage, largely incorrect, between various points on a voyage from Constantinople to Crete.

37. P. A. Jaubert, ed., *Geographie d'Edrisi* (Paris, 1836); reprinted as vols. 2 and 3 of Fuat Sezgin, ed., *Islamic Geography* (Frankfurt-am-Main: Institute for History of Arabic-Islamic Science, 1992).

38. Gautier Dalché, *Carte Marine*, 64. On Arab cartography, see chapters by Gerald Tibbetts and S. Maqbul Ahmad, in *Cartography in the Traditional Islamic and South Asian Societies*, ed. J. B. Harley and David Woodward, vol. 2 of *The History of Cartography* (Chicago: University of Chicago Press, 1992): "Cartography of al-Sharīf al-Idrīsī" by S. Maqbul Ahmad; and "The Beginnings of Cartographic Tradition," "The Balkhi School of Geographers," "Later Cartographic Developments," and "The Role of Charts in Islamic Navigation in the Indian Ocean" by G. Tibbetts.

39. Fuat Sezgin, "Arabischer Ursprung europäischer Karten," *Cartographica Helvetica* 24 (July 2001): 21–28.

40. Tibbetts, "Later Cartographic Developments," 153, fig. 6.14.

41. Campbell, "Portolan Charts," 390.

42. MS Cotton Domitian A.13, fols. 114r–129v, British Library, London. Bound with other works of history and divinity, *Liber* is an independent foliation and unrelated to them. Gautier Dalché, *Carte Marine*, 5–6.

43. Gautier Dalché, *Carte Marine*, 116, 126.

44. "Hoc Nili flumen originem habet a monte inferioris Mauritanie Affrice quod occeano appropinquat. Hoc affirmant Punici libri, hoc Iubam regem accipimus tradidisse. Mox a terra absorbetur, per quam occultomeatu procedens in littore Rubri maris denuo funditur, Ethiopiam circumiens per Egyptum labitur, et sic in .vi. ostia divisus hoc mare, ut supra diximus, iuxta Alexandriam Mediterraneum ingreditur." Ibid., 125, lines 494–501.

45. "A Cesarea vero ad montem Carmelum .xx., caput sinus Acconis, de quo monte in Canticis: 'Collum tuum ut Carmelus,' et in quo conversari voluit Elias et Eliseus." Ibid., 128, lines 611–13.

46. Ibid., 116, line 193.

47. Patrick Gautier Dalché, *Du Yorkshire à l'Inde: Une "Geographie" Urbaine et Maritime de la Fin du XIIe Siècle (Roger de Howden?)* (Geneva: Droz, 2005).

48. Gautier Dalché, *Carte Marine*, 53. See also his appendix 2, pp. 183–203.

49. A brief summary in Taylor, *The Haven-Finding Art*, 78–85.

50. *Konungs Skuggsjá*, or *King's Mirror*, trans. L. M. Larson (New York: American Scandinanvian Fdn., 1917).

51. Arne E. Christensen, "Ships and Navigation," in *Vikings: The North Atlantic Saga*, ed. William Fitzhugh and Elisabeth Ward (Washington, DC: Smithsonian Institution, 2000), 97.

52. *De Expugnatione Lyxbonensi*, ed. Charles W. David (New York: Columbia University Press, 1936), 121.

53. Gautier Dalché, *Carte Marine*, 46.

54. Goro Dati, *La Sfera*, ed. Enrico Narducci (1865; reprint, Milan: G. Daelli, 1865), 3.4. Here the work is attributed to his brother Leonardo Dati.

55. "Les croisés découvraient un monde qui sans doute les fascinait et par là suscitait leur curiosité, bouleversant ainsi les schémas mentaux précédement acquis." Gautier Dalché, *Carte Marine*, 54.

56. See Conterio, "L'Arte del Navegar," 199–204, for the case of Michele da Rodi.

57. Scott D. Westrem, *The Hereford Map: A Transcription and Translation of the Legends with Commentary* (Turnhout, Belgium: Brepols, 2001), 17.

58. Barbara Obrist, "Wind Diagrams and Medieval Cosmology," *Speculum* 72 (1997): 33–84.

59. Example from *Compasso*, given by Taylor, *The Haven-Finding Art*, 106.

60. Cotton MS Nero D.5, last folio, British Library, London; E. G. R. Taylor, "The *De Ventis* of Matthew Paris," *Imago Mundi* 2 (1937): 23–26.

61. Dati, *La Sfera*, 3.3.

62. *The Log of Christopher Columbus*, trans. Robert H. Fuson (Camden, ME: International Marine Publishing Company, 1987), 44; James E. Kelley Jr., "Columbus's Navigation: Fifteenth-Century Technology in Search of Contemporary Understanding," in *Columbus and the New World*, ed. Joseph C. Schnaubelt and Frederick Van Fleteren (New York: Peter Lang, 1998), 122.

63. Patrick Gautier Dalché, "D'une Technique à une Culture: Carte Nautique et Portulan au XIIe et au XIIIe siècle," in *L'Uomo e il Mare nella Civiltà Occidentale: Da Ulisse a Cristoforo Colombo*, Atti della Società Ligure di Storia Patria, n.s., 32 (1992): fasc. 2, p. 293n31.

64. Jean de Joinville, *The Life of Saint Louis*, in *Chronicles of the Crusades*, trans. M. R. B. Shaw (Baltimore: Penguin, 1963), 319–20.

65. "Qui ergo munitiam vult habere navem . . . habet etiam acum jaculo superpositam; rotabitur enim et circumvolvetur, donec cuspis acus respiciat septentrionem, sicque comprehendent quo tendere debeant nautae, cum Cynosura latet in aeris turbatione, quamvis ea occasum numquam teneat propter circuli brevitatem" (Alexander Neckham, *De Nominibus Utensilium* [c. 1175–83], quoted in Nordenskiold, *Periplus*, 49–50).

66. Taylor, *The Haven-Finding Art*, 116; Frederic C. Lane, "The Economic Meaning of the Invention of the Compass," *American Historical Review* 68, no. 3 (April 1963): 605–17.

67. MS. fr. 2810 (1403), Bibliothèque Nationale, Paris.

68. Taylor, *The Haven-Finding Art*, 172–74, 181.

69. "Q: Marinarii quomodo mensurant millaria in mari? R: Marinarii considerant IIII ventos generales, videlicet ventum orientalem, occidentalem, meriodionalem et septentrionalem; similter alios IIII ventos qui ex primis exeunt, viz. grecum, exalochum, lebeig et maestre. Et centrum circuli considerant in quo venti angulos faciunt; de inde per ventum orientalem navem euntem centum miliaria (sic) a centro, quot sunt miliaria usque ad ventum de exaloch. Et miliaria duplicant usque ad ducenta miliaria, et cognoscunt quot miliaria sunt multiplicata [que sunt ducenta] a vento orientali usque ad ventum de exaloch per multiplicationem miliarium que sunt de termini centenario orientis usque ad terminum de exaloch. Et ad hoc instrumentum habent chartam, compassum, acum et stella maris." Quotation from Motzo, *Il compasso*, lv; see also Taylor, *The Haven-Finding Art*, 118.

70. MS It. Z 76, fol. 1, Biblioteca Marciana, Venice; Conterio, "L'Art del Navegar," 196. A clear description of the use of the *marteloio* and an illustration is in Taylor, *The Haven-Finding Art*, 117–21, and plate 8.

71. Kelley, "Columbus's Navigation," 120–21.

72. Tullia Gasparrini Leporace, *Il Mappamondo di Fra Mauro* (Rome: Istituto Poligrafico dello Stato, 1956), 25.

73. Claudio de Polo Saibanti, "Arte del Navigare: Manuscritto Inedito Datato 1464–65," in *Imago et Mensura Mundi*, ed. Carla Clivio Marzoli, Atti del IX Congresso Internazionale di Storia della Cartografia 1 (Rome: Enciclopedia Italiana, 1981): 71–79; Conterio, "L'Arte del Navegar;" Falchetta, "Marinai," 16–18. The complete text of one of these has been edited by Conterio: *Pietro di Versi, Raxion de' Marineri: Taccuino Nautico del XV Secolo*, Fonti per la Storia di Venezia, sez. v (Venice: Comitato per la Pubblicazione delle Fonti Relative alla Storia di Venezia, 1991).

74. Saibanti, "Arte del Navigare," 77; the manuscript is in a private collection. The same theme, even in the same words, is apparent in the work of Benedetto Cotrugli, "Della carta da navigare," *De Navigatione Liber* (1464–65), MS 557, fols. 61v–62r, Taylor Collection, Beinecke Library, Yale University,

75. Campbell, "Portolan Charts," 427–28. The Beccari portolan is in the Beinecke Library at Yale University (MS 1980.158).

76. Leporace, *Il Mappamondo*, 25–27, and throughout. This work contains a facsimile of the map and a complete transcription of the legends.

77. Geoffrey Chaucer, *A Treatise on the Astrolabe* (1391), ed. Walter W. Skeat (1872; reprint, London: Early English Text Society, 1968).

78. On quadrants, see A. J. Turner, "Astrolabes" in *The Time Museum Catalogue: Time Measuring Instruments* (Rockford, IL: The Time Museum, 1985), 202–28; on mariners' astrolabes, see Roderick Webster and Marjorie Webster, *Western Astrolabes* (Chicago: Adler Planetarium, 1998), 148–51.

79. Kelley, "Columbus's Navigation," 135–37.

80. MS Ayer 746 in the Newberry Library, Chicago, contains a mid-fifteenth-century Catalan treatise on the astrolabe. See chapters 34 and 35 for surveying techniques. It is the opinion of Derek J. Price that the astrolabe was not used for surveying in the Middle Ages because the process was too complicated. He also believes that the passages on surveying in medieval astrolabe texts were introduced "by teachers of geometry in order to make their subject seem of practical use" ("Medieval Land Surveying and Topographical Maps," *The Geographical Journal* 121, no. 1 (March 1955): 1–10, esp. 2.

81. For examples, see Mollat du Jourdin and de la Roncière, *Sea Charts*. Tony Campbell says that no surviving chart with a latitude scale can be dated before the sixteenth century ("Portolan Charts," 386). Alison Sandman describes the resistance of pilots to abandoning the rhumbline chart; see, e.g., "An Apologia for the Pilots' Charts: Politics, Projections, and Pilots' Reports in Early Modern Spain," *Imago Mundi* 56, no. 1 (2004): 7–22.

82. Campbell, "Portolan Charts," 372–73, 415–24.

83. Ibid., 407–9.

84. Mollat du Jourdin and de la Roncière, *Sea Charts*, plate 7 and p. 201; Rés. Ge B 696, Cartes et Plans, Bibliothèque Nationale, Paris.

85. Add. MS 15760, fols. 68v–69, British Library, London; see fig. 8.2.

86. Campbell, "Portolan Charts," 415.

87. Opicinus's work was first noticed by Robert Almagià, "Intorno all più Antica Carto-grafica Nautica Catalan," *Bollettino della Reale Società Geografica Italian*, 7th ser., 10 (1945): 20–27; also Richard G. Salomon, "A Newly Discovered Manuscript of Opicinus de Canistris: A Preliminary Report," *Journal of the Warburg and Courtauld Institute*, 16 (1953): 45–57; and "Aftermath to Opicinus de Canistris," *Journal of the Warburg and Courtauld Institute*, 25 (1963): 137–46.

88. Falchetta, "Marinai," throughout and p. 68.

89. Campbell, "Portolan Charts," 430, 434.

90. Guillaume de Nangis, *Gesta sancti Ludovici*, in *Recueil des Historiens des Gaules et de la France* 20 (Paris: Imprimerie Royale, 1840): 444.

91. Campbell, "Portolan Charts," 439.

92. Cited by Campbell, "Portolan Charts," 430, and note: "Non modo utile verum etiam necessarium sit Januensibus navigantibus."

93. Falchetta, "Marinai," 27.

94. Frederic C. Lane, *Venice: A Maritime Republic* (Baltimore: Johns Hopkins University Press, 1973), 216–17.

95. *Summa Theologica*, part 2:2, "On Cheating in Buying and Selling," question 77, 4th article: Whether in trading it is lawful to sell a thing at a higher price than was paid for it? Trans. Fathers of the English Dominican Province (Benziger Brothers, 1947). Available at www.ccel.org/a/Aquinas/summa/home.html (accessed September 20, 2006).

96. John K. Wright, *Geographical Lore of the Time of the Crusades* (New York: American Geographical Society, 1925), 70–71, but see also 292–93.

Three • Sea Chart and Mappamundi in the Fourteenth Century

1. David Jacoby, "L'Expansion Occidentale dans le Levant: Les Vénetiens à Acre dans le second moietié du treizième siècle," *Journal of Medieval History* 3 (1977): 225–64, esp. 250.

2. Dante, *The Divine Comedy: Paradiso*, trans. John D. Sinclair (New York: Oxford University Press, 1961), canto 15, lines 139–44; A. S. Atiya, *The Crusade in the Later Middle Ages* (London: Butler and Tanner, 1938); Antony Leopold, *How to Recover the Holy Land: The Crusade Proposals of the Late 13th and Early 14th Centuries* (Aldershot: Ashgate, 2000).

3. The title changes in various manuscripts: Fausta Gualdi, "Marin Sanudo Illustrato," *Commentari*, n.s., 20 (July–September 1969): fasc. 3, 162–98, esp. 166. The most accessible edition of Sanudo's book is a reprint of the 1621 Hanover edition with a foreword by Joshua Prawer: Marino Sanutus (Sanudo) dictus Torsellus, *Liber Secretorum Fidelium Crucis*, ed. Joshua Prawer (Toronto: University of Toronto Press, 1972). This edition includes fine color prints of the maps from the British Library manuscript. For more information on the content of Sanudo's book, see Evelyn Edson, "Reviving the Crusade: Sanudo's Schemes and Vesconte's Maps," in *Eastward Bound: Travel and Travellers, 1050–1550*, ed. Rosamund Allen (Manchester: Manchester University Press, 2004), 131–55.

4. MS Vat. lat. 2972, Biblioteca Apostolica Vaticana, Vatican City.

5. This is MS Pal. lat. 1362A, Biblioteca Apostolica Vaticana, Vatican City. See Bernhard Degenhart and Annegrit Schmitt, "Marino Sanudo und Paolino Veneto," *Römisches Jahrbuch für Kunstgeschichte* 14 (1973): 1–137, esp. 23 and 64.

6. Christopher J. Tyerman, "Marino Sanudo Torsello and the Lost Crusade: Lobbying in the 14th Century," *Transactions of the Royal Historical Society,* 5th ser., 32 (1982): 57–73.

7. The letter appears in F. Kunstmann, "Studien über Marino Sanudo den Älteren mit einem Anhangseiner ungegedruckten Briefen," *Abhandlungen, Phil.-Historische Classe, Königliche Bayerische Akademie der Wissenschaften* 7 (1853): 794. The translation is by Frank Frankfort, "Marino Sanudo Torsello: A Social Biography" (unpublished Ph.D. diss., University of Cincinnati, 1974), 223.

8. Reprinted and translated in Frankfort, "Marino Sanudo Torsello," app. B.

9. Sanudo, *Liber Secretorum,* II.iv.25, pp. 85–90.

10. Rés Ge DD 687, Cartes et Plans, Bibliothèque Nationale, Paris.

11. MS Pal. Lat. 1362A, Biblioteca Apostolica Vaticana, Vatican City.

12. Piero Falchetta, "Marinai, Mercanti, Cartografi, Pittori: Ricerche sulla Cartografia Nautica a Venezia (sec. 14–15)," *Ateneo Veneto: Rivista di Scienze, Lettere ed Arti,* n.s., 33 (1995): 7–109, esp. 34–37.

13. This was the work of Konrad Kretschmer, who established the similarity between the maps in Sanudo's works and other, signed maps by Vesconte. Konrad Kretschmer, *Die italienischen Portolane des Mittelalters: Ein Beitrag zur Geschichte der Kartographie und Nautik* (Reinheim: Lokay, 1909; reprint, Hildesheim: G. Olms, 1962), 113–16.

14. Sanudo, *Liber Secretorum,* III.xiv.3.

15. Roberto Almagià, *Monumenta Cartographica Vaticana* (Vatican City: Biblioteca Apostolica Vaticana, 1944), 18, says that the text was added later and composed from the map.

16. Kenneth Nebenzahl, *Maps of the Holy Land: Images of Terra Sancta Through Two Millennia* (New York: Abbeville, 1986), 43, 58–59.

17. Burchard, *A Description of the Holy Land,* ed. by Aubrey Stewart (London: Palestine Pilgrims' Text Society, 1896; reprint, New York: AMS Press, 1971), 12.

18. Rehav Rubin, *Image and Reality: Jerusalem in Maps and Views* (Jerusalem: Magnes Press, 1999), 36.

19. See Alessandro Scafi, "Paradise: The Essence of a Mappamundi," in Paul Harvey, ed., *The Hereford World Map: Medieval World Maps and Their Context* (London: British Library, 2006). Scafi, *Mapping Paradise: A History of Heaven on Earth* (Chicago: University of Chicago Press, 2006).

20. Add. MS 27,376, fols. 8v–9, British Library, London.

21. Nathalie Bouloux, *Culture et Savoirs Géographiques en Italie au XIVe Siècle* (Turnhout: Brepols, 2002), 66.

22. Reprinted in Sanudo, *Liber secretorum,* 285–87.

23. "Moderni Sitiam et Yrcaniam, partes que adiacentes, aliter dividunt, & nominant: ponentes, ubi Sitia, regnum Catay quod ab Oriente, habet Oceanum; a Meridie, Insulas Oceani; ab Occidente, regnum Tarsae; a Septentrione, desertum de Belina" (Sanudo, *Liber secretorum,* 285).

24. Hayton (Hetoum), *La Flor des Estoires de la Terre d'Orient* in *Recueil des Historiens des Croisades: Documents Arméniens,* vol. 2 (1906; reprint, Paris: Imprimerie Nationale, 1967), 121–35.

25. "Sciendum quod huiusmodi Mappa mundi non ut cuncta sigillatim contineat cum sit impossible, est descripta: sed ut qua in libro Secreta Fidelium Crucis intitulato supra ultramarino negotio edito, inferuntur, Orbis situs ignaris, per eam, quadam sensitiva demonstratione lucescant" (Sanudo, *Liber secretorum,* 285); Add. MS, fols. 187v–188r, British Library, London.

26. Konrad Miller, *Mappaemundi: Die ältesten Weltkarten* (Stuttgart: J. Roth, 1895–98) 3:132–36, has a list of all the legends on the world maps. See Degenhart and Schmitt, "Marino Sanudo," 21–27, for an annotated list of the manuscripts, including those of Paolino.

27. Sanudo, *Liber Secretorum*, I.v.1, pp. 31–32.

28. Tadeusz Lewicki, "Marino Sanudos Mappa Mundi (1321) and die Runde Weltkarte von Idrisi," *Rocznik Orietalistyczny* 37 (1976): 169–96; J. Lelewel, *Géographie du Moyen Age*, 5 vols. (Brussels: J. Pilliet, 1852–57).

29. The copy dates from the early thirteenth century and is MS Arab c.90 at the Bodleian Library, Oxford. See Jeremy Johns and Emilie Savage-Smith, "The Book of Curiosities: A Newly Discovered Series of Islamic Maps," *Imago Mundi* 55 (2003): 7–24, esp. 12, and fig. 6, an illustration of the map of the Indian Ocean. More images from this manuscript can be found in Evelyn Edson and Emilie Savage-Smith, *Medieval Views of the Cosmos* (Oxford: Bodleian Library, 2004).

30. Book I, chapters 121–24 in Brunetto Latini, *Brunetto Latini: The Book of the Treasure*, ed. Paul Barrette and Sturgeon Baldwin (New York and London: Garland Publishing Co., 1993), 85–98. This edition does not include a reproduction of the map.

31. MS Douce 319, fol. 8, Bodleian Library, Oxford; on this map, see Anna-Dorothee von den Brincken, *Fines Terrae* (Hannover: Hahnsche Buchhandlung, 1992), 96–97, 171. She suggests that its wordlessness indicates that the copyist had an Arabic map and could not understand the writing on it. Fuat Sezgin, "Arabischer Ursprung europäischer Karten," *Cartographica Helvetica* 24 (July 2001): 21–28, includes a reproduction of the map, fig. 6, p. 24. The map is also catalogued in Marcel Destombes, *Mappemondes, A.D. 1200–1500* (Amsterdam: N. Israel, 1964), 175–76.

32. Latini, *Brunetto Latini*, 97.

33. Bouloux, *Culture*, 49–51; Degenhart and Schmitt, "Marino Sanudo," deal extensively with the relationship between the two.

34. Latin text in Anna-Dorothee von den Brincken, " 'Ut describeretur universis orbis' zur Universalkartographie des Mittelalters," in *Methoden in Wissenschaft und Kunst des Mittelalters*, ed. Albert Zimmermann, Miscellanea Mediaevalia 7 (Berlin, 1970): 249–78, esp. 261. "Incipit prologus in mapa mundi cum trifaria orbis divisione. Sine mapa mundi ea, que dicuntur de filiis ac filiis filiorum Noe et que de IIIIor monarchiis ceterisque regnis atque provinciis tam in divinis quam in humanis scripturis, non tam difficile quam impossibile dixerim ymaginari aut mente posse concipere. Requiritur autem mapa duplex, picture et scripture. Nec unum sine altero putes sufficere, quia pictura sine scriptura provincias seu regna confuse demonstrat, scriptura vero non tamen sufficienter sine adminiculo picture provinciarum confinia per varias partes celi sic determinat, ut quasi ad oculum conspici valeant. Pictura autem hic posita ex mapis variis est composita sumptis de exemplaribus, que scripturis actorum concordant illustrium, quos imitamur, videlicet: Ysi[dori] in libro Eth[imologiarum], J[er]o[nimi] de distantia locorum et hebraicarum questionum, Hug[onis] de S. Vic[tore] et Hug[onis] Floriacensis in sua ecclesiastica ystoria, Orosii de ormesta mundi, Solini de mirabilibus mundi, G[er]vasii de mirabilibus terrarum, Pomponii Mela de situ orbis, Ho[no]rii de ymagine mundi, Eusebii, Bede, Justini, Balderici Dolensis episcopi in itinerario transmarino et aliorum plurium scribentium maxime de situ Terre Sancte et circumstantium regnorum Syrie et Egyptii, que ad multos passus intelligendos Sacre Scripture necessaria sunt; in quibus studiosissimum doctorem J[er]o[nimum] plurimum laborasse qui legit, intelligit. Quod vero per pictores non vicietur pictura, magna est cautio adhibenda.

Orbis autem primaria et generali sua divisione partitur in Asyam, que ad numerum partium mundi tertia dicitur, magnitudine vero medietas invenitur, Europam et Africam. Explicit prologus."

35. Almagià, *Monumenta Cartographica*, 3–4. He believes that Paolino was the author of the original map.

36. Paolino's two world maps are MS Vat. lat.1960, fol. 264, Biblioteca Apostolica Vaticana, Vatican City; and MS lat. 4939, fol. 9r, Bibliothèque Nationale, Paris.

37. Quoted in Miller, *Mappaemundi*, 3:135: "Istud dicitur mare de Sara propter civitatem in qua imperator moratur, et dicitur etiam Caspium propter vicinitates ad montes Caspios et Georgianie eadem causa. In eo erat vorago, ubi descendebat aqua maris, sed propter terremotum obturata fuit. Ideo mare tumescit per palmam omni anno, et jam plures bone civitates destructe sunt. Tandem videtur quod debeat intrare mare Tane non absque multorum periculo. Habet in circuitu MMD millia, et de Sara usque Norgacium ponunt millia ??. Circum mare est regio arenosa (et invia?) in magna parte."

38. See map, "Medieval Commerce (Asia)," in William R. Shepherd, *Historical Atlas* (New York: C. S. Hammond, 1956), no. 104-C.

39. On the dates of Paolino's manuscripts, see Degenhart and Schmitt, "Marino Sanudo," 25–27.

40. William Rubruck, *Journey to the Eastern Parts of the World, 1253–55*, trans. William W. Rockhill (London: Hakluyt Society, 1900; reprint, New Delhi: Asian Educational Services, 1998), 119–20; Roger Bacon, *Opus Majus*, ed. John Henry Bridges (Oxford: Clarendon, 1897; reprint, Frankfurt-am-Main: Minerva, 1964). 354–55.

41. Marco Polo, *The Travels of Marco Polo*, ed. Henry Yule and Henri Cordier (London: John Murray, 1903 and 1920; reprint, New York: Dover, 1993), 2:494. Elsewhere, he calls it the Sea of Baku, Ghel, or Ghelan and insists on its inland position, twelve days' journey from the Ocean. He estimates its circumference as 2,800 miles (ibid., 1:52).

42. S. E. Banks and J. W. Binns, eds., *Gervase of Tilbury, Otia Imperialia: Recreation for an Emperor* (Oxford: Clarendon, 2002), 2.7, p. 242.

43. This is the opinion of Almagià, *Monumenta Cartographica*, 3–4, and Marcel Destombes, *Mappemondes*, 246.

44. Bouloux, *Culture*, 49.

45. The atlas is MS Esp. 30, Bibliothèque Nationale, Paris. The question of its identity and origin is treated by (among others) Jean-Michel Massing, "Observations and Beliefs: The World of the Catalan Atlas," in *Circa 1492*, ed. Jay A. Levenson (Washington, DC: National Gallery of Art, 1991), 27–33. A fine facsimile edition is Georges Grosjean, ed., *Mappamundi: The Catalan Atlas for the Year 1375* (Zurich: Urs Graf, 1978). The complete text of all the legends with translations into Spanish can be found, along with interpretive essays, in *El Atlas Catalàn de Cresques Abraham* (Barcelona: Díafora, 1975). See also the CD-ROM produced by the Bibliothèque Nationale de France: Monique Pelletier, Danielle Le Coq, and Jean-Paul Saint Aubin, *Mapamondi: Une Carte du Monde au XIVe Siècle* (Paris: Montparnasse Multimedia, 1998). The CD-ROM includes translations of all the inscriptions into several languages, plus brief essays and images of related maps.

46. Pelletier et al., *Mapamondi*.

47. Grosjean, 13–14.

48. Eva G. R. Taylor, *The Haven-Finding Art*, rev. ed. (New York: American Elsevier Publishing Co. 1971), 137–38, with diagram.

49. Honorius Augustodunensis, *Imago Mundi*, Valerie Flint, ed., *Archives d'histoire doctrinale du Moyen Age*, ann. 57 (1982), 49. The Atlas draws from book 1, chapters 1, 2, 3, 5, 7, 26, 30, 33, 34, 50, 51, 48, 21, 22, 23, 40, and 41. The usual division of the "ages of the world" was (1) Adam to Noah, (2) Noah to Abraham, (3) Abraham to David, (4) David to the Babylonian Captivity, (5) the Babylonian Captivity to the birth of Christ, and (6) from the birth of Christ to the present.

50. There is a good color reproduction in J. B. Harley and David Woodward, eds., *History of Cartography*, vol. 1 (Chicago: University of Chicago Press, 1987), plate 32.

51. See Evelyn Edson, *Mapping Time and Space: How Medieval Mapmakers Viewed Their World* (London: British Library, 1997), chap. 4, "Space and Time in the Computus Manuscript."

52. David Woodward, "Medieval *Mappaemundi*," in Harley and Woodward, *History of Cartography*, 1:314.

53. Ibid., 315.

54. Sandra Sáenz-López Pérez, "La Representación de Gog y Magog y la Imagen del Antichristo en leas Cartas Náuticas Bajomedievales," *Archivo Español de Arte* 78 (2005): 263–76.

55. Grosjean, *Mappamundi*, 90.

56. This is the opinion of François Avril, *Manuscrits Enluminées de la Péninsule Ibérique* (Paris: Bibliothèque Nationale, 1982), 97.

57. Daniel 8:23–25; see the interesting footnote in *El Atlas Catalàn*, 138. Also, see James M. Scott, *Geography in Early Judaism and Christianity: The Book of the Jubilees* (Cambridge: Cambridge University Press, 2002), 62.

58. MS C.G.A.5a., Biblioteca Estense e Universitaria, Modena. For a reproduction, see Guglielmo Cavallo, ed., *Cristoforo Columbo e l'Apertura degli Spazi*, 2 vols. (Rome: Libreria dello Stato, 1992), 1:511–14, plate 3.25b.

59. *El Libro del Conoscimiento de Todos los Reinos*, ed. and trans. Nancy F. Marino (Tempe, AZ: Center for Medieval and Renaissance Studies, 1999), 45, describes the flag of the king of Brischan as a six-pointed star, although the illustration (fig. 62) shows it with eight points. The same flag is on the British Library Dulcert map (Add. MS 25,691) but not on the Dulcert map in Paris (Rés. Ge B 696, Cartes et Plans, Bibliothèque Nationale). The *Libro del Conoscimiento* is a late fourteenth-century travel account, supposedly by a Spanish Franciscan. It is believed to have been written, at least in part, with the use of a mappamundi. For more on this work, see chapter 5. As for the star of David, it was only beginning to emerge as a symbol of Judaism in the fourteenth century. In 1354 Emperor Charles IV gave the Jews of Prague the right to have their own flag, bearing the star of David. Before this time the star was a mystical symbol, used by both Christians and Jews. See Marc M. Epstein, *Dreams of Subversion in Medieval Jewish Art and Literature* (University Park: Pennsylvania State University Press, 1997), 61.

60. Jacob Oliel, *Les Juifs au Sahara: Le Touat au Moyen Age* (Paris: CNRS, 1994), 82–85, has a good section on cartography; on Majorcan Jews and the Africa trade, see David Abulafia, *A Mediterranean Emporium: The Catalan Kingdom of Majorca* (Cambridge: University Press, 1994), 79.

61. Felipe Fernández-Armesto, *Before Columbus: Exploration and Civilization from the Mediterranean to the Atlantic* (Philadelphia: University of Pennsylvania, 1987), 136.

62. Ibid., 146.

63. *El Atlas Catalàn*, 119.

64. E. Denison Ross, "Prester John and the Empire of Ethiopia," in *Travel and Travellers of*

the Middle Ages, ed. Arthur P. Newton (New York: Barnes and Noble, 1968), 174–94, including a copy of the letter. For more on Prester John, see chapter 5.

65. Yoro K. Fall, *L'Afrique à la Naissance de la Cartographie Moderne XIVe-XVe Siècles: Les Cartes Majorquines* (Paris: Éditions Karthala, 1982), 211.

66. Felipe Fernández-Armesto, *Before Columbus*, makes an interesting case for the Canaries as a sort of dress rehearsal for the colonial exploitation of the New World that was to follow. See his last chapter, "The Mental Horizon," esp. 231–34.

67. Pliny, *Natural History*, trans. H. Rackham (Cambridge, MA: Harvard University Press, 1942), book VI.37.202.

68. Fernández-Armesto, *Before Columbus*, 160–62, asserts that this is evidence of their four-teenth-century discovery. These are the names of the islands today. See also Susan Ludmer-Glebe, "Visions of Madeira," *Mercator's World* 8, no. 3 (May–June 2002): 38–43.

69. Marco Polo, *The Travels*, trans. by Ronald Latham (Baltimore: Penguin, 1958), 84–85; Polo, *The Travels of Marco Polo*, ed. Yule and Cordier, 1:196–97.

70. Janet Abu-Lughod says that these troops were the source of the Black Plague (*Before European Hegemony: The World System 1250–1350* [New York: Oxford University Press, 1989], 126).

71. John Larner points out that these names appear to come from an edition of Marco Polo made in 1377 (*Marco Polo and the Discovery of the World* [New Haven, CT: Yale University Press, 1999], 135).

72. This number varies in the manuscripts; see Larner, *Marco Polo*, 136; Polo, *The Travels of Marco Polo*, ed. Yule and Cordier, 2:264.

73. Ibn Baṭṭūṭah also reports on an island governed by a warrior-princess somewhere in the Indian Ocean (*Travels, 1325–54*, trans. C. Defrémery, B. R. Sanguinetti, and C. F. Beckingham; ed. H. R. Gibb, 5 vols. (Cambridge: Cambridge University Press, 1994), 4:884. See also G. R. Tibbetts, *A Study of Arabic Texts Containing Material on Southeast Asia* (Leiden: Brill, 1979), 98, 182.

74. M. J. Gambin, "L'Île Taprobane: Problèmes de Cartographie dans l'Océan Indien," in *Géographie du Monde au Moyen Age et à la Renaissance*, ed. Monique Pelletier (Paris: Interna-tional Cartographic Association, 1989), 191–200.

75. This is the opinion of Fall, *L'Afrique*, 207. Ibn Baṭṭūṭah, *Travels*, 4:969, mentions the tomb but does not describe its form.

76. The best reproduction is in Marcel Destombes, *Mappemondes*, plate 28. The map is de-scribed by R. Hervé on 205–7. Arthur Dürst, "Die Weltkarte von Albertin de Virga von 1411 oder 1415," *Cartographica Helvetica* 13 (1996): 18–21, draws heavily on Hervé. The map belonged to a Jewish family in Heidelberg and disappeared, along with its owners, in the late 1930s. The sea chart is Rés. Ge D 7900, Cartes et Plans, Bibliothèque Nationale, Paris.

77. See Gunnar Thompson, *America's Oldest Map—1414 A.D.* (Seattle: Misty Isles Press—The Argonauts, 1995). His theory about the northern peninsula is thoroughly criticized by Kirsten A. Seaver, "Albertin de Virga and the Far North," in *Mercator's World* 2, no. 6 (Novem-ber–December 1997): 58–63.

78. Homer, *Iliad*, trans. Robert Fagles (New York: Viking Penguin, 1990), 3.6, p. 128.

Four • Merchants, Missionaries, and Travel Writers

1. A recent edition is *The Alexandreis of Walter of Chatillon*, David Townsend, trans. (Philadelphia: University of Pennsylvania Press, 1996). For Alexander's letter, see *Epistola Alexandri ad Aristotelem*, ed. W. Walter Boer (Meisenheim: A. Hain, 1973).

2. Accounts by these pilgrims have been collected in the dozen volumes of the Library of the Palestine Pilgrims' Text Society, originally published in London (1887–97) and reissued by AMS Press (New York, 1971).

3. Francesco Petrarca, *Itinerario in Terra Santa*, ed. Francesco Lo Monaco (Bergamo: Pierluigi Lubrina, 1990); it is discussed in Nathalie Bouloux, *Cultures et Savoirs Géographiques en Italie au XIVe siècle* (Turnhout: Brepols, 2002), 134–37.

4. "Viginti, nisi fallor, passuum milia emensus extentum in undas promontorium, Caput Montis ipsi vocant, obvium habebis et Delphini sive, ut naute nuncupant, Alphini portum, perexiguum sed tranquillum et apricis collibus abditum, inde Rapallam . . ." Petrarch, *Itinerario in Terra Santa*, no. 18, p. 46.

5. "Iam vero non longe hinc, mare quod Sodomorum dicitur, Iordanis influit, ubi consumptarum urbium vindicteque celestis aperta vestigia apparent. His deserti solitudo proxima est. Durum iter fateor, sed ad salutem tendenti nulla difficilis via videri debet. Multas ubique difficultates, multa tibi tedia vel hominum vel locorum hostis noster obiciet, quibus te ab incepto vel retrahat vel retardet vel, si neutrum possit, saltem in sacra peregrinatione hac minus alacrem efficiat." Petrarch, nos. 65–66, p. 78.

6. John Mandeville, *Travels*, ed. and trans. C. W. R. D. Moseley (Baltimore: Penguin, 1983), pp. 99–104.

7. Leonardo Olschki, *Marco Polo's Precursors* (Baltimore: Johns Hopkins University Press, 1943), 3–4.

8. Matthew Paris, *English History*, trans. John A. Giles, 3 vols. (London: Henry Bohn, 1852), 1:312. Matthew's first reference to the Tartars is in 1238 (ibid., 1:131).

9. Suzanne Lewis, *The Art of Matthew Paris in the Chronica Majora* (Berkeley: University of California Press,1987), 282–88.

10. Giovanni di Pian di Carpini, *Historia Mongolorum*, in Anastasius van den Wyngaert, *Sinica Franciscana*, 3 vols. (Florence: Ad Claras Aquas, 1929), 1:27–130; English translation in Christopher Dawson, *Mission to Asia* (Toronto: University of Toronto Press, 1980), 60. His name is spelled and translated variously, from Iohannes de Plano Carpini to John of Pian de Carpine. The Europeans frequently referred to the Mongols as Tartars (from Tatars), making use of a pun on "Tartarus," the word for hell. The Tatars, though related to the Mongols, were a distinct people.

11. Letters in Dawson, *Mission*, 73–76, 85–86.

12. Guillelmus de Rubruc, *Itinerarium*, in van den Wyngaert, *Sinica Franciscana*, 1:164–332; William of Rubruck, *Journey to the Eastern Parts of the World*, ed. and trans. William W. Rockhill (London: Hakluyt Society, 1900; reprint, Madras: Asian Educational Services, 1998), 40–282.

13. Guillelmus de Rubruc, *Itinerarium*, 285; William of Rubruck, *Journey*, 220.

14. Examples in Guillelmus de Rubruc, *Itinerarium*, 237–38; and William of Rubruck, *Journey*, 157–59.

15. Odoric's phrase in Latin is "Nequissimi sunt heretici"; see p. 442 in Fr. Odoricus de Portu

Naonis, *Relatio,* in van den Wyngaert, *Sinica Franciscana,* 1:413–95; Blessed Odoric of Pordenone, *Travels,* trans. Henry Yule (Grand Rapids, MI: William Eerdmans, 2002), 101. He is in India on the Malabar coast, where Syriac Christianity still thrives. (Hereafter he is referred to as Odoric.)

16. MS 350A, Beinecke Library, Yale University; R. A. Skelton, Thomas E. Marston, and George D. Painter, *The Vinland Map and the Tartar Relation* (New Haven, CT: Yale University Press, 1965), 54–102.

17. Roger Bacon, *Opus Majus,* ed. by Robert Belle Burke (Philadelphia: University of Pennsylvania Press, 1928), 1:381–82.

18. Vincentius Bellovacensis, *Speculum Historiale,* vol. 4 of *Speculum Quadruplex* (1624; reprint, Graz: Akademische Druck-und Verlagsanstalt, 1965) book 31, pp. 1286–1303. John Larner says that Vincent skipped over the geographical sections, including none of books 1 or 9, but this is not correct (*Marco Polo and the Discovery of the World* [New Haven, CT: Yale University Press, 1999], 26). Nearly all of book 1 appears in Vincent, Vincentius Bellovacensis, *Speculum Historiale,* book 31, chap. 3, pp. 1286–87, and the opening paragraph on p. 1292, chap. 19. This passage gives information about the location and terrain of the land of the Tartars, as well as their horrific weather. Excerpts from book 9 can be found on pp. 1292–94, chaps. 19–25; p. 1296, chap. 30; p. 1297, chap. 33; p. 1297–99, chap. 35–39, including the return journey. This is a little hard to follow, as Vincent uses his privilege as an editor to chop and arrange his texts to suit his own purposes.

19. Guillelmus de Rubruc, *Itinerarium,* 210; William of Rubruck, *Journey,* 118; and Bacon, *Opus Majus,* 1:373.

20. William of Rubruck, *Journey,* 165 and note; Guillelmus de Rubruc, *Itinerarium,* 242–43.

21. Pian di Carpini in Dawson, *Mission,* 41; Guillelmus de Rubruc, *Itinerarium,* 88–91 and 234.

22. Guillelmus de Rubruc, *Itinerarium,* 269; William of Rubruck, *Journey,* 198–99.

23. Skelton, Marston, and Painter, *The Vinland Map,* 72.

24. For example, read the essay "Excursus A: The Identification of Ornas" by George D. Painter on the identity of "Ornas" in Skelton, Marston, and Painter, *The Vinland Map,* 102–4.

25. Guillelmus de Rubruc, *Itinerarium,* 255; William of Rubruck, *Journey,* 180–81.

26. Guillelmus de Rubruc, *Itinerarium,* 331; William of Rubruck, *Journey,* 281–82.

27. Scott D. Westrem, *The Hereford Map* (Turnhout: Brepols, 2001), xxxiv–xxxvii. The recent critical edition is Patrick Gautier Dalché, *Du Yorkshire à l'Inde: Une "Géographie" Urbaine et Maritime de la Fin du XIIe Siècle (Roger de Howden?)* (Geneva: Droz, 2005), 143–64 and 233–51.

28. Ibid., item nos. 213–14.

29. See Konrad Miller, *Mappaemundi: die Ältesten Weltkarten* (Stuttgart: J. Roth, 1895–98), 5:31–51, for list of place-names and 68–71 for sources.

30. MS Roy. 14 C.VII, fol. 4v, British Library, London. Suzanne Lewis, *The Art of Matthew Paris,* 349, gives a reproduction of this map.

31. For examples, see Pian di Carpini, in Dawson, *Mission* p. 59; and *Historia Mongolorum,* 113. Tadeusz Lewicki proposes that "Carab" is derived from "Harab," an Arabic word for a ruined city: "Marino Sanudos Mappamundi [1321] und die Runde Weltkarte von Idrisi [1154]," *Rocznik Orientalistyczny* 37 [1976]: 169–96, esp. 194.

32. William of Rubruck, *Journey,* 12 note; Guillelmus de Rubruc, *Itinerarium,* 89 and note.

33. Marino Sanudo, *Liber Secretorum Fidelium Crucis,* ed. Joshua Prawer (Hanover, 1621; reprint, Toronto: University of Toronto Press, 1972), 285.

34. Sanudo, *Liber Secretorum*, books 2 and 3, part 13, chaps. 7–8; Charles Kohler, introduction to Hayton, *La Flor des Estoires de la Terre d'Orient*, in *Recueil des Historiens des Croisades: Documents Armeniens*, ed. Charles Kohler (1967; Paris: Imprimerie Nationale, 1906), 2:xxxix. For the geographical text, see *Flor*, 121–35 (for the French version) and pp. 261–73 (for the Latin version). Also see *A Lytell Cronycle: Richard Pynson's Translation (c. 1520) of La Fleur des Histoires de la terre d'Orient (c. 1307)*, ed. Glenn Burger (Toronto: University of Toronto Press, 1988). This has an excellent introduction that gives useful historical background.

35. There are numerous works on Prester John. See, e.g., Vsevolod Slessarev, *Prester John: The Letter and the Legend* (Minneapolis: University of Minnesota Press, 1959).

36. An English translation of the letter is in E. Denison Ross, "Prester John and the Empire of Ethiopia," in *Travel and Travellers of the Middle Ages*, ed. Arthur P. Newton (New York: Barnes and Noble, 1968), 174–94, esp. 174–78. For more detail about the manuscript tradition and numerous versions in Latin and German, see Bettina von Wagner, *Die Epistola Presbiteri Johannis Lateinisch und Deutsch* (Tübingen: Max Niemeyer, 2000), 345–466.

37. Umberto Eco, *Baudolino* (New York: Harcourt, 2000), 132–44. They never send the letter, but when the real one turns up, it is almost identical to their creation.

38. Rés Ge.B 696, Bibliothèque Nationale, Paris. For an account of Prester John's career in Africa, see Francesc Relaño, *The Shaping of Africa* (Aldershot: Ashgate, 2002), chap. 3, pp. 51–74.

39. Relaño, *The Shaping of Africa*, 60.

40. Entry for 1241, Paris, *English History*, 1:348.

41. Pian di Carpini, *Historia Mongolorum*, 129; Dawson, *Mission*, 71.

42. The journey of Niccolò and Maffeo is treated in summary fashion in the opening chapters of Marco Polo, *The Travels of Marco Polo*, ed. Henry Yule and Henri Cordier (London: John Murray, 1903 and 1920; reprint, New York: Dover, 1993), 1:1–19. See also Larner, *Marco Polo*, 33–35, for an excellent analysis and some speculation on the dates. His opinion is that Marco's book consistently exaggerates the travel time.

43. Luciano Petech suggests that Milione was possibly his real last name. See "Les Marchands Italiens dans l'Empire Mongol," *Journal Asiatique* 249–50 (1961–62): 549–74, esp. 557.

44. Larner discusses the Asian sources of his work, suggesting that they were itineraries, geographies and possibly maps (*Marco Polo*, 84–86, 90).

45. Polo, *Travels*, ed. Yule and Cordier, 2:185–218.

46. Translation is by Ronald Latham in Marco Polo, *The Travels* (Baltimore: Penguin, 1958), 248.

47. Polo, *Travels*, ed. Yule and Cordier, 2:331.

48. Introduction to Polo, *Travels*, ed. Yule and Cordier, 4. Yule is quoting Ramusio (c. 1553).

49. Larner, *Marco Polo*, 97.

50. Polo, *Travels*, ed. Yule and Cordier, 2:253–65.

51. On China, see Francesco Balducci Pegolotti, *La Pratica della Mercatura*, ed. Allan Evans (Cambridge, MA: Medieval Academy of America, 1936), 21–23.

52. Petech, "Les Marchands," 551–52.

53. This is one version; see Polo, *Travels*, ed. Yule and Cordier, 1:220, 223n5.

54. For examples, see Denis Sinor, "The Mongols and West Europe," in *The Fourteenth and Fifteenth Centuries*, ed. Harry W. Hazard, vol. 3 of *A History of the Crusades*, ed. Kenneth

M. Setton (Madison: University of Wisconsin Press, 1975), 533–36; Petech, "Les Marchands," 560–67.

55. Odoric, *Travels*, 64; Odoric, *Relatio*, 413: "ad partes infidelium volens ire ut fructus aliquos lucrifacerem animarum."

56. This city on the south China coast has been variously identified as Amoy or as Chüanchou, now Quanzhou, ninety miles south of Fuzhou. It is relatively small today but was one of the greatest Chinese ports of the thirteenth century.

57. Odoric, *Travels*, 69, 98–99, 106–7, 121–22; Odoric, *Relatio*, 418, 439, 447, 459.

58. Odoric, *Travels*, 126; this sentence does not appear in all manuscripts. Van den Wyngaert does not include it in Odoric, *Relatio*, 463.

59. For Old Man, see Odoric, *Travels*, 155; Polo, *Travels*, ed. Yule and Cordier, 1:139–46. For Quinsay, see Odoric, *Travels*, 126–27; Polo, *Travels*, ed. Yule and Cordier, 2:185. Odoric, *Relatio*, 488–89, 463ff.

60. Larner, *Marco Polo*, 128.

61. Odoric, *Travels*, 139; Odoric, *Relatio*, 474: "Nam nos fratres Minores in hac curia sua habemus locum deputatum, et nos sic semper opportet ire et dare sibi benedictionem nostram."

62. Ibn Baṭṭūṭah, *Travels of Ibn Baṭṭūṭah 1325–54*, trans. C. Defrémery and B. R. Sanguinetti, completed by C. F. Beckingham, 5 vols. (Cambridge: Cambridge University Press for the Hakluyt Society, 1958–2000). A summary account can be found in Ross E. Dunn, *The Adventures of Ibn Baṭṭūṭah* (Berkeley: University of California Press, 1986).

63. Ibn Baṭṭūṭah, *Travels*, 4:918–20.

64. Janet Abu-Lughod, *Before European Hegemony: The World System 1250–1350* (New York: Oxford University Press, 1989), 236–39.

65. An account of the martyrdom of the Christians at Almalyk can be found in van den Wyngaert, 510–11. Those killed included the bishop, a group of Franciscan friars, and a Genoese merchant.

66. Larner, *Marco Polo*, 116–24; Abu-Lughod, *Before European Hegemony*, 356–61; Sinor, "The Mongols," 543–44.

67. Percy G. Adams, *Travelers and Travel Liars* (Berkeley: University of California Press, 1962).

68. Pian di Carpini, in Dawson, *Mission*, 3–4; *Historia Mongolorum*, 28. His Christian sources were captives of the Mongols.

69. There are many editions and translations of Mandeville. My English quotations are from *Travels of Sir John Mandeville*, trans. C. W. R. D. Moseley (Baltimore: Penguin, 1983). For his sources, see Christiane Deluz, *Le Livre de Jehan de Mandeville* (Louvain: Institute d'Etudes Médiévales de l'Université Catholique de Louvain, 1988), 57–58, and annexe 6, 428–92.

70. Mandeville, *Travels*, 43–44.

71. Ibid., 45.

72. Ibid., 100–104.

73. Ibid., 111.

74. Ibid., 182.

75. Deluz, *Le Livre*, 399–401. For more on Mandeville's cartographic sense, see Evelyn Edson, "Travelling on the Mappamundi: The World of John Mandeville," in *The Hereford World Map: Medieval Maps and Their Context*, ed. Paul D. A. Harvey (London: British Library, 2006), 389–403.

76. Mandeville, *Travels*, 137.

77. An early printed edition (Augsburg: Anton Sorg, 1478) included 150 woodcuts, many of which featured the monstrous races. MS Harley 3954, British Library, London (early fifteenth century) also has wonderful monster illuminations.

78. Deluz, *Le Livre*, 59–60; examples in Mandeville, *Travels*, 81–85.

79. William de Boldensele, *Itinerarius Guilielmi de Boldensele*, in *Zeitschrift des historischen Vereins für Niedersachsen*, ed. C. E. Grotefend, Jahr. 1852 (Hannover: Hahn, 1855), 236–86.

80. Iain Macleod Higgins, *Writing East: The "Travels" of Sir John Mandeville* (Philadelphia: University of Pennsylvania Press, 1997), 9.

81. Mandeville, *Travels*, 108.

82. William de Boldensele, *Itinerarius*, 246: "in deserto Arabiae hominibus bestialibus et indoctis, legemque diabolicam ipsis imposuit."

83. Odoric, *Travels*, 100; similar remarks are found on pp. 104, 106, 107, 111, 114, 116, 124; Odoric, *Relatio*, 441.

84. Odoric, *Travels*, 99–100; Odoric, *Relatio*, 440.

85. Isidore, *Etymologiarum sive Originum*, ed. W. M. Lindsay (Oxford: Clarendon, 1911), VIII.xi.4–8; Mandeville, *Travels*, 121, 188.

86. Mandeville, *Travels*, 121, 186–87.

87. Ibid., 188.

88. Ibid., 46–47, 132, 166. Higgins's theory is that Mandeville constantly treated other races as the "Other," but I think this theory, while currently fashionable, stretches the text. See Higgins, *Writing East*, 80–81. He points out that the anti-Semitic diatribes do not appear in all versions of the text (44–45).

89. Odoric, *Travels*, 105; Odoric, *Relatio*, 445; Mandeville, *Travels*, 127; Deluz, *Le Livre*, 472, lists Mandeville's source as Sacrobosco's *De Sphaera*, which discusses the two poles and the Arctic and Antarctic Circles, but Sacrobosco does not propose a southern polestar. See Sacrobosco, *De Sphaera*, in *The Sphere of Sacrobosco and its Commentators*, ed. Lynn Thorndike (Chicago: University of Chicago Press, 1949), 76–117 (Latin text) and 118–42 (English translation), esp. 92–93 or 127–28. A more likely source for the southern polestar is Brunetto Latini, *The Book of the Treasure*, trans. Paul Barrett and Spurgeon Baldwin (New York: Garland, 1993), book 1, chap. 119, p. 84.

90. Mandeville, *Travels*, 128.

91. Nicholas Crane, *Mercator: The Man Who Mapped the Planet* (New York: Henry Holt, 2002), 226.

92. "Auctor licet alioqui fabulosus." The inscription appears on a reproduction of the world map in Marcel Watelet, ed., *Gerard Mercator, Cosmographe, le Temps et l'Espace* (Anvers: Fonds Mercator, 1994), 202. See also Moseley's introduction to Mandeville, *Travels*, 29–35.

93. See Higgins, *Writing East*, for the amazing variety of manuscript versions of Mandeville.

94. Nancy F. Marino, ed. and trans., *El Libro del Conoscimiento de Todos los Reinos*, (Tempe: Arizona Center for Medieval and Renaissance Studies, 1999); an earlier edition is Clements R. Markham, trans., *Knowledge of the World*, (London: Hakluyt Society, 1912).

95. Marino, *El Libro*, 47.

96. Ibid., 107, 93.

97. Ibid., 83.

98. Ibid., 73, 63; Mandeville, *Travels*, 185.

99. Marino, *El Libro*, 101.

100. Ibid., xxxvii. The latest events described in the book take place around 1350. Marino says it was written after 1378 because the author described the Pope as being in Avignon, but of course the Pope was installed at Avignon from 1305. All other indications point to the earlier date (1350–75).

101. Ibid., xl.

102. Ibid., 49.

103. Relaño, *The Shaping of Africa*, 149–50. The Pillars of Hercules and the island of Cadiz appear off Cape Verde in the Catalan world map in Modena (c. 1450).

104. Marino, *El Libro*, xvi. A doubtful term as the book has neither plot nor characters!

105. Ibid., 49. See Peter Russell, *Prince Henry the Navigator: A Life* (New Haven, CT: Yale University Press, 2000), 111–115, for more on the rounding of the Cape and its identity, and pp. 99–101 on the Azores; see also Felipe Fernández-Armesto, *Before Columbus* (Philadelphia: University of Pennsylvania Press, 1987), chap. 6.

106. Marino, *El Libro*, 57.

107. On the theory that the author was a herald, see Marino, *El Libro*, xli; for illustrations of shields, see lxi–lxxiii. For flags on sea charts, see Tony Campbell, "Portolan Charts from the Late Thirteenth Century to 1500," in *Cartography in Prehistoric, Ancient, and Medieval Europe and the Mediterranean*, ed. J. B. Harley and David Woodward, vol. 1 of *History of Cartography* (Chicago: University of Chicago Press, 1987), 398–401. Patrick Gautier Dalché also mentions the role of heralds in the growing use of maps in the early fourteenth century: "Pour une Histoire de Regard Géographique: Conception et Usage de la Carte au XVe Siècle," *Micrologus* 4 (1996): 102–3.

108. Marino, *El Libro*, xxi; George H. T. Kimble, foreword to *Catalan World Map of the R. Biblioteca Estense at Modena* (London: Royal Geographical Society, 1934), 8.

109. Christine de Pisan, *Le Livre du Chemin de Long Estude* (Berlin, 1887; reprint, Geneva: Slatkine Reprints, 1974).

110. R. A. Skelton has produced a facsimile of the 1482 printed edition of Berlinghieri's Ptolemy (Amsterdam: Theatrum Orbis Terrarum, 1966). On Berlinghieri's introduction, see Roberto Almagià, "Osservazioni sull'Opera Geografica di Francesco Berlinghieri," *Deuptazione Romana di Storia Patria, Archivio* 68 (1945): 211–55.

111. See Abu-Lughod, *Before European Hegemony*, esp. chap. 1.

Five • The Recovery of Ptolemy's Geography

1. There is some confusion over whether this was a book or a map. Strozzi says "la pictura." Sebastiano Gentile, ed., *Firenze e la Scoperta dell'America: Umanesimo e Geografia nel '400 Fiorentino* (Florence: Leo Olschki, 1992), catalogue item no. 43 (Palla Strozzi's will), 88. This excellent work catalogues 116 items from an exhibit on the role of Florence in the changing of geographical consciousness in the fifteenth century. It has many well-produced illustrations and a detailed description for each entry.

2. There is a manuscript tradition with a dedication to Pope Gregory XII, who was elected in 1406, but the oldest surviving manusripts are MS Vat. lat. 2974 (1409) and MS Otto. lat. 1771 (1411) in the Biblioteca Apostolica Vaticana, Vatican City, both with the dedication to Alexander, who succeeded Gregory in 1409.

3. Patrick Gautier Dalché gives a collection of references to Ptolemy in medieval works in "Le Souvenir de la Géographie de Ptolémée dans le Monde Latin Médiéval (Vie–XIVe siècles)," *Euphrosyne* 27 (1999): 79–106. On the survival and use of Ptolemy in the Arab world, see Gerald R. Tibbetts, "Islamic Cartography: The Beginnings of a Cartographic Tradition," in *Cartography in Traditional Islamic and South Asian Societies*, vol. 2, part 1, *History of Cartography*, ed. J. B. Harley and David Woodward (Chicago: University of Chicago Press, 1992), 95; and Emilie Savage-Smith and Jeremy Johns, "*The Book of Curiosities:* A Newly Discovered Series of Islamic Maps," *Imago Mundi* 55 (2003): 7–24, esp. 9.

4. This is believed to be MS Urb. gr. 82, Biblioteca Apostolica Vaticana, Vatican City. A facsimile was published in *Claudii Ptolemaei Geographiae: Codex Urb. gr. 82*, ed. Joseph Fischer, 2 vols. in 4 (Leiden: Brill, 1932); Gentile, *Firenze*, cat. item no. 37 (Vespasiano da Bisticci's lives of the Strozzi family), p. 77, and cat. item no. 42 (*Geographia*, MS Paris lat. 17542), p. 86.

5. O. A. W. Dilke,"Cartography in the Byzantine Empire," in *Cartography in Prehistoric, Ancient, and Medieval Europe and the Mediterranean*, ed. J. B. Harley and David Woodward, vol. 1 of *History of Cartography* (Chicago: University of Chicago Press, 1987), 268.

6. These are MS lat. 441, Bibliothèque Municipale, Nancy; and MS Vat. lat. 5698, Biblioteca Apostolica Vaticana, Vatican City. See Gentile, *Firenze*, cat. item no. 40 (MS Vat. lat. 5698), p. 83. Here it is billed as "Il più antico codice delle tavole latine della "Geographia."

7. Ibid., cat. item no. 22 (Petrarch's copy of Pliny), p. 50.

8. Ibid., cat. item no. 23 (tenth century manuscript of Pliny), p. 54. This manuscript was acquired by Cosimo de' Medici.

9. Giovanni Boccaccio, *Zibaldone*, in *Tutte le Opere di Giovanni Boccaccio*, ed. M. Pastore Stocchi, vol. 5 (Milan; Mondadori, 1998). See Gentile, *Firenze*, cat. item no. 31, p. 68. Nathalie Bouloux asserts that he added these comments some years after the initial copying, when his geographical understanding had become more sophisticated (*Cultures et Savoirs Géographiques en Italie au XIVe Siècle* [Turnhout: Brepols, 2002]), 125–34.

10. An example is in Bouloux, *Cultures*, 126–27.

11. Boccaccio, *Tutte le Opere*, vols. 7, 8.

12. The astrolabe's spherical projection is centered on the South Pole, which would not make much sense for a map of the land in the Northern Hemisphere. For a description and some good diagrams, see A. J. Turner, *Astrolabes and Astrolabe-Related Instruments*, vol. 1, part 1 of *The Time Museum: Catalogue of the Collection* (Rockford, IL: Time Museum, 1985), 3.

13. For a good description, see Ptolemy, *Ptolemy's Geography: An Annotated Translation of the Theoretical Chapters*, trans. and ed. J. Lennart Berggren and Alexander Jones (Princeton, NJ: Princeton University Press, 2000), 35–40; see also Ptolemy, book 1, chap. 20, and book 7, chap. 6 in this edition. Samuel Edgerton discusses this projection as an exercise in perspective in *The Heritage of Giotto's Geometry: Art and Science on the Eve of the Scientific Revolution* (Ithaca, NY: Cornell University Press, 1991), 153.

14. See Ptolemy, *Geography*, book 1, chap. 20, on Marinos; book 8, chap. 1 on regional maps.

15. This is on his world map in MS 26, p. 284, Corpus Christi College, Cambridge.

16. *Liber Floridus*, MS 92, fols. 92v–93r, Rijksuniversiteit Centrale Bibliothek, Ghent. For a facsimile, see Albert Derolez, ed., *The Autograph Manuscript of the Liber Floridus* (Turnhout: Brepols, 1998).

17. Tony Campbell, "Portolan Charts from the Late Thirteenth Century to 1500," in Harley and Woodward, *History of Cartography*, 1:385–86.

18. Michel Mollat du Jourdin and Monique de la Roncière, with Marie-Madeleine Azard, Isabelle Raynaud-Nguyen, and Marie Antoinette Vannereau, *Sea Charts of the Early Explorers*, trans. L. leR. Dethan (London: Thames and Hudson, 1984), 15.

19. Roger Bacon, *Opus Majus*, ed. and trans. Robert Belle Burke (Philadelphia: University of Pennsylvania Press, 1928) 1:315–16; Gautier Dalché, "Le Souvenir," 102.

20. Noel Swerdlow, "The Recovery of the Exact Sciences of Antiquity: Mathematics, Astronomy, Geography," *Rome Reborn: The Vatican Library and Renaissance Culture*, ed. Anthony Grafton (Washington, DC: Library of Congress, 1993), 125–168, esp. 160: he says the modern length of the ecumene is 130° but this is an error for 150°. The western coast of Spain is 10° west of the Greenwich meridian, and the Chinese coast is 140° east. See also Edward L. Stevenson, *Portolan Charts: Their Origin and Characteristics* (New York: Hispanic Society of America, 1911), 19–20. For Ptolemy's errors in mapping Italy, see Oswald A. W. Dilke and Margaret S. Dilke, "Italy in Ptolemy's Manual of Geography," in *Imago et Mensura Mundi*, ed. Carla Clivio Marzoli (Rome: Istituto della Enciclopedia Italiana, 1981), 2:353–60.

21. Dana B. Durand, *The Vienna-Klosterneuberg Map Corpus* (Leiden: E. J. Brill, 1952). Patrick Gautier Dalché takes a dim view of Durand's assessment of the accomplishments of this school in "Pour une Histoire du Regard Géographique: Conception et Usage de la Carte au XVe Siècle," *Micrologus* 4 (Turnhout: Brepols, 1996), 77–103, esp. 86. For more on the Klosneuberg School, see chap. 7.

22. Occasionally, mapmakers could not resist dressing up the Ptolemy maps. In the Bologna edition of Ptolemy (1477), Noah's ark is shown perched on a mountain in Greater Armenia. Map reproduced by R. V. Tooley, *Maps and Map-makers* (New York: Dorset, 1987), plate 7, between pp. 8 and 9.

23. Edgerton, *The Heritage of Giotto's Geometry*, 151.

24. Ricardo Padrón, *The Spacious Word: Cartography, Literature, and Empire in Early Modern Spain* (Chicago: University of Chicago Press, 2004), 35–36.

25. David Buisseret gives a good account of the relationship between painting and mapmaking in *The Mapmakers' Quest: Depicting New Worlds in Renaissance Europe* (Oxford: Oxford University Press, 2003), chap. 2. See also Edgerton, *The Heritage of Giotto's Geometry*.

26. Recently his responsibility for this globe has been questioned. See Jacques Paviot, "La Mappemonde Attribuée à Jan van Eyck," *Revue des Archéologues et Historiens d'Art du Louvain* 24 (1991): 57–62.

27. Joan Kelly Gadol, *Leon Battista Alberti: Universal Man of the Early Renaissance* (Chicago: University of Chicago Press, 1969), esp. chap. 1.

28. An excellent collection of articles may be found in Didier Marcotte, ed., *Humanisme et Culture Géographique à l'Époque du Concile de Constance* (Turnhout: Brepols, 2002). For the comment by Jean-Patrice Boudet, see p. 138.

29. C. M. Gormley, M. A. Rouse, and Richard H. Rouse, "The Medieval Circulation of Pomponius Mela," *Medieval Studies* 46 (1984): 266–320. Petrarch was the first to recover a copy of Mela; see Gentile, *Firenze*, cat. item no. 17 (a copy of Mela belonging to Coluccio Salutati), pp. 44–46.

30. Colette Jeudy, "La Bibliothèque de Guillaume Fillastre," in Marcotte, *Humanisme*, 245–92.

31. Fillastre's introduction is preserved in four manuscripts: MS 1321, Bibliothèque Municipale, Reims; MS H.31, Archivio di San Pietro, Biblioteca Apostolica Vaticana, Rome; MS 91 inf. 7, Biblioteca Medicea Laurenziana, Florence; MS 256, Bibliothèque Municipale, Rennes.

Patrick Gautier Dalché has published a critical edition, "Guillelmi Fillastri Introductio in Pomponii Melae Cosmographiam," which appears in Marcotte, *Humanisme*, 319–44.

32. Pomponius Mela, *Description of the World*, ed. and trans. F. E. Romer (Ann Arbor, Michigan: University of Michigan Press, 1998); on the Caspian, see 3.38; Indian Ocean, 1.9; and Africa, 1.20, 3.90. For Fillastre's discussion of the waters of the earth, see Gautier Dalché, "Guillelmi Fillastri Introductio," para. 15, p. 332.

33. "Que omnia Deus fecit ut ostendat mirabilia potestatis eius." Gautier Dalché, "Guillelmi Fillastri Introductio," 331–32.

34. On the Christian basis of French humanism at this time, see Patrick Gilli, "L'Humanisme Français au Temps du Concile de Constance," in Marcotte, *Humanisme*, 1–62, esp. 53.

35. "De oriente ad occidentem est planus et facilis transitus quantum ad situm terre, nisi malicia vel diversitas gentium impediret . . ." Gautier Dalché, "Guillelmi Fillastri Introductio," para. 17, 332–33.

36. Ibid., para. 17, p. 333.

37. Ibid., chap. 18, p. 333.

38. This manuscript (MS H.31, Arch. San Pietro, Biblioteca Apostolica Vaticana, Vatican City) was made for Cardinal Giordano Orsini during the council.

39. This manuscript is described by Jeudy in "La Bibliothèque," 252. The Clavus map is reproduced in Marcotte, *Humanisme*, 302–3.

40. Adam of Bremen, *Hamburgische Kirchengeschichte*, ed. Bernhard Schmeidler (Hannover: Hahnsche, 1917), book 4, chaps. 36 and 38, pp. 274–75.

41. "Item continet ultra quod ponit Tholomeus Norvegiam, Suessiam, Rossiam utramque et sinum Codanum dividens Germaniam a Norvegia et Suessia, item alium sinum ultra ad septentrionem qui omni anno congelatur in tercia parte anni. Et ultra illum sinum est Grolandia que est versus insulam Tyle magis ad orientem. Et ita tenet totam illam plagam septentrionalem usque ad terram incognitam. De quibus Tholomeus nullam facit mencionem et creditur de illis non habuisse noticiam. Ideo hec viiia tabula est multo amplior describenda, propter quod quidam Claudius Cymbricus illas septentrionales partes descripsit et fecit de illis tabulam que iungitur Europe et ita erunt xi." Gautier Dalché, "Guillelmi Fillastri Introductio," 346.

42. Christiane Deluz, "L'Europe selon Pierre d'Ailly ou selon Guillaume Fillastre?" in Marcotte, *Humanisme*, 151–60.

43. Gautier Dalché comments on the increasing use of maps for practical purposes as a sign of the changing attitude toward cartography in "Pour une histoire du regard cartographique," 102.

44. MS 441, fols. 184v–185, Bibliothèque Municipale, Nancy.

45. A good color reproduction of the map is in Marcotte, *Humanisme*, 302–3.

46. A good color reproduction can be found in Harley and Woodward, eds., *History of Cartography*, vol. 1, plate 19.

47. See an account of this embassy in Peter Russell, *Prince Henry "the Navigator": A Life* (New Haven, CT: Yale University Press, 2000), 125–26.

48. Pierre d'Ailly, *Ymago Mundi*, 3 vols., ed. Edmond Buron (Paris: Maisonneuve Frères, 1930), 2:356–58 and 2:404–5, has reproductions of the map from two different manuscripts.

49. Gerald R. Tibbetts, "Islamic Cartography: Later Cartographic Developments," in Harley and Woodward, *History of Cartography*, 2:1, 147–48. Konrad Miller, *Mappaemundi: Die Ältesten Weltkarten* (Stuttgart: J. Roth, 1895–98), 3:127, has a sketch of Petrus Alfonsus's map, which he pairs with d'Ailly's.

50. In d'Ailly, *Compendium Cosmographiae*, in *Ymago Mundi*, 3:659.

51. MS 927, Bibliothèque Municipale, Cambrai. See Jean-Patrice Boudet, "Un Prélat et son Équipe de Travail à la Fin du Moyen Age," in Marcotte, *Humanisme*, 127–150, esp. 137.

52. The *Compendium* can be found in d'Ailly, *Ymago Mundi*, vol. 3.

53. Boudet, "Un Prélat," 139.

54. D'Ailly, *Ymago Mundi*, 2:454–55.

55. John Larner, *Marco Polo and the Discovery of the World* (New Haven, CT: Yale University Press, 1999), 140. Buron's edition of d'Ailly includes Columbus's marginal notes and underlinings. He calls the idea that Columbus did not read this book until after his first voyage an "absurdité" (d'Ailly, *Ymago Mundi*, 1:27).

56. Buron describes this as the purpose of *Imago Mundi* in the introduction to the *Compendium* (d'Ailly, (d'Ailly, *Ymago Mundi*, 3:557).

57. Charles L. Stinger, *Humanism and the Church Fathers: Ambrogio Traversari (1386–1439) and Christian Antiquity in the Italian Renaissance* (Albany: State University of New York, 1977), 13.

58. His dedication is reprinted in James Hankins, "Ptolemy's *Geography* in the Renaissance," in *The Marks in the Fields: Essays on the Uses of Manuscripts*, ed. Rodney G. Dennis (Cambridge, MA: Harvard University Press, 1992), 125–27.

59. "Ut mos est." From Poggio's dialogue, "De Infelicitate Principium," Gentile, *Firenze*, cat. item no. 51, p. 101.

60. Hans Baron, "Dati's *Istoria* of Florence," in *The Crisis of the Early Italian Renaissance* (Princeton, NJ: Princeton University Press, 1955), 1:149.

61. Thomas Goldstein, "Geography in 15th Century Florence," in *Merchants and Scholars: Essays in the History of Exploration and Trade*, ed. John Parker (Minneapolis: University of Minnesota Press, 1965), 11–32.

62. Gentile, *Firenze*, cat. item no. 81 (Letter from the Ethiopians to the Pope) and cat. item no. 82 (*De Varietate Fortunae* of Poggio Bracciolini), pp. 168–73.

63. Nicolò de' Conti, *L'India di Nicolò de' Conti*, ed. Alessandro Grossato (Padova: Studio Editoriale Programma, 1994), contains a facsimile of MS Marc. 2560, Biblioteca Marciana, Venice, as well as a transcription of the Latin and an Italian translation.

64. Grossato, *L'India*, 40.

65. Portolano 1, Biblioteca Nazionale Centrale, Florence.

66. For a classic account of the traditional view, see Boies Penrose, *Travel and Discovery in the Renaissance, 1420–1620* (New York: Atheneum, 1962), 42–61. For a more balanced modern view, see Russell, *Prince Henry*, esp. the intro., pp. 1–9.

67. Russell, *Prince Henry*, 341, and map, plate 8. An English translation of Cadamosto's report is in *Voyages of Cadamosto and Other Documents on West Africa in the Second Half of the 15th Century*, ed. G. R. Crone, Hakluyt Society Series, 2d ser., 80 (London: Hakluyt Society, 1937).

68. MSS Urb. lat. 274 and 275, Biblioteca Apostolica Vaticana, Vatican City; MS lat. 4802, Bibliothèque Nationale, Paris, by Piero del Massaio, originally thought to be earlier, has now been redated to 1475/80. The oldest manuscript in this group is MS Vat. lat. 5699, Apostolica Vaticana Biblioteca, Vatican City, dated 1469. See Louis Duval-Arnould, "Les Manuscrits de la *Géographie* de Ptolémée Issus de l'Atelier de Piero del Massaio," in Marcotte, *Humanisme*, 227–44.

69. R. A. Skelton, Thomas E. Marston, and George D. Painter, *The Vinland Map and the*

Tartar Relation (New Haven, CT: Yale University Press, 1965), 190, reconstructed map, fig. 9. The so-called towns on Greenland are apparently nonsense, words that Clavus took from a popular song. Fridtjob Nansen, *In Northern Mists*, trans. Arthur G. Chater (London: W. Heineman, 1911), 252–53.

70. A color reproduction of this map is plate 27 in Gentile, *Firenze*, cat. item no. 102 (MS 30.3, Biblioteca Medicea Laurenziana, Florence).

71. A list of printed editions of Ptolemy's *Geography* can be found in Tooley, *Maps*, 6–8.

72. Bruce Fetter has reported that fifteenth-century cartographers working freehand made slight, unheralded corrections in the maps, such as changing the shapes of Sicily and Scotland (communication on MapHist discussion group, December 17, 1996). The Pirrus da Noha map (MS Arch. di San Pietro H.31, fol. 8v, Biblioteca Apostolica Vaticana, Vatican City) is discussed by Anna-Dorothee von den Brincken, *Fines Terrae* (Hannover: Hahnsche Buchhandlung, 1992), 140–41.

73. R. A. Skelton, introduction to the facsimile of the printed edition of Ptolemy, *Geographia. Florence. 1482* (Amsterdam: Theatrum Orbis Terrarum, 1966); Roberto Almagià, "Osservazioni sull'Opera Geografica di Francesco Berlinghieri," *Deputazione Romana di Storia Patria, Archivio* 68 (1945): 211–55. There are also two lavish manuscripts of Berlinghieri's work.

74. Almagià, "Osservazioni," 251.

75. Margriet Hoogvliet, "The Medieval Texts of the 1486 Ptolemy Edition by Johann Reger of Ulm," *Imago Mundi* 54 (2002): 7–18.

76. Mollat du Jourdin et al., *Sea Charts*, 23; A. E. Nordenskjold, *Facsimile Atlas to the Early History of Cartography* (Stockholm, 1889; reprint, New York: Dover, 1973), 18–19.

77. Gautier Dalché, "Le Souvenir." A contrary view is expressed by Marica Milanese, "La Rinascita della Geografia dell'Europa, 1350–1480," in *Viaggiare nel Medioevo*, ed. Sergio Gensini (Pisa: Fondazione Centro, 2000), 35–59.

78. E. G. R. Taylor, *The Haven-Finding Art*, rev. ed. (New York: American Elsevier, 1971), chaps. 7 and 8. On Columbus, see *The Log of Christopher Columbus*, trans. and intro. by Robert H. Fuson (Camden, ME: International Marine Publishing Company, 1987), 42, 113 (entry for November 12, 1492). On the resistance of sailors to astronomical navigation in the sixteenth century, see Surekha Davies, "The Navigational Iconography of Diogo Ribeiro's 1529 Vatican Planisphere," *Imago Mundi* 55 (2003): 103–12.

Six • Fra Mauro

1. There is a excellent facsimile of this map; unfortunately, it is hard to find: Tullia Gasparrini Leporace, *Il Mappamondo di Fra Mauro* (Rome: Istituto Poligrafico dello Stato, 1956) has the complete text as well as color, actual-size photos of all sections of the map. The captions for the Fra Mauro map have been put on the web by the Biblioteca Marciana at http://geoweb .venezia.sbn.it/geoweb/HSL/FraMauro/FraMauroIndex.htm (accessed September 20, 2006). There are two indices, one alphabetical and one by tavola number from Leporace's work. My thanks to Angelo Cattaneo for providing me with a copy of the captions before they were posted on the web. Piero Falchetta has a new book, *Fra Mauro's Map of the World* (Turnhout: Brepols, 2006), with the inscriptions translated into English. For a brief, good description of the map, see George Kish, *La Carte: Images des Civilisations* (Paris: Seuil, 1980), 221–26.

2. Angelo Cattaneo, "Fra Mauro, 'Cosmographus Incomparabilis' and His 'Mappamundi': Documents, Sources, and Protocols for Mapping," in *La Cartografia Europea tra Primo Rinas-*

cimento e Fine dell'Illuminismo, ed. Diogo Ramada Curto, Angelo Cattaneo, and André Ferrand Almeida (Florence: Olschki, 2003), 19–48. Piero Falchetta agrees with this date; see *Andreas Biancho Me Fecit* (Venice: Arsenal Editrice, 1993), 9. Alberto Pinheiro Marques has argued for an earlier date for the Portuguese commission, on the assumption that the map was ordered by Prince Pedro of Portugal before his assassination in 1459; see "The Portuguese Prince Pedro's Purchase of the Fra Mauro Map From Venice," *Globe* 48 (1999): 1–34.

3. Ingrid Baumgärtner, "Kartographie, Reisebericht und Humanismus: Die Erfahrung in der Weltkarte des Venezianischen Kamaldulensermönchs Fra Mauro," *Das Mittelalter* 3 (1998): 2, 161–97, esp. 162–65. For the Florentine conversations, see chap. 5.

4. MS Borgia 5, Biblioteca Apostolica Vaticana, Vatican City. On this map, see Roberto Almagià, *Planisferi, Carte Nautiche e Affini dal Sec. XIV al XVII*, vol. 1 in *Monumenta Cartographica Vaticana* (Vatican City: Biblioteca Apostolica Vaticana, 1944), 32–41; and Marcel Destombes, *Mappemondes 1200–1500* (Amsterdam: N. Israel, 1964), 226–27. There are other theories.

5. "Brother Mauro of San Michele of Murano of the Venetian Order of Camaldolensians, Incomparable Cosmographer." The medal was struck in the fifteenth century, but only an eighteenth-century copy survives.

6. James Cowan, *A Mapmaker's Dream* (Boston: Shambala, 1996).

7. This beautiful little vignette is attributed to the painter Leonardo Bellini by Angelo Cattaneo, "God in His World: The Earthly Paradise in Fra Mauro's *Mappamundi* Illuminated by Leonardo Bellini," *Imago Mundi* 55 (2003): 97–102.

8. For the complete inscription, see Edward L. Stevenson, *The Genoese World Map of 1457* (New York: American Geographical Society, 1912), 63.

9. Leporace, *Il Mappamondo*, tav. 26: "Nota che in questo monte."

10. Alessandro Scafi, "Il Paradiso Terrestre di Fra Mauro," *Storia dell'Arte* 93/94 (1998): 411–19.

11. Cattaneo, "God in His World," 35.

12. Leporace, *Il Mappamondo*, tav. 22: "Queli cho sono experti." All references to the text on the map are keyed to Leporace's work, which is the same as that on the internet site.

13. Leporace, *Il Mappamondo*, tav. 33: "Alguni scrive."

14. Ibid., tav. 38: "De qui e vulgo."

15. Ibid., tav. 34: "El fiume thanai."

16. Ibid.: "El fiume Edil."

17. Ibid., tav. 23: "Perché io no ho habuto loco."

18. Wojciech Iwanczak, "Entre l'Espace Ptolémaïque et l'Empirie: Les Cartes de Fra Mauro," *Médiévales* 18 (Spring 1990): 53–68, esp. 61–62. Also on Borgia 5, see H. Winter, "The Fra Mauro Portolan Chart in the Vatican," *Imago Mundi* 16 (1962): 17–28.

19. See the interesting comments on this issue in the essay by Brian Harley and David Woodward, "Concluding Remarks" in *Cartography in the Traditional Islamic and South Asian Societies*, ed. J. B. Harley and David Woodward, vol. 2 of *History of Cartography* (Chicago: University of Chicago, 1992), 1, 510–18, esp. 518.

20. Leporace, *Il Mappamondo*, tav. 23: "Perché sono molti cosmographi."

21. Ibid., tav. 30: "Se'l parerà ad alguno."

22. Ibid., tav. 15, 27, 14.

23. Ibid., tav. 8, 46.

24. Ibid., tav. 30: "In questa insula." Aristotle's *Generation of Animals* is a more likely source for monstrosities.

25. Kish, *La Carte*, 222.

26. Leporace, *Il Mappamondo*, tav. 22: "Queli che sono experti."

27. Ibid., tav. 28: "Su queste do' cime."

28. Ibid., tav. 34: "De questa Gothia," and XLI.

29. Ibid., tav. 40: "Io non credo derogar."

30. Ibid., tav. 35: "Parme che Tolomeo."

31. Ibid., tav. 30: "In questa opera."

32. Ibid., tav. 34: "Nota che i cosmographi fano division de le mauritanie."

33. Ibid., tav. 9: "Nota che secondo el dir."

34. Gianbattista Ramusio, *Navigazioni e Viaggi*, vol. 2, ed. Marica Milanese (Turin: Einaudi, 1978–1985), fol. 17.

35. Leporace, *Il Mappamondo*, tav. 25: "Questa nobillissima città dita Cansay."

36. Ibid., tav. 11: "Molte opinion." This legend is posted in Africa. Elsewhere he refers to the map as "la più justa carta ho possudo" (the most correct map I have). Tav. 23: "Benché io habi servato."

37. Ibid., tav. 46: "Una nave de Catelani."

38. Ibid., tav. 9: "Questa region fertilissima." An account of these interviews can be found in Sebastiano Gentile, *Firenze e la Scoperta dell'America* (Florence: Olschki, 1992), cat. item no. 81, pp. 168–70. For more on Fra Mauro's depiction of Ethiopia, see Charles de la Roncière, *La Découverte de l'Afrique au Moyen Age*, vol. 2 (Cairo: Société Royale de Géographie d'Égypte, 1925), 123–139. De la Roncière calls this "le chef-d'oeuvre de son planisfere" (p. 123).

39. Leporace, *Il Mappamondo*, tav. 10: "Perché ad alguni."

40. Ibid., tav. 16: "Nota che abassini."

41. Ibid., tav. 9; "Alguni autori." Elsewhere, he cites Pomponious Mela on this issue, see tav. 11: "Molte opinion."

42. Ibid., tav. 11: "Molte opinion." In this rather confusing caption, he says that the Portuguese sailed southwest for 2,000 miles, then turned southeast and sailed to the longitude ("indromo") of Tunis and then to that of Alexandria. These points would have been the far eastern end of the Gulf of Guinea (10° east) and around the Cape of Good Hope (30° east). The determination of longitude at that time was faulty, so it is unclear where they were.

43. Ibid., tav. 10: "Circa hi anni," and tav. 11: "Molte opinion."

44. Marco Polo, *Travels of Marco Polo: The Complete Yule-Cordier Edition*, (London: John Murray, 1929; reprint, New York: Dover, 1993), 2:249–51; Nicolò de' Conti, *L'India di Nicolò de' Conti*, ed. Alessandro Grossato (Padova: Editoriale Programma, 1994), 37.

45. Gavin Menzies, *1421: The Year the Chinese Discovered America* (New York: William Morrow, 2003), 91–93, discusses this text from Fra Mauro. Ibn Baṭṭūṭah tells the same story almost exactly about his mid-fourteenth-century travels. Crone suggests it is a traditional Arab tale dating five centuries before Fra Mauro (*Maps and Their Makers* [New York: Capricorn, 1962], 62).

46. MS Borgia V, Bibliotheca Apostolica Vaticana, Vatican City.

47. *El Libro del Conoscimiento*, ed. and trans. Nancy F. Marino (Tempe: Arizona Center for Medieval and Renaissance Studies, 1999), 59.

48. Leporace, *Il Mappamondo,* tav. 17: "Questo colfo."

49. Ibid., tav. 18: "Io ho piu volte aldido."

50. Ibid., tav. 9: "Note che questo cavo de diab."

51. Crone, *Maps and Their Makers,* 62.

52. Leporace, *Il Mappamondo,* tav. 11: "Questa provincia."

53. Ibid., tav. 10: "Se dice che presto Jane," and "Questo re de abbasia."

54. Ibid., tav. 16: "El nilo nasce."

55. Ibid.: "Credo che."

56. Ibid., tav. 18: "Note che dal cavo verde," and tav. 15: "Note che queli che navegano."

57. Ibid., tav. 23: "Qui fra terra." G. R. Crone, ed. and trans., *Voyages of Cadamosto and Other Documents on West Africa,* 2d ser., no. 80 (London: Hakluyt Society, 1937). Perhaps Cadamosto picked up this detail from Fra Mauro.

58. Felipe Fernandez Armesto, "Mapping in the Eastern Atlantic," in *Before Columbus: Exploration and Civilization from the Mediterranean to the Atlantic* (Philadelphia: University of Pennsylvania Press, 1987), 152–80.

59. Leporace, *Il Mappamondo,* tav. 35: "Per questo mar."

60. Ibid., 9: "Le nave ouer çonchi."

61. Marco, *Travels,* 2:312–13.

62. Leporace, *Il Mappamondo,* tav. 14: "Nota che Tolomeo."

63. Ibid., tav. 19: "In questo mar."

64. The interesting question of Arabic place-names is discussed by Gerald R. Tibbetts, "The Role of Charts in Islamic Navigation in the Indian Ocean," in Harley and Woodward, eds., *History of Cartography,* 2:256–62, esp. 260–62, in relation to the Cantino map. This is an area that needs further study.

65. Conti, *L'India,* 20, 28.

66. Leporace, *Il Mappamondo,* tav. 14: "Questa maxima citade," and tav. 15: "Questa città."

67. He does not mention the parrots' ability to talk. Ibid., tav. 19: "Sondai," and tav. 13: "Bandan." Conti, *L'India,* 30.

68. Conti, *L'India,* 30; Leporace, *Il Mappamondo,* tav. 14: "Qui sono serpe longe."

69. Leporace, *Il Mappamondo,* tav. 21: "Nota che alguni istoriographi."

70. Ibn Baṭṭūtah, *Travels,* trans. C. Defrémery and B. R. Sanguinetti (Cambridge: Cambridge University Press, 1958–2000), 4:884, 911.

71. Kish, *La Carte,* 223, has a good analysis of his portrayal of Scandinavia and the Baltic.

72. Leporace, *Il Mappamondo,* tav. 36: "Norvega e provincia." This inscriptions on this section are not on the website.

73. Ibid., tav. 39: "In queste parte"; tav. 40: "Questi Permiani."

74. Ibid., tav. 36: "Scotia chome apar."

75. Ibid., tav. 20: "Alguni scriveno."

76. Ibid., tav. 40: "Questa opera."

Seven • The Persistence of Tradition in Fifteenth-Century World Maps

1. J. G. Edwards, "Ranulf: Monk of Chester," *English Historical Review* 47 (1932): 94. On Higden in general, see John Taylor, *The Universal Chronicle of Ranulf Higden* (Oxford: Clarendon, 1966). On his maps, see V. H. Galbraith, "An Autograph MS of Ranulph Higden's Poly-

chronicon," *Huntington Library Quarterly* 34 (1959): 1–18; R. A. Skelton, "Ranulf Higden," in *Mappemondes 1200–1500*, ed. Marcel Destombes (Amsterdam: N. Israel, 1964), 149–60.

2. Ranulf Higden, *Polychronicon Ranulphi Higden Monachi Cestrensis*, ed. by Churchill Babington, in *Chronicles and Memorials of Great Britain and Ireland in the Middle Ages* (London: Longman Green, 1865), vol. 41, 2d preface, chap. 3, p. 26. A map appears in at least one "first edition" (MS 33.4.12, fol. 13v, National Library of Scotland, Edinburgh). Possibly it was added when the manuscript was copied.

3. MS HM 132, fol. 4v, Huntington Library, San Marino, California. Illustration in J. B. Harley and David Woodward, eds., *Cartography in Prehistoric, Ancient, and Medieval Europe and the Mediterranean*, vol. 1 of *History of Cartography* (Chicago: University of Chicago Press, 1987), plate 15.

4. This is David Woodward's suggestion in "Medieval Mappaemundi," in *Cartography in Prehistoric, Ancient, and Medieval Europe and the Mediterranean*, ed. J. B. Harley and David Woodward, vol. 1 of *History of Cartography* (Chicago: University of Chicago Press, 1987), 313.

5. Skelton, *Mappemondes*, says there are twenty manuscripts, but one of these is MS Harley 3673 (1466), British Library, London. It does in fact include two chapters from Higden, but the map is many folios further on and bears no relation to the standard Higden map form.

6. Royal MS 14.C.IX, fols. 1v–2, British Library, London; 46 × 34 cm or 17 × 13.5″.

7. For more on these texts, see Evelyn Edson, *Mapping Time and Space* (London: British Library, 1997), 126–31.

8. These are MS 89, Corpus Christi College, Oxford; Add. MS 10,104, British Library, London; MS 33.4.12, National Library of Scotland, Edinburgh; MS A.4.17, Lincoln Cathedral; MS Digby 196, Bodleian Library, Oxford; and MS Reg. lat. 731, Biblioteca Apostolica Vaticana, Vatican City.

9. Peter Barber, "The Evesham World Map: A Late Medieval English View of God and the World," *Imago Mundi* 47 (1995): 13–33. This map is now at the College of Arms, Muniment Room 18/19.

10. Higden, *The Polychronicon*, chap. 10, pp. 66–78.

11. Cited by Anna-Dorothee von den Brincken, "Universalkartographie und Geographische Schulkenntnisse im Inkunabelzeitalter: Unter Besonderer Berücksightigung des Rudimentum Novitiorum und Hermann Schedels," in *Studien zum Städtischen Bildungswesen des Späten Mittelalters und der frühen Neuzeit*, Abhandlungen der Akademie der Wissenschaften zu Göttingen, Phil-Hist.-Klasse, no. 3 (Göttingen, 1983): 401. Alessandro Scafi thinks this should be interpreted as the earthly paradise: *Mapping Paradise* (Chicago: University of Chicago Press, 2006), 55–57.

12. Wesley A. Brown, "The World Image Expressed in the *Rudimentum Novitiorum*," in Phillips Society, *Occasional Paper Series*, no. 3 (Washington, DC: Library of Congress, 2000), includes reproductions of the maps and a complete list of place-names.

13. Despite the best work of Wes Brown, Testudinum, Mons Aliariorum, Mons Alpharye, Mons Calesti remain mysteries. There is a good color illustration in Peter Whitfield, *The Image of the World* (London: British Library, 1994), 34–35.

14. On technical matters, see von den Brincken, "Universalkartographie," 409–10.

15. For other theories, see Heinrich Winter, "Notes on the World Map in *Rudimentum Novitiorum*," *Imago Mundi* 9 (1952): 102.

16. At least twenty-seven, according to von den Brincken, "Universalkartographie," 402.

17. Ibid., 404.

18. Tony Campbell, *The Earliest Printed World Maps, 1472–1500* (London: British Library, 1987), 144–51.

19. Hugo Hassinger, "Deutsche Weltkarten-Inkunabeln," *Zeitschrift der Gesellschaft für Erdkunde zu Berlin* 9/10 (1927): 455–82, esp. 466–67; Leo Bagrow, "Rüst and Sporer's World Maps," *Imago Mundi* 7 (1950): 32–36; Tony Campbell, *The Earliest Printed World Maps*, 79–84.

20. Such as Edelstein, the magnetic mountain, and the Congealed Sea; see Hassinger, 469–70; (Anon.) *The Legend of Duke Ernst*, trans. J. W. Thomas and Carolyn Dussère (Lincoln: University of Nebraska Press, 1979), 83–109. The legend also includes an eerie account of the visit to the land of the evil cranes.

21. Von den Brincken, "Universalkartographie," 424.

22. MS Borgia 16, Biblioteca Apostolica Vaticana, Vatican City. The map is described by Marcel Destombes, *Mappemondes*, 239–41, and plate 29; Joachim Lelewel, *Géographie du Moyen Age* (Brussels: J. Pilliet, 1852–57), 2:96–103; A. H. L. Heeren, "Explicatio Planiglobii," *Commentationes: Societatis Regiae Scientiarum Gottingensis* (Göttingen: H. Dieterich, 1808), 16:250–84; Roberto Almagià, *Monumenta Cartografica Vaticana* (Vatican City: Biblioteca Apostolica Vaticana, 1944–52), 1:27–29; Konrad Miller, *Mappaemundi: Die Ältesten Weltkarten* (Stuttgart: J. Roth, 1895–98), 3:148–50; Aksinia Džurova, Božidar Dimitrov, José Ruysschaert, and Ivan Dujčev, *Manoscritti Slavi Documenti e Carte Riguardanti la Storia Bulgara* (Sofia: Nauka, 1979), no. 24 and tav. 57 An actual-size reproduction of the map serves as the cover of the book by Džurova et al. Also thanks to John Hamer, who shared his research and ideas about the map and sent me a copy of his master's thesis, "The Borgia Map: Europe's Rise and the Redefinition of the World" (University of Michigan, Ann Arbor, 1995).

23. A. E. Nordenskjold, "The Fifteenth-Century Map of the World on Metal in the Collection Borgia at Velletri," *Ymer* 11 (1891): fasc. 1. Almagià suggests the date 1430, seven hundred years after the Donation of Sutri.

24. John Mandeville, *The Travels of John Mandeville*, trans. C. W. R. D. Moseley (Baltimore: Penguin, 1983), p. 187.

25. "Italia nitens, pinguis, fortis, et superba; ex quibus caret domino uno, iustitia vana."

26. By the twelfth century, European Christians had conflated the story of Alexander and Gog and Magog with that of the Ten Lost Tribes of Israel. This idea was spread by Peter Comestor's popular *Historia Scholastica* (c. 1169). See Andrew Gow, "Gog and Magog on Mappaemundi and Early Printed World Maps: Orientalizing Ethnography in the Apocalyptic Tradition," *Journal of Early Modern History* 2, no. 1 (February 1993): 61–88, esp. 68n17.

27. Anton Mayer, "Mittelalterliche Weltkarten aus Olmütz," in *Kartographische Denkmäler der Sudetenländer*, vol. 8, ed. Bernhard Brandt (Prague: K. André, 1932), 1–5. The map seems to have disappeared after World War II, and the best efforts of Scott Westrem and helpful librarians in the Czech Republic have been unable to locate it. Personal communication from Scott Westrem.

28. "Mare Rubrum quod filii Israel pertransierunt sicis pedibus."

29. Scott D. Westrem, "Against Gog and Magog," in *Text and Territory: Medieval Geographical Imagination in the European Middle Ages*, ed. Sylvia Tomasch and Sealy Gilles (Philadelphia: University of Pennsylvania Press, 1998), 54–75.

30. The Walsperger map is at Biblioteca Apostolica Vaticana, Vatican City (MS Pal. Lat. 1362B). The Zeitz map is at Stiftsbibliothek, Zeitz (Saxony) (MS Lat. Hist., fol. 497, f. 48). The Bell map is in the James Ford Bell Collection, University of Minnesota, Minneapolis.

31. For instructions, see "Instrumentum de distantiis civitatum et regionum et dicitur Cosmographia," by Theodore Ruffi, CLM 11067, fols. 174 et seq., Bayerische Staatsbibliothek, Munich; the tables from which he constructed the map are in CLM 14583, fols. 236r–277v, Bayerische Staatsbibliothek, Munich. These documents are reprinted in Dana B. Durand, *The Vienna-Klosterneuberg Map Corpus of the Fifteenth Century* (Leiden: Brill, 1952), appendices 6 and 14.

32. On the Bell map, see Scott D. Westrem, "Learning from Legends on the James Ford Bell Library Mappamundi," James Ford Bell Lectures, no. 37 (Minneapolis: Associates of the J. F. Bell Library, 2000); on the Zeitz map, see Heinrich Winter, "A Circular Map in a Ptolemaic MS," *Imago Mundi* 10 (1953): 15–22; on the Walsperger map, see Konrad Kretschmer, "Eine Neue Mittelalterliche Weltkarte der Vatikanischen Bibliothek," *Zeitschrift der Gesellschaft für Erdkunde zu Berlin* 26 (1891): 371–406. The Walsperger map is also described in Almagià, *Monumenta*, 1:30–31. Konrad Miller covers Walsperger in *Mappaemundi*, 3:147–48, but his list of inscriptions is incomplete.

33. Kretschmer, "Eine Neue Mittelalterliche Weltkarte," 377.

34. Winter, "Circular Map," 15.

35. Durand, *The Vienna-Klosterneuberg Map Corpus*, 177–78.

36. "Instrumentum," CLM 11067, fol. 174r, Bayerische Staatsbibliothek, Munich.

37. Robert Almagià, "Mappemonde d'Andreas Walsperger," in Destombes, *Mappemondes*, 212–14.

38. Durand, *The Vienna-Klosterneuberg Map Corpus*, 205–206.

39. Kretschmer, "Eine Neue Mittelalterliche Weltkarte," 402.

40. On King Catolonabes, see Mandeville, *The Travels*, 171–72; Odoric, *The Travels of Friar Odoric*, trans. Henry Yule (Grand Rapids, MI: William B. Eerdmans, 2002), 155–57. For a more accurate contemporary account of the cult of the Assassins, see Jean de Joinville, *The Life of Saint Louis*, in *Chronicles of the Crusades*, trans. M. R. B. Shaw (Baltimore: Penguin, 1963), 277–80.

41. Kretschmer, "Eine Neue Mittelalterliche Weltkarte," 397.

42. Durand, *The Vienna-Klosterneuberg Map Corpus*, 192.

43. Ibid., 206.

44. Ibid., 193.

45. This can be found in CLM 14583, fols. 300–12, Bayerische Staatsbibliothek, Munich; Durand, *The Vienna-Klosterneuberg Map Corpus*, 457–76.

46. Von den Brincken, "Universalkartographie," 399; this is also the opinion of Hassinger, "Deutsche Weltkarten-Inkunabeln," 481.

47. Durand, *The Vienna-Klosterneuberg Map Corpus*, 206.

48. Scott D. Westrem, *Broader Horizons: Johannes Witte de Hese's Itinerarius and Medieval Travel Narratives* (Cambridge, MA: Medieval Academy, 2001).

49. MS 3119, Biblioteca Comunale (1442) Verona; MS 598(a) (1448), Museo Civico, Vicenza; Leardo World Map, American Geographical Society (1452), Milwaukee, WI. See John K. Wright, *The Leardo Map of the World* (New York: AGS, 1928), with facsimile; also Harley and Woodward, eds., *History of Cartography*, 1:316–18, plate 20 (Vicenza, color) and fig. 18.50 (AGS, black and white); M. de la Roncière, "Mappemondes de Giovanni Leardo" in Destombes, *Mappemondes*, 208–11, and plate 30 (AGS map).

50. Wright, *The Leardo Map*, 31–60.

51. Quoted in Wright, *The Leardo Map*, 22, from Cod. ital. 7.291, p. 542. Biblioteca Marciana, Venice.

52. Tony Campbell, *The Earliest Printed Maps*, 23–26 and 41 (illus.); Erich Woldan, "A Circular, Copper-Engraved, Medieval World Map," *Imago Mundi* 11 (1956): 13–16. The two surviving copies are in private collections.

53. Portolano 1, Biblioteca Nazionale Centrale, Florence.

54. Theobald Fischer, *Mittelalterliche Welt-und Seekarten: Italienischen Ursprungs* (Venice: F. Ongania, 1886), 155–206, esp. 157; Edward L. Stevenson, *The Genoese World Map, 1457* (New York: American Geographical Society, 1912), with facsimile; Gerald Crone, "Anonyme dîte genoise, 1457," in Destombes, *Mappemondes*, 222–23. Peter Whitfield has a color illustration in his *The Image of the World*, 40–41.

55. Gaetano Ferro, *The Genoese Cartographic Tradition and Christopher Columbus*, trans. Ann Heck and Luciano F. Farina (Rome: Libreria dello Stato, 1996), 38.

56. T. Fischer, *Mittelalterliche Welt-und Seekarten* 158.

57. This is the suggestion of David Woodward, "Medieval Mappaemundi," 313.

58. Fazio degli Uberti, *Il Dittamondo e le Rime*, ed Giuseppe Corsi (Bari: Gius, Laterza and Figli, 1952), 1:19.

59. On classical circumnavigations of Africa, see Pomponius Mela, *Description of the World*, trans. F. E. Romer (Ann Arbor: University of Michigan Press, 1998), book 3.90.

60. For a look at this not very edifying controversy, consult *Rivista Geografica Italiana* 49 (1942): 44–54, for an article by Sebastiano Crinò, "Ancora sul Mappamondo del 1457 e Sulla Carta Navigatoria di Paolo dal Pozzo Toscanelli," and a response by Renato Biasutti, "Il Mappamondo del 1457 non è la Carta Navigatoria di Paolo dal Pozzo Toscanelli."

61. C.G.A. 1 Biblioteca Estense, Modena. A facsimile and transcription of all the legends were made by Ernesto Milano and Annalisa Battini of the Biblioteca Estense of Modena: *Il Mappamondo Catalano Estense del 1450* (Dietikon, Switzerland: Urs Graf, 1995). Unfortunately, it was a very expensive, limited edition and is difficult to find in libraries. Other accounts of the map can be found in R. A. Skelton, "Mappemonde Anonyme Catalane de Modena," in Destombes, *Mappemondes*, 217–21; Arthur Dürst, "Die Katalanische Estense-Weltkarte, um 1450: Bericht zur Faksimile-Ausgabe," *Cartographica Helvetica* 14 (July 1996): 42–44; George Kimble, *Memoir: The Catalan World Map at R. Biblioteca Estense at Modena* (London: Royal Geographical Society, 1932), with facsimile; Charles de la Roncière, *La Découverte de l'Afrique au Moyen Age* (Cairo: Société Royale de Géographie d'Egypte, 1925), 1:118 and plate 10. The Biblioteca Estense has a photo of the map, which may be examined in detail, at www.cedoc.mo.it/estense (Septmeber 22, 2006).

62. Skelton cites this as Destombes's opinion, "Mappemonde Anonyme Catalane de Modena," 221.

63. "Aphrica comensa al flum de nilles en les partes degipta e fenex en Gutzola vert ponent circuit tota la barbaria e la part de mig jorn." Gutzola, variously spelt, is Gazuola on the Fra Mauro map, Gaççola on the Petrus Roselli chart of 1462, Gozol on the Catalan Atlas, and Gazula in the *Libro del conoscimiento*, ed. and trans. Nancy F. Marino (Tempe: Arizona Center for Medieval and Renaissance Studies, 1999). The author describes the city as "rich and comfortable" (*Libro*, 57).

64. "A quest cap es fi de la terra del ponent de la part de affrica aquesta linia es an la equinocsiall en la qual sta continuament lo soll . . . que fa xii hores de nit e xii de dia . . . "

65. "Esi posa ercules dues colones."

66. Milano and Battini, *Il Mappamondo Catalano*, 62.

67. "Questas illas son appelladas islandes."

68. Milano, and Battini, *Il Mappamondo Catalano*, 24.

69. The mappamundi is fol. 9 in MS It. Z, 76 (Atlas of Andrea Bianco), Biblioteca Nazionale Marciana, Venice; Piero Falchetta, *Andreas Biancho de Veneciis me Fecit M.CCCC.XXX.VJ.* (Venice: Arsenal Editrice, 1993. This facsimile, made for wealthy customers of the Banco San Marco, is now out of print and difficult to find.

70. Note that these inscriptions are difficult to make out and receive variant readings from different scholars.

71. Peter Barber, "The Maps, Town-Views and Historical Prints in the Columbus Inventory," in *The Print Collection of Ferdinand Columbus (1488–1539): A Renaissance Collector in Seville,* ed. Mark P. McDonald (London: British Museum Press, 2004), 1:251.

72. Erin C. Blake, "Where Be Dragons?" *Mercator's World* 4, no. 4 (July/August 1999): 80.

73. Examples are Mercator's world map of 1569, reproduced by Peter Whitfield, *Images of the World,* 66–67; G. Blaeu, Map of Africa (1648–65), in R. V. Tooley, *Maps and Map-Makers* (New York: Dorset, 1987), plate 72.

74. For example, Heinrich Berghaus's world map of the distribution of the races (1852), reproduced in Whitfield, *Images of the World,* 120–21.

75. Hartmann Schedel, *The Nuremberg Chronicle* (Nuremberg: Anton Koberger, 1493). A facsimile of the German edition was printed in New York by the Landmark Press in 1979.

76. This is "Missa de lo Mapamundo." Juan Cornago (fl. 1455–75) was Spanish and worked at the Aragonese court of Naples. See Allan W. Atlas, "Aggio Visto lo Mappamondo: A New Reconstruction," in *Studies in Musical Sources and Style: Essays in Honor of Jan LaRue,* ed. Eugene K. Wolf and Edward H. Roesner (Madison, WI: A-R Editions, 1990), 109–17.

Eight • *The Transformation of the World Map*

1. There was a huge variation in measures during the late Middle Ages. For a table of comparisons, see Gustavo Uzielli, *La Vita e i Tempi di Paolo dal Pozzo Toscanelli*, in Reale Commissione Colombiana, Raccolta di documenti e studi, 5:1 (Rome, 1894), 419–21.

2. Boies Penrose, *Travel and Discovery in the Renaissance, 1420–1620* (New York: Atheneum, 1962), 100.

3. A *maravedi* was a small copper coin worth about nine U.S. cents today. So the expedition would have cost about $90,000, and Columbus's salary would have been worth $12,600.

4. Washington Irving, *The Life and Voyages of Christopher Columbus,* ed. John Harmon McElroy (Boston: Twayne, 1981), 47–53. See also the comments of Samuel Eliot Morison on this issue in the introduction, p. xciv. Irving based his work on extensive reading in the relevant documents, but, as was common in the nineteenth century, there are few footnotes and the style is florid.

5. Aristotle, *On the Heavens (De Caelo),* trans. W. K. C. Guthrie (Cambridge, MA: Harvard University Press, 1953), 2.14, p. 253.

6. Richard Southern, *Robert Grosseteste: The Growth of an English Mind in Medieval Europe* (Oxford: Clarendon Press, 1986), 143–44.

7. Danielle LeCoq, "Saint Brandan, Christophe Columb et le Paradis Terrestre," *Révue de la Bibliothèque Nationale* 45 (Autumn 1992): 14–21.

8. Felipe Fernández-Armesto, *Before Columbus: Exploration and Civilization from the Mediterranean to the Atlantic* (Philadelphia: University of Pennsylvania Press, 1987), 250–51.

9. Aristotle, *On the Heavens*, 2.14, p. 253.

10. For Pierre D'Ailly's interpretation, see *Ymago Mundi*, ed. and trans. Edmond Buron (Paris: Maisonneuves Frères, 1930), 2:536–37; Pliny, *Natural History*, trans. H. Rackham (Cambridge, MA: Harvard University Press, 1942), 6.21.56–58, p. 381.

11. Jerome, *Liber de Situ et Nominibus Hebraicorum*, in Pauli de Lagarde, *Onomastica Sacra* (Göttingen: Horstmann, 1887), "Ailath" and "Sophira." The reference to 1 Kings 10:22 is unclear. The Bible says that the ships came every three years, not that this was the length of the journey.

12. Ptolemy, *Ptolemy's Geography*, ed. and trans. J. Lennart Berggren and Alexander Jones (Princeton, NJ: Princeton University Press, 2000), 1.11, pp. 71–72). In Ptolemy, *Almagest*, trans. G. J. Toomer (London: Springer, 1984), 2.1, he says that the known world is one-quarter of the globe

13. D'Ailly, chap. 8, "De quantitate terre habitabilis," in *Ymago Mundi*, 1:206–15, includes Columbus's notes. There is some dispute about which notes are by Bartholomew and which by Christopher Columbus. Apparently, their handwriting was similar.

14. D'Ailly, *Ymago Mundi*, 2:530–31; of course, a degree of longitude varies according to latitude, being sixty-nine miles at the equator (modern figure) and zero at the pole.

15. For a full account of the length of the degree supposedly measured by Columbus, see George E. Nunn, *Geographical Conceptions of Columbus* (New York: American Geographical Society, 1924), 1–17.

16. D'Ailly, *Ymago Mundi*, 1:206–9. Dias' estimate was 10° off, the modern reading being 35°. Arthur Davies argues that Bartholomew Columbus distorted this reading to support his case that the trip around Africa was too long; see his "Behaim, Martellus, and Columbus," *Geographical Journal* 143 (November 1977): 451–59. His thesis is convincingly denounced by Ilaria Luzzana Caraci, "Il Planisfero di Enrico Martello," *Rivista Geografica Italiana* 85 (1978): 132–43.

17. Examples of other significant errors are cited by Luzzana Caraci, "Il Planisfero," 140, due to "the imperfection of the instruments, the difficulty of using an unfamiliar method, and the difference of longitude from Lisbon."

18. Nunn, *Geographical Conceptions*, 23–24.

19. Sebastiano Gentile, *Firenze e la Scoperta dell'America: Umanesimo e Geografia nel '400 Fiorentino* (Florence: Olschki, 1992), cat. item no. 72 (Francesco Castellani's memo book), 146–48, and cat. item no. 83 (Genoese map of 1457), 173–75.

20. For a text of the letter, see G. Uzielli, 571–75, for all three versions; Henry Vignaud has an English translation, in *Toscanelli and Columbus: The Letter and Chart of Toscanelli* (London: Sands, 1902), 275–92. For a reconstruction of the map, see Uzielli, *La Vita*, plate 10; Vignaud, *Toscanelli and Columbus*, appendix J.

21. Nicolò de' Conti, *L'India di Nicolò de' Conti*, ed. Alessandro Grossato (Padova: Studio Editoriale Programma, 1994), 26. Poggio adds some information he got from a Nestorian Christian from "upper India" who visited the papal court (ibid., 42).

22. Vignaud's theory is that Bartholomew Columbus forged the letter that appears on the flyleaf of Columbus's copy of Pius II's *Historia Rerum Ubique Gestarum* (Venice: Ioannes de Colonia, 1477), and that the other two copies are derived from this one. See *Toscanelli and Columbus*, 153–56.

23. Patrick Gautier Dalché, "Pour une histoire du regard géographique: Conception et usage de la carte au XVe siècle," *Micrologus* 4 (1996): 77–103, esp. 77–78.

24. The map is in the Beinecke Library at Yale University (MS 350A).

25. Kirsten Seaver, *Maps, Myths and Men: The Story of the Vinland Map* (Stanford, CA: Stanford University Press, 2004). The primary proponent of the map's authenticity has been James R. Enterline. See his *Erikson, Eskimos, and Columbus* (Baltimore: Johns Hopkins University Press, 2002), esp. 61–70. The map was launched into the academic world with a tome by R. A. Skelton, Thomas E. Marston, and George D. Painter, *The Vinland Map and the Tartar Relation* (New Haven, CT: Yale University Press, 1965.) This work has been severely criticized by Seaver. A review of its reissue in 1995 is Paul Saenger, "Vinland Re-Read," *Imago Mundi* 50 (1998): 199–202, which casts doubt on the paleography and provenance of the map.

26. This is the opinion of Michael Livingstone, "More Vinland Maps and Texts: Discovering the New World in Higden's *Polychronicon*," *Journal of Medieval History* 30, no. 1 (March 2004): 25–44. For a description of Vinland in an anonymous geographical text, see Marvin L. Colker, "America Rediscovered in the 13th Century?" *Speculum* 54, no. 4 (1979): 712–26.

27. This assertion goes back to Las Casas, but see Gaetano Ferro, "Cristoforo e Bartolomeo Colombo Cartografi," in *Cristoforo Columbo e l'Apertura degli Spazi*, ed. Guglielmo Cavallo (Rome: Libreria dello Stato, 1992), 1:381–99.

28. Tony Campbell, "Portolan Charts from the Late Thirteenth Century to 1500," in *Cartography in Prehistoric, Ancient, and Medieval Europe and the Mediterranean*, vol. 1 of *History of Cartography*, ed. David Woodward and J. B. Harley (Chicago: University of Chicago Press, 1987), 430; See also Gaetano Ferro, "Cartografi e Dinastie di Cartografi a Genova," in Cavallo, ed., *Cristoforo Columbo*, 1:245–62, 252–56.

29. D'Ailly, *Ymago Mundi*, 1:306–7, 3:602–3.

30. Rés. Ge AA 562, Cartes et Plans, Bibliothèque Nationale, Paris.

31. Arguments can be found in de la Roncière, *La Carte de Christophe Colomb* (Paris: Eds. Historiques, 1924), and *La Découverte de l'Afrique au Moyen Age* (Cairo: Société Royale de Géographie d'Égypte, 1925), 2:40–63.

32. Marcel Destombes, ed., *Mappemondes, 1200–1500* (Amsterdam: N. Israel, 1964), 51.26, p. 185.

33. De la Roncière, *La Découverte de l'Afrique*, 2:49; the inscription is almost invisible on the chart.

34. *The Journal of Christopher Columbus*, trans. Cecil Jane (London: A. Blond, 1968), 25 September 25 and October 3, 1492, pp. 14–15, 18; November 14, p. 62.

35. De la Roncière, *La Découverte de l'Afrique*, 2:53. D'Ailly, *Ymago Mundi*, 1:306–7, 3:602–3. I have seen only one other "map with a sphere," that of Vesconte Maggiolo, 1535 in Turin (J.B.III.18, Biblioteca Antica, Archivio di Stato). This map is reproduced in Cavallo, ed., *Cristoforo Columbo*, vol. 1, plate 2.21. I. L. Caraci gives some other examples in "Regarding the So-Called 'Columbus Map,'" in *The Puzzling Hero* (Rome: Carocci Ed., 2002), 85–107, esp. 99–103.

36. De la Roncière, *La Découverte de l'Afrique*, 2:44. "Mélange plus or moins heureux de termes archaïques et de la nomenclature la plus moderne, de hypothèse et de la réalité, de résultats acquis et de découvertes supputées d'avance, ce type étrange, c'est celui des cartes de Christophe Colomb."

37. Monique Pelletier, "Peut-on Encore Affirmer que la BN Possède la Carte de Christophe Colomb?" *Révue de la Bibliothèque Nationale* 45 (Autumn 1992): 22–25. For a more skeptical opinion, see I. L. Caraci, "Regarding the So-Called 'Columbus Map,'" 85–107.

38. Note by Roberto Almagià on his name, "I Mappamondi di Enrico Martello e Alcuni Concetti Geografici di Cristoforo Colombo," *Bibliofilia* 42 (1940), 288–311, esp. 290. We have no reference to him other than his own signature, always in Latin, on his maps.

39. See R. A. Skelton, "Henricus Martellus," in Destombes, *Mappemondes*, 229–33, and plates 37–38. Also Alexander Vietor, "A Pre-Columbian Map of the World, ca. 1489," *Yale University Library Gazette* 37 (1962): 8–12, on the Yale map.

40. MS Vat. lat. 7289, Biblioteca Apostolica Vaticana, Vatican City. The *Insularium* at the James F. Bell Library, University of Minnesota, may be slightly earlier.

41. "Ornatissimo." Gentile, *Firenze*, cat. item no. 114, p. 241.

42. More than seventy manuscripts survive. See Giuseppe Ragone, "Il *Liber Insularum Archipelagi* di Cristoforo Buondelmonti," in *Humanisme et Culture Géographique à l'Époque du Concile de Constance Autour de Guillaume Fillastre*, ed. Didier Marcotte (Turnhout: Brepols, 2002), 177–217, esp. 181.

43. MS 29.25, Biblioteca Medicea Laurenziana, Florence. A reproduction of the world map is in Gentile, *Firenze*, tav. 45. The map of Cipangu is reproduced in Almagià, "I Mappamondi," 303, and in Gentile, *Firenze*, tav. 46.

44. Skelton, "Henricus Martellus," 230.

45. Add. MS 15,760, fols. 68v–69r, British Library, London.

46. For a thorough quashing of these claims, see William A. R. Richardson, "South America on Maps Before Columbus? Martellus's 'Dragon's Tail' Peninsula," *Imago Mundi* 55 (2003): 25–37.

47. MS 29.25 Biblioteca Medicea Laurenziana, Florence; there is a good reproduction in Gentile, *Firenze*, tav. 45.

48., MS 29.25, fol. 55, Biblioteca Medicea Laurenziana, Florence. The poem is quoted in Gentile, *Firenze*, 239: "Si vacat, ipse potes que scribimus, hospes, adire:/ Tunc quoque sit, quamvis utilis iste labor./ At si non facile est, patria tellure relicta,/ Alba procellosum, per mare vela dare,/ Me duce que multis ipsi lustravimus annis./ Si sapis, exiguo tempore disce domi" (If time allows, dear reader, you may draw near to what we write. Although it may be worthwhile, it is not easy to leave one's native soil and set a white sail upon the stormy sea. With me, who has traveled about for many years, as your guide, you, if you are wise, may learn all this at home in a short space of time). The verse also appears in MS483, Musée Condé, Chantilly, France, fol. lv.

49. The only surviving copy is in Florence in the Landau-Finaly collection, Biblioteca Nazionale Centrale.

50. Tony Campbell, *The Earliest Printed Maps, 1472–1500* (London: British Library, 1987), 70–78, plate 42; the map is also reproduced almost actual size and in color in Cavallo, ed., *Cristoforo Colombo*, 1, plate 3.28.

51. "Forma universalis totius quod oceano mari ambitur cum superioris Indiae portione post ptholomei tempus inventa cumque ea parte Aphricae quam temporibus nostris lusitani nautae perlustrarunt ita se habet." The complete inscription may be read from the reproduction of the map in Cavallo's book; Tony Campbell omits a few words (*The Earliest Printed Maps*, 70).

52. On this map, see also Robert Almagià, "On the Cartographic Work of Francesco Rosselli," *Imago Mundi* 8 (1951), 27–34; Diogo Ramada Curto, Angelo Cattaneo, and André Ferrand Almeida, *La Cartografia Europea tra Primo Rinascimento e Fine dell'Illuminismo* (Florence: Olschki, 2003), cat. 1.5, pp. 345–46; Florio Banfi, "Two Italian Maps of the Balkan Peninsula," *Imago Mundi* 11 (1954): 17–34, discusses the two cartographers, especially in connection with their regional maps.

53. Ilana Luzzana Caraci, III.8 "Francisco Rosselli, Planisfero," in Cavallo, *Cristoforo Colombo*, 1:521–24.

54. Patrick Gautier Dalché, "Jean Fusoris et la Géographie," in *Humanisme*, ed. Marcotte, 161–176, esp. 162.

55. Jacques Paviot, "La Mappemonde Attribuée à Jan van Eyck par Fàcio: Une Pièce à Retirer du Catalogue de son Oeuvre," *Revue des Archéologues et Historiens d'Art du Louvain* 24 (1991): 57–62.

56. Oswald Dreyer-Eimbcke, "Conrad Celtis: Humanist, Poet and Cosmographer," *Map Collector* 74 (1996): 18–21. In 1508, Celtis left two globes to the university in his will. It has been suggested that his globes were made by Hans Dorn (1430–1509).

57. Wolfgang Pülhorn and Peter Laub, eds., *Focus Behaim Globus*, 2 vols. (Nürnberg: Germanisches Nationalmuseum, 1992). The first volume is a collection of scholarly articles, and the second is a catalogue of the exhibition, directed by Johann K. W. Willers and mounted on the occasion of the 500th anniversary of the globe's creation.

58. Nuremberg, Germanisches Nationalmuseum. G. R. Crone, "Globe de Martin Behaim," in Destombes, *Mappemondes*, 234–35.

59. The inscriptions are printed and translated by E. G. Ravenstein, *Martin Behaim's 1492 "Erdapfel"* (London: Greaves and Thomas, 1992). This is a reissue, somewhat abridged, of a 1908 edition published in London by George Philip, accompanied by a facsimile.

60. Ravenstein, *Martin Behaim's 1492 "Erdapfel,"* 32.

61. Ibid., 15.

62. Almagià, "I Mappamondi," 311; de la Roncière, *La Carte Christophe Colomb*.

63. Felipe Fernández-Armesto, "Columbus and Maps," *Map Collector* 58 (Spring 1992): 2–5.

Conclusion • *The World Map Transformed*

1. With the possible exception of the Vinland map, see chap. 8.

2. *Liber de Existencia Riveriarum* in *Carte Marine et Portulan au XIIe Siécle*, ed. Patrick Gautier Dalché (Rome: École Française de Rome, 1995), 116.

3. Valerie Flint presents an interesting treatment of this question in *The Imaginative Landscape of Christopher Columbus* (Princeton, NJ: Princeton University Press, 1992), chap. 6.

4. Sebastiano Gentile, "Umanesimo e Cartografia: Tolomeo nel Secolo XV," in Diogo Ramada Curto, Angelo Cattaneo, and André Ferrand Almeida, *La Cartografia Europea tra Primo Rinascimento e Fine dell'Illuminismo* (Florence: Olschki, 2003), 3–18.

5. Rés Ge. DD 683, Cartes et Plans, Bibliothèque Nationale de France, Paris.

6. *The Merchant of Venice*, Act I, sc. 1. Antonio is agonizing about the fate of his ships out at sea. *Piring* means *peering.*

7. On this issue, see Walter D. Mignolo, *The Darker Side of the Renaissance: Literacy, Territoriality, and Colonization* (Ann Arbor: University of Michigan Press, 1995), esp. chap. 6; Felipe Fernández-Armesto, *Before Columbus* (Philadelphia: University of Pennsylvania Press, 1987), 244.

8. Ptolemy, *Geography*, ed. J. Lennart Berggren and Alexander Jones (Princeton, NJ: Princeton University Press, 2000), book 1, pp. 58–59.

9. The phrase is Mignolo's (*The Darker Side*, 258).

Manuscripts

Berlin, Staatsbibliothek Preussischer Kulturbesitz, MS Hamilton 396.

Burgo de Osma, Cathedral, MS 1.

Cambridge, Corpus Christi College, MS 26.

Chantilly (France), Musée Condé, MS 483.

Chicago, Newberry Library, MS Ayer 746.

Cortona, Biblioteca dell'Accademia Etrusca, inv. 100.

Edinburgh, National Library of Scotland, MS 33.4.12.

Florence, Biblioteca Medicea Laurenziana, MS 91, inf. 7; MS 29.25.

Florence, Biblioteca Nazionale Centrale, Portolano 1; Landau-Finaly Collection, Rosselli map printed; MS Magliabechiano 13.16.

Ghent, Rijksuniversiteit, Centrale Bibliotheek, MS 92.

Hereford Cathedral, mappamundi.

Lincoln, Lincoln Cathedral, MS A.4.17.

London, British Library, Add. MS 10,104; Add. MS 10,049; Add. MS 15,760; Add. MS 25,691; Add. MS 27,376; Add. MS 28,681; MS Cotton Domitian A.13; MS Cotton Nero D.5; MS Cotton Tiberius B.V-1; MS Egerton 1500; MS Harley 3954; MS Harley 3673 (1466); MS Roy.14 C.IX; MS Roy.14 C.VII.

London, The College of Arms, Muniment Room 18/19.

Milwaukee, Wisc., American Geographical Society (Leardo, 1452).

Minneapolis, University of Minnesota, James Ford Bell Collection, Bell map.

Modena, Biblioteca Estense e Universitaria, MS C.G.A. 1, MS C.G.A. 5a.

Munich, Bayerische Staatsbibliothek, CLM 11067, CLM 14583, CLM 14731.

Nancy, Bibliothèque Municipale, MS 441.

New Haven, Conn., Yale University, Beinecke Library, MS 506, MS 1980.158; Taylor Collection, MS 30, MS 557; Vinland map, MS 350A; World map of Henricus Martellus Germanus.

Nuremberg, Germanisches Nationalmuseum, Behaim globe.

Oxford, Bodleian Library, MS Douce 319, MS e Mus. 223, MS Tanner 190.

Oxford, Corpus Christi College, MS 2*, MS 89.

Paris, Bibliothèque Nationale de France, Cartes et Plans, Rés. Ge. AA 562, B 1118, D 7900, DD 687, B 696; MS Esp. 30, MS fr. 1116, MS fr. 2810, MS lat. 3123, MS lat. 4802, MS lat. 4939, MS lat. 17542.

Reims, Bibliothèque Municipale, MS 1321.

Rennes, Bibliothèque Municipale, MS 256.

San Marino, California, Huntington Library, MS HM 132.

Turin, Archivio di Stato, Biblioteca Antica, MS J.B.III.18.

Valenciennes, Bibliothèque Municipale, MS 344.

Vatican City, Biblioteca Apostolica Vaticana, Arch. di S. Pietro, MS H.31; MS Borgia 5, MS Borgia 16, MS Otto.lat. 1771, MS Pal.lat. 1362a, MS Pal.lat. 1362b, MS Reg.lat. 731, MS Urb.gr. 82, MS Urb.lat. 274, MS Urb.lat. 275, MS Vat.lat. 1960, MS Vat.lat. 2972, MS Vat.lat. 2974, MS Vat.lat. 5698, MS Vat.lat. 5699, MS Vat. lat. 7289.

Venice, Biblioteca Marciana, MS Ital. 7.291, MS Ital. Z.76, Fra Mauro world map.

Verona, Biblioteca Comunale, Leardo, 1442.

Vicenza, Museo Civico, Portulano I (1448).

Vienna, Nationalbibliothek, Cod. 324.

Zeitz (Saxony), Stiftsbibliothek, MS Lat. Hist., fol.497.

Primary Sources

Adam of Bremen. *Hamburgische Kirchengeschichte*. Edited by Bernhard Schmeidler. Hannover: Hahnsche, 1917.

Alexander. *Epistola ad Aristotelem*. Edited by W. Walter Boer. Meisenheim: Anton Hain, 1973.

Anonymous. "Arte del Navigare: Manuscritto Inedito Datato 1464–65." Edited by Claudio de Polo Saibanti. In Marzoli 1981.

———. *Il Compasso da Navigare*. Edited by Bacchisio Motzo. In *Annali della Facoltà di Lettere e Filosofia della Università di Cagliari*, 8 (1947): 1–137.

———. *Konungs Skuggsjá*, or *King's Mirror*. Translated by L. M. Larson. New York: American Scandinanvian Foundation, 1917.

———. *De Expugnatione Lyxbonensi*. Edited by Charles W. David. New York: Columbia University Press, 1936.

———. *Knowledge of the World*. Edited by Clements R. Markham. London: Hakluyt Society, 1912.

———. *The Legend of Duke Ernst*. Translated by J. W. Thomas and Carolyn Dussère. Lincoln: University of Nebraska Press, 1979.

———. *El Libro del Conoscimiento de Todos los Reinos*. Edited and translated by Nancy F. Marino. Tempe: Arizona Center for Medieval and Renaissance Studies, 1999.

———. *Periplus Maris Erythraei*. Edited by Lionel Casson. Princeton, NJ: Princeton University Press, 1989.

———. *The Tartar Relation*. Translated by George D. Painter. In Skelton, Marston, and Painter 1965.

Aristotle. *On the Heavens* (*De Caelo*). Translated by W. K. C. Guthrie. Cambridge, MA: Harvard University Press, 1953.

Bacon, Roger. *Opus Majus*. Edited by John Henry Bridges. Oxford: Clarendon, 1897. Reprint, Frankfurt-am-Main: Minerva, 1964.

———. *Opus Majus of Roger Bacon*. Edited by Robert Belle Burke. 2 vols. Philadelphia: University of Pennsylvania, 1928.

Burchard of Mt. Sion. *A Description of the Holy Land*. Edited by Aubrey Stewart. Vol. 12. London: Palestine Pilgrims' Text Society, 1896. Reprint, New York: AMS Press, 1971.

Cadamosto, Alvise. *Voyages of Cadamosto and Other Documents on West Africa in the Second Half of the 15th Century*. Edited by G. R. Crone. Hakluyt Society Series. 2d ser. Vol. 80. London: Hakluyt Society, 1937.

Chaucer, Geoffrey. *A Treatise on the Astrolabe* 1391. Edited by Walter W. Skeat. 1872. Reprint, London: Early English Text Society, 1968.

Christine de Pisan. *Le Livre du Chemin de Long Estude*. Berlin, 1887. Reprint, Geneva: Slatkine Reprints, 1974.

Columbus, Christopher. *The Journal of Christopher Columbus*. Translated by Cecil Jane. London: A. Blond, 1968.

———. *The Log of Christopher Columbus*. Translated by Robert H. Fuson. Camden, ME: International Marine Publishing Company, 1987.

Comnena, Anna. *The Alexiad*. Translated by E. R. A. Sewter. Baltimore: Penguin, 1969.

Conti, Nicolò de'. *L'India di Nicolò de' Conti*. Edited by Alessandro Grossato. Padova: Studio Editoriale Programma, 1994.

Cresques, Abraham. *Atlas Catalan*. In *Mapamondi: Une Carte du Monde au XIVe Siècle* by Monique Pelletier, Danielle Le Coq, and Jean-Paul Saint Aubin. Paris: Montparnasse Multimedia, 1998. CD-ROM produced by the Bibliothèque Nationale de France.

D'Ailly, Pierre. *Ymago Mundi*. Edited by Edmond Buron. 3 vols. Paris: Maisonneuve Frères, 1930.

Dati, Goro. *La Sfera*. Edited by Enrico Narducci. Milan: G. Daelli, 1865. Reprint, 1975.

Fazio degli Uberti. *Il Dittamondo e le Rime*. Edited by Giuseppe Corsi. 2 vols. Bari: Gius, Laterza and Figli, 1952.

Fillastre, Guillaume. "L'Oeuvre Géographique du Cardinal Fillastre." Edited by Patrick Gautier Dalché. In Marcotte 2002.

Gautier de Châtillon. *The Alexandreis of Walter of Châtillon*. Translated by David Townsend. Philadelphia: University of Pennsylvania Press, 1996.

Gervase of Tilbury. *Otia Imperialia*. Edited and translated by S. E. Banks and J. W. Binns. Oxford: Clarendon, 2002.

Hayton (Hetoum). *La Flor des Estoires de la Terre d'Orient*. In *Recueil des Historiens des Croisades: Documents Armeniens*, edited by Charles Kohler. Vol. 2. Paris: Imprimerie Nationale, 1906. Reprint, 1967.

———. *A Lytell Cronycle: Richard Pynson's Translation c.1520 of La Fleur des Histoires de la Terre d'Orient c. 1307*. Edited by Glenn Burger. Toronto: University of Toronto Press, 1988.

Higden, Ranulf. *Polychronicon Ranulphi Higden Monachi Cestrensis*. Edited by Churchill Babington. Vols. 41 and 42 of *Chronicles and Memorials of Great Britain and Ireland in the Middle Ages*. London: Longman Green, 1865.

Honorius Augustodunensis. *Imago Mundi*. Edited by Valerie Flint. *Archives d'histoire doctrinale du Moyen Age* 57 (1982): 48–93.

Ibn Baṭṭūṭah. *Travels of Ibn Baṭṭūṭah, 1325–54*. Translated by C. Defrémery and B. R. Sanguinetti. Completed by C. F. Beckingham. 5 vols. Cambridge: Cambridge University Press for the Hakluyt Society, 1958–2000.

Al-Idrisi. *Géographie d'Edrisi*. Edited by P. A. Jaubert. Paris, 1836. Reprinted as volumes 2 and 3 of *Islamic Geography*, edited by Fuat Sezgin. Frankfurt-am-Main: Institute for History of Arabic-Islamic Science, 1992.

Isidore. *Etymologiarum sive Originum Libri xx*, Edited by W. M. Lindsay. 2 vols. Oxford: Clarendon, 1911.

Jerome. *Liber de Situ et Nominibus Locorum Hebraicorum.* Edited by Paul Lagarde. *Onomastica Sacra: Studia et Sumptibus Alterum Edita.* Göttingen: Horstmann, 1887.

Joinville, Jean de. *The Life of Saint Louis.* In *Chronicles of the Crusades,* translated by M. R. B. Shaw. Baltimore: Penguin, 1963.

Jordanes, Friar. *Mirabilia Descripta: Wonders of the East, c. 1330.* Edited by Henry Yule. London: Hakluyt Society, 1863.

Lambert of St. Omer. *The Autograph Manuscript of the Liber Floridus.* Edited by Albert Derolez. Turnhout: Brepols, 1998.

Latini, Brunetto. *Brunetto Latini: The Book of the Treasure.* Edited by Paul Barrette and Sturgeon Baldwin. New York: Garland, 1993.

Macrobius. *Commentary on the Dream of Scipio.* Translated by William H. Stahl. New York: Columbia University Press, 1952.

Mandeville, John. *Travels.* Edited by C. W. R. D. Moseley. Baltimore: Penguin, 1983.

Marignolli, Iohannes. *Relatio.* In van den Wyngaert 1929.

Marco Polo. *The Travels.* Translated by Ronald Latham. Baltimore: Penguin, 1958.

———. *The Travels of Marco Polo: The Complete Yule-Cordier Edition.* 2 vols. London: John Murray, 1929. Reprint, New York; Dover, 1993.

Mela, Pomponius. *Description of the World.* Edited and translated by F. E. Romer. Ann Arbor: University of Michigan Press, 1998.

Monte Corvino, Iohannes. *Epistolae Fr. Iohannis.* In van den Wyngaert 1929.

Monte Corvino, John. *Letters of John of Monte Corvino.* In Dawson 1980.

Nangis, Guillaume de. *Gesta sancti Ludovici.* Vol. 20 of *Recueil des Historiens des Gaules et de la France.* 23 vols. Edited by Leopold Delisle. Paris: Imprimerie Royale, 1840.

Odoric of Pordenone, Blessed. *Travels.* Translated by Henry Yule. Grand Rapids, MI: William Eerdmans, 2002.

Odoricus de Portu Naonis. *Relatio.* In van den Wyngaert 1929.

Orosius, Paulus. *Seven Books of History Against the Pagans.* Edited and translated by Roy J. Deferrari. Washington, DC: Catholic University Press, 1964.

Paris, Matthew. *English History.* Translated by John A. Giles. 3 vols. London: Henry Bohn, 1852.

Pegolotti, Francesco Balducci. *La Pratica della Mercatura.* Edited by Allan Evans. Cambridge, MA: Medieval Academy of America, 1936.

Petrarca, Francesco. *Itinerario in Terra Santa.* Edited and translated by Francesco Lo Monaco. Bergamo: Pierluigi Lubrina, 1990.

Pian di Carpini, Giovanni di. *Historia Mongolorum.* In van den Wyngaert 1929.

Plano Carpini, John. *History of the Mongols.* In Dawson 1980.

Piccolomini, Aeneas Sylvius (Pope Pius II). *Historia Rerum Ubique Gestarum.* Venice: Ioannes de Colonia, 1477.

Pliny, *Natural History.* Translated by H. Rackham. 10 vols. Cambridge, MA: Harvard University Press, 1942.

Ptolemy, *Claudii Ptolemaei Geographiae: Codex Urb. gr. 82.* 2 vols. in 4 bks. Edited by Joseph Fischer. Leiden: Brill, 1932.

Ptolemy, Claudius. *Geographia Florence 1482.* Printed edition of Berlinghieri's Ptolemy. Edited by R. A. Skelton. Amsterdam: Theatrum Orbis Terrarum, 1966.

———. *The Geography.* Edited and translated by Edward L. Stevenson. New York: New York Public Library, 1932. Reprint, New York: Dover, 1991.

————. *Ptolemy's Geography: An Annotated Translation of the Theoretical Chapters.* Edited by J. Lennart Berggren and Alexander Jones. Princeton, NJ: Princeton University Press, 2000.

Ramusio, Gianbattista. *Navigazioni e Viaggi.* Edited by Marica Milanese. 6 vols. Turin: Einaudi, 1978–85.

Sacrobosco, John. *De Sphaera.* In *The Sphere of Sacrobosco and its Commentators.* Edited by Lynn Thorndike. Chicago: University of Chicago Press, 1949.

Sanudo, Marino. *Liber Secretorum Fidelium Crucis.* Edited by Joshua Prawer. Hanover, 1621. Reprint, Toronto: University of Toronto Press, 1972.

Schedel, Hartmann. *The Nuremberg Chronicle.* Nuremberg: Anton Koberger, 1493. Facsimile of the German edition, New York: Landmark Press, 1979.

Solinus. *The Excellent and Pleasant Worke: Collectanea Rerum Memorabilium.* Translated by Arthur Golding. 1587. Gainesville, FL.: Scholars' Facsimiles and Reprints, 1955.

Versi, Pietro di. *Pietro di Versi, Raxion de' Marineri: Taccuino Nautico del XV Secolo.* Fonti per la Storia di Venezia. Edited by Annalisa Conterio, ser. 5. Venice: Comitato per la Pubblicazione delle Fonti Relative alla Storia di Venezia, 1991.

Vincentius Bellovacensis (Vincent of Beauvais). *Speculum Historiale.* In *Speculum Quadruplex.* 1624. Reprint, Graz: Akademische Druck-und Verlagsanstalt, 1965.

William de Boldensele. *Itinerarius Guilielmi de Boldensele.* In *Zeitschrift des historischen Vereins für Niedersachsen.* Edited by C. E. Grotefend. Jahr 1852. Hannover: Hahn, 1855.

William of Rubruck. *Journey to the Eastern Parts of the World.* Edited and translated by by William W. Rockhill. London: Hakluyt Society, 1900. Reprint, Madras: Asian Educational Services, 1998.

Secondary Sources

Abulafia, David. *A Mediterranean Emporium: The Catalan Kingdom of Majorca.* Cambridge: Cambridge University Press, 1994.

Abulafia, David, and Nora Berends, eds. *Medieval Frontiers: Concepts and Practices.* Aldershot: Ashgate, 2002.

Abu-Lughod, Janet. *Before European Hegemony: The World System 1250–1350.* New York: Oxford University Press, 1989.

Adams, Percy G. *Travelers and Travel Liars.* Berkeley: University of California Press, 1962.

Ahmad, S. Maqbul. "Al-Sharīf al-Idrīsī as a Mapmaker." In Harley and Woodward 1992.

Albu, Emily. "Imperial Geography and the Medieval Peutinger Map." *Imago Mundi* 57, no. 2 (2005): 136–48.

Allen, Rosamund, ed. *Eastward Bound: Travel and Travellers, 1050–1550.* Manchester: Manchester University Press, 2004.

Almagià, Roberto. "I Mappamondi di Enrico Martello e Alcuni Concetti Geografici di Cristoforo Colombo." *Bibliofilia* 42 (1940): 288–311.

————. "Intorno al più Antica Cartografica Nautica Catalan." *Bollettino della Reale Società Geografica Italiana,* 7th ser., 10 (1945): 20–27.

————. "On the Cartographic Work of Francesco Rosselli." *Imago Mundi* 8 (1951): 27–34.

————. "Osservazioni sull'Opera Geografica di Francesco Berlinghieri." *Deputazione Romana di Storia Patria: Archivio* 68 (1945): 211–55.

————. *Planisferi, Carte Nautiche e Affini dal Secoli XIV al XVII.* Vol. 1 in *Monumenta Carto-graphica Vaticana.* 4 vols. Vatican City: Biblioteca Apostolica Vaticana, 1944–52.

Atiya, A. S. *The Crusade in the Later Middle Ages.* London: Butler and Tanner, 1938.

Atlas, Allan W. "Aggio Visto lo Mappamondo: A New Reconstruction." In *Studies in Musical Sources and Style: Essays in Honor of Jan LaRue.* Edited by Eugene K. Wolf and Edward H. Roesner. Madison, WI: A-R Editions, 1990.

Aujac, Germaine. *Claude Ptolémée: Astronome, Astrologue, Géographe: Connaissance et Représentation du Monde Habité.* Paris: Comm. des Travaux Historiques et Scientifiques, 1998.

Avril, François. *Manuscrits Enluminées de la Péninsule Ibérique.* Paris: Bibliothèque Nationale, 1982.

Bagrow, Leo. "Rüst and Sporer's World Maps." *Imago Mundi* 7 (1950): 32–36.

Banfi, Florio. "Two Italian Maps of the Balkan Peninsula." *Imago Mundi* 11 (1954): 17–34.

Barber, Peter. "The Evesham World Map: A Late Medieval English View of God and the World." *Imago Mundi* 47 (1995): 13–33.

————. "The Maps, Town-Views and Historical Prints in the Columbus Inventory." In vol. 1 of *The Print Collection of Ferdinand Columbus, 1488–1539: A Renaissance Collector in Seville.* Edited by Mark P. McDonald. London: The British Museum Press, 2004.

Baron, Hans. "Dati's *Istoria* of Florence." In *The Crisis of the Early Italian Renaissance.* Princeton, NJ: Princeton University Press, 1955.

Baumgärtner, Ingrid. "Die Wahrnehmung Jerusalems auf Mittelalterlichen Weltkarten." In *Jerusalem im Hoch-und Spätmittelalter.* Edited by Dieter Bauer, Klaus Herbers, and Nikolas Jaspert. Frankfort: Campus, 2001.

————. "Kartographie, Reisebericht und Humanismus: Die Erfahrung in der Weltkarte des Venezianischen Kamaldulensermönchs Fra Mauro." *Das Mittelalter* 3, no. 2 (1998): 161–97.

Biasutti, Renato. "Replica a Sebastiano Crinò." *Rivista Geografica Italiana* 49 (1942): 44–54.

Blake, Erin C. "Where Be Dragons?" *Mercator's World* 4, no. 4 (July/August 1999): 80.

Boudet, Jean-Patrice, "Un Prélat et son Équipe de Travail à la Fin du Moyen Age: Remarques sur l'Oeuvre de Pierre d'Ailly." In Marcotte 2002.

Bouloux, Nathalie, *Cultures et Savoirs Géographiques en Italie au XIVe siècle.* Turnhout: Brepols, 2002.

Brincken, Anna-Dorothee von den. *Fines Terrae.* Hannover: Hahnsche Buchhandlung, 1992.

————. "Jerusalem on Medieval Mappaemundi: A Site Both Historical and Eschatological." In Harvey 2006.

————. "Universalkartographie und Geographische Schulkenntnisse im Inkunabelzeitalter: Unter Besonderer Berücksightigung des 'Rudimentum Noviciorum' und Hartmann Schedels." In *Studien zum Städtischen Bildungswesen des Späten Mittelalters und der Frühen Neuzeit. Abhandlungen der Akademie der Wissenschaften zu Göttingen, Phil-Hist.-Klasse* 3 (Göttingen, 1983): 398–428.

————. " 'Ut describeretur universus orbis': Zur Universalkartographie des Mittelalters." In *Methoden in Wissenschaft und Kunst des Mittelalters.* Edited by Albert Zimmermann. Miscellanea Mediaevalia 7. Berlin: Walter de Gruyter, 1970.

Brown, Wesley A. "The World Image Expressed in the *Rudimentum Novitiorum.*" Phillips Society Occasional Paper Series 3. Washington, DC: Library of Congress, 2000.

Buisseret, David. "The Cartographic Background to the Voyages of Columbus." In Schnaubelt and Van Fleteren 1998.

————. *The Mapmakers' Quest: Depicting New Worlds in Renaissance Europe.* Oxford: Oxford University Press, 2003.

Campbell, Tony. *The Earliest Printed Maps, 1472–1500.* London: British Library, 1987.

————. "Portolan Charts From the Late Thirteenth Century to 1500." In Harley and Woodward 1987.

Caraci, Ilaria Luzzana. "Il Planisfero di Enrico Martello." *Rivista Geografica Italiana* 85 (1978): 132–43.

————. *The Puzzling Hero: Christopher Columbus and the Culture of his Age.* Rome: Carocci, 2002.

Cattaneo, Angelo. "Fra Mauro, 'Cosmographus Incomparabilis' and His 'Mappamundi': Documents, Sources, and Protocols for Mapping." In Curto et al. 2003.

————. "God in His World: The Earthly Paradise in Fra Mauro's *Mappamundi* Illuminated by Leonardo Bellini." *Imago Mundi* 55 (2003): 97–102.

Cavallo, Guglielmo, ed. *Cristoforo Colombo e l'Apertura degli Spazi.* 2 vols. Rome: Libreria dello Stato, 1992.

Chekin, Leonid S. *Northern Eurasia in Medieval Cartography: Inventory, Text, Translation and Commentary.* Turnhout: Brepols, 2006.

Christensen, Arne E. "Ships and Navigation," In *Vikings: The North Atlantic Saga.* Edited by William Fitzhugh and Elisabeth Ward. Washington, DC: Smithsonian Institution, 2000.

Colker, Marvin L. "America Rediscovered in the 13th Century?" *Speculum* 54, no. 4 (1979): 712–26.

Conterio, Annalisa. "L'Arte del Navegar: Cultura, Formazione Professionale ed Esperienza dell'Uomo di Mari Veneziano nel XV Secolo." In *L'Uomo e il Mare nella Civiltà Occidentale: da Ulisse a Cristoforo Colombo.* Genoa: Atti della Società Ligure di Storia Patria, new ser., 32 (1992): 187–225.

Conti, Simonetta. "Portolano e Carta Nautica: Confronto Toponomastica." In Marzoli 1981.

Cowan, James. *A Mapmaker's Dream.* Boston: Shambala, 1996.

Crane, Nicholas. *Mercator: The Man Who Mapped the Planet.* New York: Henry Holt, 2002.

Crinò, Sebastiano. "Ancora sul Mappamondo del 1457 e sulla Carta Navigatoria di Paolo dal Pozzo Toscanelli." *Rivista Geographica Italiana* 49 (1942): 35–43.

Crone, G. R. *Maps and Their Makers.* New York: Capricorn, 1962.

————. "New Light on the Hereford Map." *Geographical Journal* 131, no. 4 (December 1965): 447–62.

Crosby, Alfred W. *Measure of Reality: Quantification and Western Society, 1250–1600.* Cambridge: Cambridge University Press, 1997.

Curto, Diogo Ramada, Angelo Cattaneo, and André Ferrand Almeida. *La Cartografia Europea tra Primo Rinascimento e Fine dell'Illuminismo.* Florence: Leo S. Olschki, 2003.

Davies, Arthur. "Behaim, Martellus, and Columbus." *Geographical Journal* 143 (November 1977): 451–59.

Davies, Surekha. "The Navigational Iconography of Diogo Ribeiro's 1529 Vatican Planisphere." *Imago Mundi* 55 (2003): 103–12.

Dawson, Christopher, ed. *Mission to Asia.* Toronto: University of Toronto Press, 1980.

Degenhart, Bernhard, and Annegrit Schmitt. "Marino Sanudo und Paolino Veneto." *Römisches Jahrbuch für Kunstgeschichte* 14 (1973): 1–137.

Delano-Smith, Catherine, and Roger Kain. *English Maps.* London: British Library, 1999.

Deluz, Christiane. "L'Europe selon Pierre d'Ailly ou selon Guillaume Fillastre? De l'*Ymago Mundi* aux Légendes de la Carte de Nancy." In Marcotte 2002.

———. *Le Livre de Jehan de Mandeville*. Louvain: Institute d'Etudes Médiévales de l'Université Catholique de Louvain, 1988.

Destombes, Marcel, ed. *Mappemondes, A.D. 1200–1500*. Vol. 1. Catalogue Préparé par la Commission des Cartes Anciennes de l'Union Géographique Internationale. Amsterdam: N. Israel, 1964.

Dilke, O. A. W. "Cartography in the Byzantine Empire." In Harley and Woodward 1987.

Dilke, O. A. W., and Margaret Dilke. "Italy in Ptolemy's Manual of Geography." In Marzoli 1981.

Dreyer-Eimbcke, Oswald. "Conrad Celtis: Humanist, Poet and Cosmographer." *Map Collector* 74 (1996): 18–21.

Dunn, Ross E., *The Adventures of Ibn Baṭṭūṭah*. Berkeley and Los Angeles: University of California Press, 1986.

Durand, Dana B. *The Vienna-Klosterneuberg Map Corpus*. Leiden: E. J. Brill, 1952.

Dürst, Arthur. "Die Katalanische Estense-Weltkarte, um 1450: Bericht zur Faksimile-Ausgabe." *Cartographica Helvetica* 14 (July 1996): 42–44.

———. "Die Weltkarte von Albertin de Virga von 1411 oder 1415." *Cartographica Helvetica* 13 (January 1996): 18–21.

Džurova, Aksinia, Božidar Dimitrov, José Ruysschaert, and Ivan Dujčev. *Manoscritti Slavi Documenti e Carte Riguardanti la Storia Bulgara*. Sofia: Nauka, 1979.

Eco, Umberto. *Baudolino*. New York: Harcourt, 2000.

Edgerton, Samuel. *The Heritage of Giotto's Geometry: Art and Science on the Eve of the Scientific Revolution*. Ithaca, NY: Cornell University Press, 1991.

Edson, Evelyn. *Mapping Time and Space: How Medieval Mapmakers Viewed Their World*. London: British Library, 1997.

———. "Travelling on the Mappamundi: The World of John Mandeville." In Harvey 2006.

Edson, Evelyn, and Emilie Savage-Smith. *Medieval Views of the Cosmos: Picturing the Universe in the Christian and Islamic Middle Ages*. Oxford: Bodleian Library, 2004.

Edwards, J. G. "Ranulf: Monk of Chester." *English Historical Review* 47 (1932): 94.

El Atlas Català de Cresques Abraham. Barcelona, 1975.

Englisch, Brigitte. *Ordo Orbis Terrae: Die Weltsicht in den Mappaemundi des Frühen und Hohen Mittelalters*. Berlin: Akademie, 2002.

Enterline, James R. *Erikson, Eskimos, and Columbus*. Baltimore: Johns Hopkins University Press, 2002.

Epstein, Marc M. *Dreams of Subversion in Medieval Jewish Art and Literature*. University Park: Pennsylvania State University Press, 1997.

Falchetta, Piero. *Andreas Biancho de Veneciis me fecit M.CCCC.XXX.VJ*. Venice: Arsenal Editrice, 1993.

———. "Marinai, Mercanti, Cartografi, Pittori: Ricerche sulla Cartografia Nautica a Venezia sec. XIV–XV." *Ateneo Veneto: Rivista di Scienze, Lettere ed Arti*, new ser., 183, no. 33 (1995).

Fall, Yoro K. *L'Afrique à la Naissance de la Cartographie Moderne XIVe–XVe Siècles: Les Cartes Majorquines*. Paris: Éditions Karthala, 1982.

Fernández-Armesto, Felipe. *Before Columbus*. Philadelphia: University of Pennsylvania Press, 1987.

Ferro, Gaetano. "Cartografi e Dinastie di Cartografi a Genova." In Cavallo 1992.

———. "Cristoforo e Bartolomeo Colombo Cartografi." In Cavallo 1992.

———. *The Genoese Cartographic Tradition and Christopher Columbus*. Translated by Ann Heck and Luciano F. Farina. Rome: Libreria dello Stato, 1996.

Fischer, Theobald. *Sammlung Mittelalterlicher Welt-und Seekarten Italienischen Ursprungs aus Italienischen Bibliotheken und Archiven*. Venice: F. Ongania, 1886.

Flint, Valerie I. J. "The Hereford Map: Two Scenes and a Border." *Transactions of the Royal Historical Society*, 6th ser., 8 (1998): 19–44.

———. *The Imaginative Landscape of Christopher Columbus*. Princeton, NJ: Princeton University Press, 1992.

Fox-Friedman, Jeanne. "Vision of the World: Romanesque Art of Northern Italy and of the Hereford Mappamundi." In Harvey 2006.

Frankfort, Frank. "Marino Sanudo Torsello: A Social Biography." PhD diss., University of Cincinnati, 1974.

Friedman, John Block. *The Monstrous Races in Medieval Art and Thought*. Cambridge, MA: Harvard University Press, 1981.

Friedman, John Block, and Kristen M. Figg, eds. *Trade, Travel, and Exploration in the Middle Ages*. New York: Garland, 2000.

Gadol, Joan Kelly. *Leon Battista Alberti: Universal Man of the Early Renaissance*. Chicago: University of Chicago Press, 1969.

Galbraith, V. K. "An Autograph MS of Ranulph Higden's Polychronicon." *Huntington Library Quarterly* 34 (1959): 1–18.

Gautier Dalché, Patrick. *Carte Marine et Portulan au XIIe Siècle: Le 'Liber de Existencia Riveriarum et Forma Maris Nostri Mediterranei' (Pise c. 1200)*. Rome: École Française de Rome, 1995.

———. "Décrire le Monde et Situer les Lieux au XIIe Siècle: L'Expositio Mappe Mundi et la Généalogie de la Mappemonde de Hereford." *Mélanges de l'Ecole Française de Rome: Moyen Age* 113 (2001): 343–409.

———. *La "Descriptio Mappe Mundi" de Hugues de Saint-Victor*. Paris: Etudes Augustiniennes, 1988.

———. "Jean Fusoris et la Géographie: Un Astronome, Auteur d'un Globe Terrestre, à la Découverte de Ptolémée." In Marcotte 2002.

———. "L'Oeuvre Géographique du Cardinal Fillastre." In Marcotte 2002.

———. "Portulans and the Byzantine World." In *Travel in the Byzantine World*. Edited by Ruth Macrides. Aldershot: Ashgate, 2002.

———. "Pour une Histoire de Regard Géographique: Conception et Usage de la Carte au XVe Siècle." *Micrologus* 4 (1996): 77–104.

———. "Le Souvenir de la Géographie de Ptolémée dans le Monde Latin Médiéval Ve–XIVe siècles." *Euphrosyne* 27 (1999): 79–106.

———. "D'une Technique à une Culture: Carte Nautique et Portulan au XIIe et au XIIIe siècle." In *L'Uomo e il Mare nella Civiltà Occidentale: da Ulisse a Cristoforo Colombo*. Atti della Società Ligure di Storia Patria, new ser., 32 (1992): fasc. 2, 283–312.

———. "La Trasmissione Medievale e Rinascimentale della Tabula Peutingeriana." In Prontera 2003.

———. *Du Yorkshire à l'Inde: Une "Géographie" Urbaine et Maritime de la Fin du XIIe Siècle (Roger de Howden?)*. Geneva: Droz, 2005.

Gentile, Sebastiano. "Umanesimo e Cartografia: Tolomeo nel Secolo XV." In Curto et al. 2003.

————, ed. *Firenze e la Scoperta dell'America: Umanesimo e Geografia nel '400 Fiorentino*. Florence: Leo Olschki, 1992.

Goldstein, Thomas, "Geography in 15th Century Florence." In *Merchants and Scholars: Essays in the History of Exploration and Trade*. Edited by John Parker. Minneapolis: University of Minnesota Press, 1965.

Gormley, C. M., M. A. Rouse, and Richard H. Rouse. "The Medieval Circulation of Pomponius Mela." *Medieval Studies* 46 (1984): 266–320.

Gow, Andrew. "Gog and Magog on Mappaemundi and Early Printed World Maps: Orientalizing Ethnography in the Apocalyptic Tradition." *Journal of Early Modern History* 2, no. 1 (February 1993): 61–88.

Greenblatt, Stephen. *Marvelous Possessions: The Wonder of the New World*. Oxford: Clarendon, 1991.

Grosjean, Georges, ed. *Mappamundi: The Catalan Atlas for the Year 1375*. Zurich: Urs Graf, 1978.

Gualdi, Fausta. "Marin Sanudo Illustrato." *Commentari*, new ser., 20 (July–September 1969): fasc. 3, pp. 162–98.

Hage, Rushika February. "The Island Book of Henricus Martellus." *Portolan* 56 (Spring 2003): 7–23.

Hamer, John. "The Borgia Map: Europe's Rise and the Redefinition of the World." Master's thesis, University of Michigan, Ann Arbor, 1994.

Hankins, James. "Ptolemy's *Geography* in the Renaissance." In *The Marks in the Fields: Essays on the Uses of Manuscripts*. Edited by Rodney G. Dennis. Cambridge, MA: Harvard University Press, 1992.

Harley, J. B., and David Woodward, eds. *History of Cartography*. Vol. 1, *Cartography in Prehistoric, Ancient, and Medieval Europe and the Mediterranean*. Chicago: University of Chicago, 1987.

————. *History of Cartography*. Vol. 2, *Cartography in Traditional Islamic and South Asian Societies*. Chicago: University of Chicago Press, 1992.

Harvey, Paul D. A. *Mappa Mundi: The Hereford World Map*. London: British Library, 1996.

————. *Medieval Maps*. London: British Library, 1991.

————. "Medieval Maps: An Introduction." In Harley and Woodward 1987.

————, ed. *The Hereford World Map: Medieval World Maps and Their Context*. London: British Library, 2006.

Haslam, Graham. "The Duchy of Cornwall Map Fragment." In Pelletier 1989.

Hassinger, H. "Deutsche Weltkarten-Inkunabeln." *Zeitschrift der Gesellschaft für Erdkunde zu Berlin* 9/10 (1927): 455–82.

Heeren, A. H. "Explicatio Planiglobii." *Commentationes: Societatis Regiae Scientiarum Gottingensis*. Vol. 16. Göttingen: H. Dieterich, 1808.

Higgins, Iain Macleod. *Writing East: The "Travels" of Sir John Mandeville*. Philadelphia: University of Pennsylvania Press, 1997.

Hoogvliet, Margriet. "Animals in Context: Beasts on the Hereford Map and Medieval Natural History." In Harvey 2006.

————. "The Medieval Texts of the 1486 Ptolemy Edition by Johann Reger of Ulm." *Imago Mundi* 54 (2002): 7–18.

Huxley, G. L. "A Porphyrogenitan Portolan." *Greek, Roman and Byzantine Studies* 17 (1976): 295–300.

Irving, Washington. *The Life and Voyages of Christopher Columbus*. Edited by John Harmon McElroy. Boston: Twayne, 1981.

Iwanczak, Wojciech. "Entre l'Espace Ptolémaïque et l'Empirie: Les Cartes de Fra Mauro." *Médiévales* 18 (Spring 1990): 53–68.

Jacob, Christian. *L'Empire des Cartes: Approache Théorique de la Cartographie à Travers l'Histoire.* Paris: Albin Michel, 1992.

Jacoby, David. "L'Expansion Occidentale dans le Levant: Les Vénitiens à Acre dans le Second Moietié du Treizième Siècle." *Journal of Medieval History* 3 (1977): 225–64.

Jeudy, Colette. "La Bibliothèque de Guillaume Fillastre." In Marcotte 2002.

Johns, Jeremy, and Emilie Savage-Smith. "*The Book of Curiosities:* A Newly Discovered Series of Islamic Maps," *Imago Mundi* 55 (2003): 7–24.

Kelley, James E., Jr. "Columbus's Navigation: Fifteenth-Century Technology in Search of Contemporary Understanding." In Schnaubelt and Van Fleteren 1998.

Kimble, George H. T. Foreword to *Catalan World Map of the R. Biblioteca Estense at Modena.* London: Royal Geographical Society, 1934.

———. *Memoir: The Catalan World Map at R. Biblioteca Estense at Modena.* London: Royal Geographical Society, 1932.

Kish, George. *La Carte: Image des Civilisations.* Paris: Seuil, 1980.

Kline, Naomi Reed. *Maps of Medieval Thought.* Woodbridge: Boydell, 2001.

Kretschmer, Konrad. *Die Italienischen Portolane des Mittelalters: Ein Beitrag zur Geschichte der Kartographie und Nautik.* 1909. Reprint, Hildesheim: G. Olms, 1962.

———. "Eine Neue Mittelalterliche Weltkarte der Vatikanischen Bibliothek," *Zeitschrift der Gesellschaft für Erdkunde zu Berlin* 26 (1891): 371–406.

Kugler, Hartmut, and Eckhard Michael, eds. *Ein Weltbild vor Columbus: Die Ebstorfer Weltkarte.* Weinheim: VCH, Acta Humaniora, 1991.

Kunstmann, F. "Studien über Marino Sanudo den Älteren mit einem Anhangseiner Ungedruckten Briefen." *Abhandlungen, Phil.-Historische Classe, Königliche Bayerische Akademie der Wissenschaften* 7 (1853): 794.

Kupfer, Marcia. "The Lost Mappamundi at Chalivoy-Milon." *Speculum* 66, no. 2 (June 1996): 286–310.

———. "Medieval World Maps: Embedded Images, Interpretive Frames." *Word and Image* 10, no. 3 (July–Sept. 1994): 262–88.

Lane, Frederic C. "The Economic Meaning of the Invention of the Compass." *American Historical Review* 68, no. 3 (April 1963): 605–17.

———. *Venice: A Maritime Republic.* Baltimore: Johns Hopkins University Press, 1973.

Lanman, Jonathan T. *On the Origin of Portolan Charts.* Chicago: Newberry Library, 1987.

Larner, John. *Marco Polo and the Discovery of the World.* New Haven, CT: Yale University Press, 1999.

LeCoq, Danielle. "Place et Fonction du Désert dans la Représentation du Monde au Moyen Age." *Revue des Sciences Humaines* 2 (Avril/Juin 2000): 15–112.

———. "Saint Brandan, Christophe Columb et le Paradis Terrestre." *Revue de la Bibliothèque Nationale* 45 (Autumn 1992): 14–21.

Lelewel, Joachim. *Géographie du Moyen Age.* 5 vols. Brussels: J. Pilliet, 1852–57.

Leopold, Antony. *How to Recover the Holy Land: The Crusade Proposals of the Late 13th and Early 14th Centuries.* Aldershot: Ashgate, 2000.

Leporace, Tullia Gasparrini. *Il Mappamondo di Fra Mauro.* Rome: Istituto Poligrafico dello Stato, 1956.

Lewicki, Tadeusz. "Marino Sanudos Mappamundi 1321 und die Runde Weltkarte von Idrisi 1154." *Rocznik Orientalistyczny* 37 (1976): 169–96.

Lewis, C. S. *Discarded Image.* Cambridge: Cambridge University Press, 1967.

Lewis, Martin W., and Kären E. Wigen. *The Myth of Continents: A Critique of Metageography.* Berkeley and Los Angeles: University of California Press, 1997.

Lewis, Suzanne. *The Art of Matthew Paris in the Chronica Majora.* Berkeley and Los Angeles: University of California Press, 1987.

Livingstone, Michael. "More Vinland Maps and Texts: Discovering the New World in Higden's *Polychronicon.*" *Journal of Medieval History* 30, no. 1 (March 2004): 25–44.

Losovsky, Natalia. *The Earth is Our Book: Geographical Knowledge in the Latin West ca. 400–1000.* Ann Arbor: University of Michigan Press, 2000.

Ludmer-Glebe, Susan. "Visions of Madeira." *Mercator's World* 8, no. 3 (May/June 2002): 38–43.

Marcotte, Didier, ed. *Humanisme et Culture Géographique à l'Époque du Concile de Constance.* Turnhout: Brepols, 2002.

Marzoli, Carla Clivio, ed. *Imago et Mensura Mundi, Atti del IX Congresso Internazionale di Storia della Cartografia.* Vol. 1. Rome: Enciclopedia Italiana, 1981.

Massing, Jean-Michel. "Observations and Beliefs: The World of the Catalan Atlas." In *Circa 1492.* Edited by Jay A. Levenson. Washington, DC: National Gallery of Art, 1991.

Mayer, Anton. "Mittelalterliche Weltkarten aus Olmütz." In *Kartographische Denkmäler der Sudetenländer.* Edited by Bernhard Brandt. Vol. 8. Prague: K. André, 1932.

McKenzie, Stephen. "Conquest Landmarks and the Medieval World Image." PhD diss., University of Adelaide, 2000.

Menzies, Gavin. *1421: The Year the Chinese Discovered America.* New York: William Morrow, 2003.

Mignolo, Walter D. *The Darker Side of the Renaissance: Literacy, Territoriality, and Colonization.* Ann Arbor: University of Michigan Press, 1995.

Milanese, Marica. "La Rinascita della Geografia dell'Europa, 1350–1480." In *Viaggiare nel Medioevo.* Edited by Sergio Gensini. Pisa: Fondazione Centro, 2000.

Milano, Ernesto, and Annalisa Battini. *Il Mappamondo Catalano Estense del 1450.* Dietikon, Switzerland: Urs Graf Verlag, 1995.

Miller, Konrad. *Itineraria Romana: Römische Reisewege an der Hand der Tabula Peutingeriana.* 1916. Reprint, Rome: Bretschneider, 1964.

Miller, Konrad. *Mappaemundi: Die Ältesten Weltkarten.* 6 vols. Stuttgart: J. Roth, 1895–98.

Mollat du Jourdin, Michel, and Monique de la Roncière, with Marie-Madeleine Azard, Isabelle Raynaud-Nguyen, and Marie Antoinette Vannereau. *Sea Charts of the Early Explorers.* Translated by L. leR. Dethan. London: Thames and Hudson, 1984.

Nebenzahl, Kenneth. *Maps of the Holy Land: Images of Terra Sancta through Two Millennia.* New York: Abbeville, 1986.

Newton, Arthur Percival. *Travel and Travellers of the Middle Ages.* New York: Barnes and Noble, 1968.

Nordenskjold, A. E. *Facsimile Atlas to the Early History of Cartography.* Stockholm, 1889. Reprint, New York: Dover, 1973.

———. "The Fifteenth-Century Map of the World on Metal in the Collection Borgia at Velletri." *Ymer* 11 (1891): fasc. 1.

———. *Periplus: An Essay on the Early History of Charts and Sailing Directions*. Translated by Francis A. Bather. Stockholm: P. O. Norstedt & Söner, 1897.

Nunn, George E. *Geographical Conceptions of Columbus*. New York: American Geographical Society, 1924.

Obrist, Barbara. "Wind Diagrams and Medieval Cosmology." *Speculum* 72 (1997): 33–84.

Oliel, Jacob. *Les Juifs au Sahara: Le Touat au Moyen Age*. Paris: CNRS, 1994.

Olschki, Leonardo. *Marco Polo's Precursors*. Baltimore: Johns Hopkins University Press, 1943.

Padrón, Ricardo. *The Spacious Word: Cartography, Literature, and Empire in Early Modern Spain*. Chicago: University of Chicago Press, 2004.

Paviot, Jacques. "La Mappemonde Attribuée à Jan van Eyck." *Revue des Archéologues et Historiens d'Art du Louvain* 24 (1991): 57–62.

Pelletier, Monique, "Peut-on Encore Affirmer que la Bibliothèque Nationale Possède la Carte de Christophe Colomb?" *Revue de la Bibliothèque Nationale* 45 (Autumn 1992): 22–25.

———, ed. *Couleurs de la Terre: Des Mappemondes Médiévales aux Imags Satellitales*. Paris: Seuil, 1998.

———, ed. *Géographie du Monde au Moyen Âge et à la Renaissance*. Paris: Éditions du Comité des Travaux Historiques et Scientifiques, 1989.

Penrose, Boies. *Travel and Discovery in the Renaissance, 1420–1620*. New York: Atheneum, 1962.

Petech, Luciano. "Les Marchands Italiens dans l'Empire Mongol." *Journal Asiatique* 249–50 (1961–62): 549–74.

Phillips, J. R. S. *The Medieval Expansion of Europe*. Oxford: Oxford University Press, 1988.

Phillips, William D., and Carla Rahn Phillips. *The World of Christopher Columbus*. Cambridge: Cambridge University Press, 1992.

Pinheiro Marques, Alberto. "The Portuguese Prince Pedro's Purchase of the Fra Mauro Map From Venice." *Globe* 48 (1999): 1–34.

Price, Derek J. de Solla. "Medieval Land Surveying and Topographical Maps." *Geographical Journal* 121, no. 1 (March 1955): 1–10.

Prontera, Francesco. "PerÍploi: Sulla Tradizione della Geografica Nautica presso i Greci." In *L'Uomo e il Mare nella Civiltà Occidentale: da Ulisse a Cristoforo Colombo Atti della Società Ligure di Storia Patria*, new ser., 32 (1992): fasc. 2., 25–44.

———, ed. *Tabula Peutingeriana: Le Antiche Vie del Mondo*. Florence: Leo S. Olschki, 2003.

Ragone, Giuseppe. "Il *Liber Insularum Archipelagi* di Cristoforo Buondelmonti." In Marcotte 2002.

Randles, W. G. L. "De la Carte-Portulan Méditerranéenne à la Carte Marine du Monde des Grandes Découvertes: La Crise de la Cartographie au XVIe Siècle." In Pelletier 1989.

Ravenstein, E. G. *Martin Behaim's 1492 'Erdapfel.'* London: G. Philip, 1908. Rev. ed., London: Greaves and Thomas, 1992.

Relaño, Francesc. *The Shaping of Africa*. Burlington, VT: Ashgate, 2001.

Richard, Jean. *Croisés, Missionaires et Voyageurs: Les Perspectives Orientales du Monde Latin Médiéval*. London: Variorum, 1983.

Richardson, William A. R. "South America on Maps Before Columbus? Martellus's 'Dragon's Tail' Peninsula." *Imago Mundi* 55 (2003): 25–37.

Roncière, Charles de la. *La Carte de Christophe Colomb*. Paris: Eds. Historiques, 1924.

———. *La Découverte de l'Afrique au Moyen Age*. 3 vols. Cairo: Société Royale de Géographie d'Égypte, 1925.

Ross, E. Denison. "Prester John and the Empire of Ethiopia." In Newton 1968.

Russell, Jeffrey B. *Inventing the Flat Earth*. New York: Praeger, 1997.

Rubiés, Joan-Pau. *Travel and Ethnology in the Renaissance: South India Through European Eyes, 1250–1625*. Cambridge: Cambridge University Press, 2000.

Rubin, Rehav. *Image and Reality: Jerusalem in Maps and Views*. Jerusalem: Magnes Press, 1999.

Russell, Peter. *Prince Henry "The Navigator": A Life*. New Haven, CT: Yale University Press, 2000.

Saenger, Paul. "Vinland Re-Read." *Imago Mundi* 50 (1998): 199–202.

Saibanti, Claudio de Polo. In Marzoli 1981.

Salomon, Richard G. "Aftermath to Opicinus de Canistris." *Journal of the Warburg and Courtauld Institute* 25 (1963): 137–46.

———. "A Newly Discovered Manuscript of Opicinus de Canistris: A Preliminary Report." *Journal of the Warburg and Courtauld Institute* 16 (1953): 45–57.

Sandman, Alison. "An Apologia for the Pilots' Charts: Politics, Projections, and Pilots' Reports in Early Modern Spain." *Imago Mundi* 56, no. 1 (2004): 7–22.

Scafi, Alessandro. "Mapping Eden: Cartographies of Earthly Paradise." In *Mappings*. Edited by Denis E. Cosgrove. London: Reaktion, 1999.

———. *Mapping Paradise: A History of Heaven on Earth*. Chicago: University of Chicago, 2006.

———. "Defining Mappaemundi." In Harvey 2006.

———. "Il Paradiso Terrestre di Fra Mauro." *Storia dell'Arte* 93/94 (1998): 411–19.

Schnaubelt, Joseph C., and Frederick Van Fleteren, eds. *Columbus and the New World*. New York: Peter Lang, 1998.

Scott, James M. *Geography in Early Judaism and Christianity: The Book of Jubilees*. Cambridge: Cambridge University Press, 2002.

Seaver, Kirsten A. "Albertin de Virga and the Far North." *Mercator's World* 2, no. 6 (November/December 1997): 58–63.

———. *The Frozen Echo: Greenland and the Exploration of North America, ca. A.D. 1000–1500*. Stanford, CA: Stanford University Press, 1996.

———. *Maps, Myths, and Men: The Story of the Vinland Map*. Stanford, CA: Stanford University Press, 2004.

Sezgin, Fuat. "Arabischer Ursprung Europäischer Karten." *Cartographica Helvetica* 24 (July 2001): 21–28.

Simek, Rudolf. *Heaven and Earth in the Middle Ages*. Translated by Angela Hall. Woodbridge: Boydell, 1996.

Sinor, Denis. "The Mongols and Western Europe." In *The Fourteenth and Fifteenth Centuries*. Edited by Harry W. Hazard, vol. 3 of *A History of the Crusades*, edited by Kenneth M. Setton. Madison: University of Wisconsin Press, 1975.

Skelton, R. A., Thomas E. Marston, and George D. Painter. *The Vinland Map and the Tartar Relation*. New Haven, CT: Yale University Press, 1965.

Slessarev, Vsevolod. *Prester John: The Letter and the Legend*. Minneapolis: University of Minnesota Press, 1959.

Southern, Robert. *Grosseteste: The Growth of an English Mind in Medieval Europe*. Oxford: Clarendon, 1986.

Stevens, Wesley M. "Figure of the Earth in Isidore's De Natura Rerum." *Isis* 71 (1980): 268–77.

Stevenson, Edward L. *The Genoese World Map, 1457*. New York: American Geographical Society, 1912.

————. *Portolan Charts: Their Origin and Characteristics.* New York: Hispanic Society of America, 1911.

Stinger, Charles L. *Humanism and the Church Fathers: Ambrogio Traversari, 1386–1439, and Christian Antiquity in the Italian Renaissance.* Albany: State University of New York, 1977.

Suárez, Thomas. *Early Mapping of Southeast Asia.* Singapore: Periplus, 1999.

Swerdlow, Noel. "The Recovery of the Exact Sciences of Antiquity: Mathematics, Astronomy, Geography." In *Rome Reborn: The Vatican Library and Renaissance Culture.* Edited by Anthony Grafton. Washington, DC, Library of Congress, 1993.

Taylor, Eva G. R. "The *De Ventis* of Matthew Paris." *Imago Mundi* 2 (1937): 23–26.

————. *The Haven-Finding Art.* 1956. Reprint, New York: American Elsevier, 1971.

Taylor, John. *The Universal Chronicle of Ranulf Higden.* Oxford: Clarendon, 1966.

Terkla, Dan. "The Original Placement of the Hereford Mappa Mundi." *Imago Mundi* 56, no. 2 (2004): 131–51.

Thompson, Gunnar. *America's Oldest Map—141 A.D.* Seattle: Misty Isles Press—the Argonauts, 1995.

Thrower, Norman J. W. *Maps and Civilization: Cartography in Culture and Society.* Chicago: University of Chicago Press, 1996.

Tibbetts, Gerald R. "Islamic Cartography: The Beginnings of a Cartographic Tradition." In Harley and Woodward 1992.

————. "Later Cartographic Developments." In Harley and Woodward 1992.

————. *A Study of Arabic Texts Containing Material on Southeast Asia.* Leiden: Brill, 1979.

Tooley, R. V. *Maps and Map-makers.* New York: Dorset, 1987.

Turner, A. J. *The Time Museum: Catalogue of the Collection.* Vol. 1, book 1, *Astrolabes and Astrolabe-Related Instruments.* Rockford, IL: Time Museum, 1985.

Tyerman, Christopher J. "Marino Sanudo Torsello and the Lost Crusade: Lobbying in the 14th Century." *Transactions of the Royal Historical Society* 5th ser., 32 (1982): 57–73.

Uzielli, Gustavo. *La Vita e i Tempi di Paolo dal Pozzo Toscanelli.* In Reale Commissione Colombiana, Raccolta di Documenti e Studi, parte 5, vol. 1. Rome, 1894.

Van den Wyngaert, Anastasius, ed. *Sinica Franciscana.* Vol. 1 of 3. Florence: Ad Claras Aquas, 1929.

Vietor, Alexander. "A Pre-Columbian Map of the World, ca. 1489." *Yale University Library Gazette* 37 (1962): 8–12.

Vignaud, Henry. *Toscanelli and Columbus: The Letter and Chart of Toscanelli.* London: Sands, 1902.

Von Wagner, Bettina. *Die Epistola Presbiteri Johannis Lateinisch und Deutsch.* Tübingen: Max Niemeyer, 2000.

Watelet, Marcel, ed. *Gerard Mercator, Cosmographe: Le Temps et l'Espace.* Anvers: Fonds Mercator, 1994.

Weber, Ekkehard. *Tabula Peutingeriana: Codex Vindobonensis 324.* 2 vols. Graz: Akademisch Druck-und Verlagsanst, 1976.

Webster, Roderick, and Marjorie Webster. *Western Astrolabes.* Chicago: Adler Planetarium, 1998.

Westrem, Scott D. "Against Gog and Magog." In *Text and Territory: Medieval Geographical Imagination in the European Middle Ages.* Edited by Sylvia Tomasch and Sealy Gilles. Philadelphia: University of Pennsylvania Press, 1998.

————. *Broader Horizons: Johannes Witte de Hese's Itinerarius and Medieval Travel Narratives.* Cambridge, MA: Medieval Academy, 2001.

————. *The Hereford Map: A Transcription and Translation of the Legends with Commentary.* Turnhout: Brepols, 2001.

————. "Learning from Legends on the James Ford Bell Library Mappamundi." James Ford Bell Lectures, no. 37. Minneapolis: Associates of the J. F. Bell Library, 2000.

Whitfield, Peter. *The Image of the World*. London: British Library, 1994.

Pülhorn, Wolfgang, and Peter Laub, eds. *Focus Behaim Globus*. 2 vols. Nürnberg: Germanisches Nationalmuseum, 1992.

Williams, John W. *The Illustrated Beatus*. 5 vols. London: Harvey Miller, 1994–1998.

————. "Isidore, Orosius, and the Beatus Map." *Imago Mundi* 49 (1997): 7–32.

Winter, Heinrich. "A Circular Map in a Ptolemaic MS." *Imago Mundi* 10 (1953): 15–22.

————. "The Fra Mauro Portolan Chart in the Vatican." *Imago Mundi* 16 (1962): 17–28.

————. "Notes on the World Map in *Rudimentum Novitiorum*." *Imago Mundi* 11 (1952): 102.

Wolf, Armin. "News on the Ebstorf World Map." In Pelletier 1989.

Woodward, David. "Medieval Mappaemundi." In Harley and Woodward 1987.Woodward, David, and Herbert Howe. "Roger Bacon on Geography and Cartography." In *Roger Bacon and the Sciences: Commemorative Essays*. Edited by Jeremiah Hackett. Leiden: E. J. Brill, 1997.

Wright, John K. *Geographical Lore of the Time of the Crusades*. New York: American Geographical Society, 1925.

————. *The Leardo Map of the World*. New York: AGS, 1928.

Zinner, Ernst. *Regiomontanus: His Life and Work*. Translated by Ezra Brown. Amsterdam: North-Holland, 1990.

Lightning Source UK Ltd.
Milton Keynes UK
UKHW011824080223
416704UK00002B/57